Building Services Engineering

Third Edition

Building Services Engineering Third edition

David V Chadderton
MSc, CPEng, CEng, MIEAust, MCIBSE, MAIRAH

London and New York

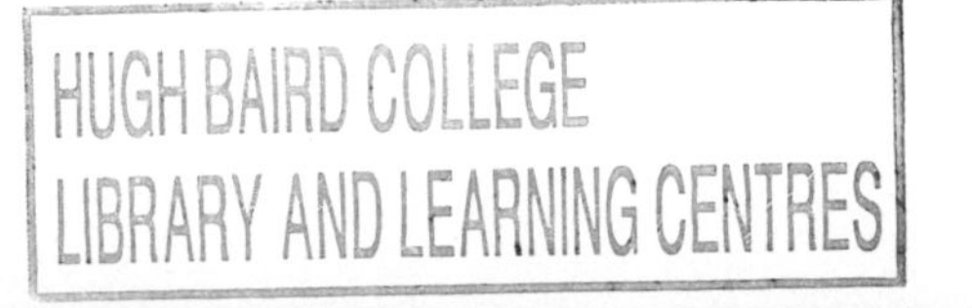

First published 2000 by E & FN Spon
11 New Fetter Lane, London EC4P 4EE

Simultaneously published in the USA and Canada
by E & FN Spon
29 West 35th Street, New York, NY 10001

E & FN Spon is an imprint of the Taylor & Francis Group

Typeset in 10/12pt Sabon by Mathematical Composition Setters Ltd, Salisbury, Wiltshire.
Printed and bound in Great Britain by TJ International Ltd, Padstow, Cornwall.

British Library Cataloguing in Publication Data
A catalogue record for this book is available from the British Library

Library of Congress Cataloging in Publication Data
Chadderton, David V. (David Vincent), 1944-
Building services engineering/David V. Chadderton.–3rd ed.
p. cm.
Includes index.
1. Buildings–Mechanical equipment. 2. Buildings–Environment engineering I. Title.

TH6010.C4867 2000
696–dc21

99-086245

ISBN 0–419–25730–6 (hbk)
ISBN 0–419–25740–3 (pbk)

Contents

Preface to third edition

Building Services Engineering third edition is an update and expands the scope of the book as a whole. Many of the numerical examples and questions have been changed and new data incorporated to increase the quantity of different work that is available, a selection of the graphics has been redrawn, additional practical information is provided in the management of energy, and the development of the need to air condition a building is explored along with further insight into how such a decision may be made. Chapters 14, 16 and 18 in the second edition have been removed and a new Chapter 14 on the design of acoustics for rooms has been included. This new chapter has the same format as those in *Building Services Engineering Spreadsheets* (Chadderton, 1997b).

The spreadsheet software file for Chapter 14, 'Room Accoustics', can be downloaded from the website *http://www.efnspon.com/spon/features/chadderton.html*. It has been produced with As-Easy-As software and bridges the gap between using paper with a calculator and higher cost dedicated software. The program is published by, and can be purchased from, Atlantic Coast PLC, The Shareware Village, Colyton, Devon EX13 6HA, UK; tel. +44 (0)1297 552222; fax. +44 (0)1297 553366. It is expected that the reader can use the spreadsheet software that is available on their own computer, or is provided on a network system for their use. If this is not the case, introductory training in spreadsheet software use is needed. The reader can make use of Chapter 1, Computer and Spreadsheet Use, in *Building Services Engineering Spreadsheets* (Chadderton, 1997b) where sufficient introductory training in computer and spreadsheet use is provided.

Building Services Engineering third edition is intended to be a broad introduction to the range of subjects involved. The engineering content and calculation methods are sufficiently rigorous to match most of what is done within the industry during the design of many building services applications. The subjects covered and the depth to which they are analysed and calculated are more than sufficient to meet the syllabus requirements of higher technician, undergraduate and some postgraduate courses in building services engineering, heating, ventilating and air conditioning, energy management, architecture, building and quantity surveying, housing management, estate management and property facility management. Those preparing for clerk of works examinations will also find the book useful. The advanced user will need to progress to specialized text books and the standard references (Kut, 1993).

The reader is challenged to become actively engaged in the design calculations carried out by design engineers, through step-by-step introduction of each stage. A standard of numerical competence is expected that some lecturers may consider to be higher than is necessary for some courses. This was deemed appropriate in order to broaden the potential readership and provide an adequate basis for a deeper design study.

Acknowledgements

I am particularly grateful to the publishers for their investment in much of my life's work. Such a production only becomes possible through the efforts of a team of highly professional people. An enthusiastic, harmonious and efficient working relationship has always existed, in my experience, with E & FN Spon. All those involved are sincerely thanked for their efforts and the result. My wife Maureen is thanked for her encouragement and understanding while I have been engrossed in keyboard work, on the drawing board and shuffling through piles of proofs. I would specifically like to thank those who have refereed this work. Their efforts to ensure that the book has comprehensive coverage, introductory work, adequate depth of study, valid examples of design, good structured worked examples and exercises are all appreciated. Users and recommenders of the book are all thanked for their support, without them, it would not exist.

The psychrometric chart, Figure 10.1, has been reproduced by permission of the Chartered Institution of Building Services Engineers. Pads of charts, for calculation purposes, may be obtained from CIBSE, Delta House, 222 Balham High Road, London SW12 9BS, UK.

Figures 1.4, 1.7, 1.8, 1.10 and 1.13 are reproduced by kind permission from Oxford Polytechnic, now Oxford Brookes University. Figures 6.5 and 6.6 are reproduced by permission from British Gas Corporation. Figure 12.7 is reproduced by permission from Jeavons Engineering Company.

Units and constants

Système International units are used and Table 1 gives the basic and derived units employed, their symbols and some common equalities.

Table 1 Units

Quantity	Unit	Symbol	Equality
mass	kilogram	kg	
	tonne	tonne	1 tonne = 10^3 kg
length	metre	m	
time	second	s	
	hour	h	1 h = 3600 s
energy, work, heat	joule	J	1 J = 1 Nm
force	newton	N	1 N = 1 kg m/s^2
power, heat flow	watt	W	1 W = 1 J/s
			1 W = 1 Nm/s
			1 W = 1 VA
pressure	pascal	Pa	1 Pa = 1 N/m^2
	newton/m^2	N/m^2	1 b = 10^5 N/m^2
	bar	b	1 b = 10^3 mb
frequency	hertz	Hz	1 Hz = 1 cycle/s
electrical resistance	ohm	R, Ω	
electrical potential	volt	V	
electrical current	ampere	I, A	$I = V/R$
absolute temperature	kelvin	K	K = (°C + 273)
temperature	degree Celsius	°C	
luminous flux	lumen	lm	
illuminance	lux	lx	1 lx = 1 lm/m^2
area	square metre	m^2	
volume	litre	l	
	cubic metre	m^3	1 m^3 = 10^3 l

Table 2 Multiples and submultiples

Quantity	Name	Symbol
10^{12}	tera	T
10^{9}	giga	G
10^{6}	mega	M
10^{3}	kilo	k
10^{-3}	milli	m
10^{-6}	micro	μ

Table 3 Physical constants

gravitational acceleration	g	9.807 m/s^2
specific heat capacity of air	SHC	1.012 kJ/kg K
specific heat capacity of water	SHC	4.186 kJ/kg K
Stefan–Boltzmann constant	σ	5.67×10^{-8} $W/m^2\ K^4$
density of air at 20 °C, 1013.25 mb	ρ	1.205 kg/m^3
density of water at 4 °C	ρ	10^3 kg/m^3
exponential	e	2.718

Symbols

Symbol	*Description*	*Units*
A	area	m^2
	electrical current	A
A_f	floor area	m^2
	physical constant	dB
A_g	cross-sectional area of gutter	mm^2
A_o	area of water flow at gutter outlet	mm^2
A_r	roof area	m^2
A_w	walling area	m^2
α (alpha)	electrical temperature coefficient of resistance	Ω/Ω °C
	percentage depreciation and interest charge	%
	absorption coefficient	dimensionless
$\overline{\alpha}$	mean absorption coefficient	dimensionless
AET	allowed exposure time	min
B	building envelope number	
	sound reduction index	dB
B_f	physical constant	dB
b	barometric pressure	bar
β (beta)	angle	degree
C	fuel cost per appropriate unit	
C_1, C_2	constant	
C_i	interior air pollution	decipol
C_o	outdoor air pollution	decipol
C_r	room concentration	%
C_s	supply air concentration	%
C_T	concentration after time T	%
C_3	electrical load	W/m^2
clo	clothing thermal insulation	
C_v	ventilation coefficient	
CO_2	carbon dioxide	%, ppm
D	gutter depth	mm
DI	directivity index	dB
DU	demand or discharge unit	
d	pipe diameter	m or mm
	distance	m

Δt (delta)	difference of temperature	°C
Δp	difference of pressure	N/m^2
d.b.	dry-bulb air temperature	°C
decipol	air pollution from one standard person	
E	emissivity	
E_{max}	maximum available evaporative cooling	W
E_{req}	required evaporative cooling	W
EWCT	equivalent wind chill temperature	°C
EL	equivalent length	m
EUPF	energy use performance factor	
e	exponential	
η (eta)	efficiency	%
F	radiation configuration factor	
F_G	fractional area	
F_m	Marston bedding factor	
F_s	factor of safety	
f	frequency	Hz
G	gradient	
	moisture mass flow rate	kg/m^2 s
	pollution load	olf/m^2
GCV	gross calorific value	MJ/kg
GJ	energy	gigajoule
g	gravitational acceleration	m/s^2
	air moisture content	kg water/kg dry air
H	height	m
	body internal heat generation	W/m^2
HSI	heat stress index	
Hz	frequency	cycle/s
h	time	hour
I	cost of installed thermal insulation	$£/m^3$
	electrical current	ampere
J	energy	joule
K	absolute temperature	kelvin
K_1, K_2	constant	
kg	mass	kilogram
kJ	energy	kilojoule
kW	power	kilowatt
kWh	energy	kilowatt-hour
L	load factor	
LDL	lighting design lumens	lumen
LH	latent heat of evaporation	kW
l	length	m
λ (lambda)	thermal conductivity	W/m K
LPG	liquefied petroleum gas	
M	metabolic rate	W/m^2
MF	maintenance factor	
MJ	energy	megajoule
MW	power	megawatt

Symbol	Quantity	Unit
m	length	metre
mm	length	millimetre
μ (mu)	diffusion resistance factor	
N	air change rate	h^{-1}
	force	newton
	number of occupants	
NR	noise rating	dimensionless
n_f	number of storeys	
olf	concentration of odorous pollutants	
Ω (omega)	electrical resistance	ohm
P	pressure	pascal
	permeance	kg/N s
P	carbon dioxide production	
Pa	pressure	pascal
P_1, P_2	area fraction	
PD	percentage of occupants dissatisfied	%
p_s	vapour pressure	pascal
ϕ (phi)	angle	degree
Q	fluid flow rate	m^3/s or l/s
	power	kW
	geometric directivity factor	dimensionless
Q_c	convection heat transfer	W
Q_e	extract air flow rate	m^3/s
Q_{ex}	exhaust air flow rate	m^3/s
Q_f	fresh air flow rate	m^3/s
Q_f	fabric heat loss	W
Q_{HWS}	hot water service power	kW
Q_L	leakage air flow rate	m^3/s
Q_p	total heat requirement	W
Q_r	radiation heat transfer	W
	recirculation air flow rate	m^3/s
Q_u	heat flow through fabric	W
Q_v	ventilation heat loss	W
q	water flow rate	kg/s
R	resistance, electrical	Ω
	thermal	m^2 K/W
	room sound absorption constant	m^2
R_A	combined resistance of pitched roof	m^2 K/W
R_a	air space thermal resistance	m^2 K/W
R_B	ceiling thermal resistance	m^2 K/W
R_n	new thermal resistance	m^2 K/W
R_{si}	internal surface thermal resistance	m^2 K/W
R_{so}	outside surface thermal resistance	m^2 K/W
R_R	thermal resistance of roof void	m^2 K/W
R_v	vapour resistance	N s/kg
r	distance	m
r_v	vapour resistivity	GN s/kg m, MN s/g m
ρ (rho)	density	kg/m^3

	specific electrical resistance	Ωm
	soil electrical resistivity	Ωm
S	spacing	m
	length of heating season	days
	surface area	m^2
s	time	second
SC	quarterly standing charge	
SE	specific enthalpy	kJ/kg
SG	specific gravity	
SH	sensible heat transfer	kW
SPL	sound pressure level	dB
SWL	sound power level	dB
SRI	sound reduction index	dB
SHC	specific heat capacity	kJ/kg K
Σ (sigma)	summation	
T	absolute temperature	kelvin
	total demand target	
	reverberation time	s
τ (tau)	time interval	
T_E	electrical demand target	
T_T	thermal demand target	
t_a	air temperature	°C
t_{ai}	inside air temperature	°C
t_{ao}	outside air temperature	°C
t_b	base temperature	°C
t_c	comfort temperature	°C
t_{dp}	dew-point temperature	°C
t_e	environmental temperature	°C
t_{ei}	internal environmental temperature	°C
t_{eo}	outside environmental temperature	°C
t_f	water flow temperature	°C
t_g	globe temperature	°C
t_{HWS}	hot water storage temperature	°C
t_m	mean water temperature	°C
	area-weighted average room surface temperature	°C
t_{max}	maximum air temperature	°C
t_{min}	minimum air temperature	°C
t_{15}	air temperature at time $\theta = 15$ h	°C
t_r	mean radiant temperature	°C
	return water temperature	°C
t_{res}	resultant temperature	°C
t_s	supply air temperature	°C
	surface temperature	°C
θ (theta)	angle	degree
θ	time	h
U	thermal transmittance	W/m^2 K
U_e	economic thermal transmittance	W/m^2 K
U_n	new thermal transmittance	W/m^2 K
U_w	wall thermal transmittance	W/m^2 K

UC	useful cost of a fuel	£/GJ, p/kWh
UF	utilization factor	
V	volume	m^3
V	electrical potential	volt
v	velocity	m/s
v_s	specific volume	m^3/kg
W	width	m or mm
W	power	watt
W_e	total vertical load	kN/m
W_t	standard test crushing strength	kN/m
w.b.	wet-bulb air temperature	°C
WCI	wind chill index	
Y	admittance factor	
	annual degree days	

1 The built environment

Learning objectives

Study of this chapter will enable the reader to:

1. relate human physiological needs to the internal and external environment;
2. understand the ways in which heat exchange between the body and its surroundings takes place;
3. calculate indoor and outdoor thermal comfort equations;
4. identify essential instruments for measuring the environment;
5. understand the thermal environment terminology used by design engineers;
6. recognize the problems of experimental work;
7. make reliable technical reports based on his or her own work and not to copy the work of others;
8. understand and use the factors that influence indoor air quality;
9. calculate fresh air ventilation rate;
10. know the instrumentation used for indoor environmental monitoring.

Key terms and concepts

Introduction

The building is an enclosure for the benefit of human habitation, work or recreation; in some cases, the ruling criteria are those demanded by an industrial need, such as machinery, products or computers. Much construction work is undertaken outdoors, where the climate moderates human working effectiveness.

The way in which the design engineer calculates measurement scales and interaction with human requirements is investigated.

World energy supply and demand

Highly developed countries have become rich in material possessions by rapidly exploiting technological advances in energy supply and use. The human ability to extract enormous amounts of natural energy from the earth has enabled rapid communication with all parts of the globe, and space, and has kept habitable structures warm or cool as required. Large parts of the globe are much poorer in terms of their standard of living, and it is incumbent upon those nations possessing knowledge to act responsibly with regard to the resources they use now, how they plan for future development, and what technology they sell to the developing nations. The building industry as a whole can learn from our own history of energy usage to promote only those systems making the best use of available power. Economical use of energy with services installations is crucially important in this respect.

The building as an environmental filter

One of people's basic needs is to maintain a constant body temperature, and the metabolism regulates heat flows from the body to compensate for changes in the environment. We have become expert in fine-tuning the environmental conditions produced by the climate in relation to the properties of the building envelope to avoid discomfort. A simple tent or cave may be sufficient to filter out the worst of adverse weather conditions, but the ability of this type of shelter to respond to favourable heat gains or cooling breezes may be too fast or too slow to maintain comfort.

Outside the tropics, houses may be advantageously oriented towards the sun to take advantage of solar heat gains, which will be stored in the dense parts of the structure and later released into the rooms to help offset heat losses to the cool external air during winter. Buildings within the tropical zone require large overhanging roofs and shutters over the windows to exclude as much solar radiation as possible and to shade the rooms. Thus the building envelope acts to moderate extremes of climate, and by suitable design of illumination and ventilation openings, together with heating, cooling and humidity controls, a stable internal environment can be matched to the use of the building.

Basic needs for human comfort

The building services engineer is involved with every part of the interface between the building and its occupant. Visually, colours rendered by natural and artificial illumination are produced by combinations of decor and windows. The acoustic environment is largely attributed to the success achieved in producing the required temperatures with quiet services equipment, all of which is part of the thermal control and transportation arrangements. Energy consumption for thermally based systems is the main concern, and close coordination between client, architect and engineer is vitally important.

Heat transfer between the human body and its surroundings can be summarized as follows.

Conduction

Points of contact with the structure are made with furniture. Clothing normally having substantial thermal insulation value and discomfort should be avoided.

Convection

Heat removed from the body by natural convection currents in the room air, or fast-moving airstreams produced by ventilation fans or external wind pressure, is a major source of cooling. The body's response to a cool air environment is to restrict blood circulation to the skin to conserve deep tissue temperature, involuntary reflex action (shivering) if necessary, and in extreme cases inevitable lowering of body temperature. This last state of hypothermia can lead to loss of life and is a particular concern in relation to elderly people.

Radiation

Radiation heat transfer takes place between the body and its surroundings. The direction of heat transfer may be either way, but normally a minor part of the total body heat loss takes place by this method. Radiation between skin and clothing surfaces and the room depends on the fourth power of the absolute surface temperature, the emissivity, the surface area and the geometric configuration of the emitting and receiving areas. Thus a moving person will experience changes in comfort level depending on the location of the hot and cold surfaces in the room, even though air temperature and speed may be constant.

Some source of radiant heat is essential for comfort, particularly for sedentary occupations, and hot-water central heating radiators, direct fuel-fired appliances and most electrical heaters provide this. The elderly find particular difficulty in keeping warm when they are relatively immobile, and convective heating alone is unlikely to be satisfactory. A source of radiant heat provides rapid heat transfer and a focal point, easy manual control and quick heat-up periods. Severe cases of underheating can be counteracted by placing aluminium foil screens in positions where they can reflect radiation onto the rear of the chair.

Overheating from sunshine can also cause discomfort and glare, and tolerance levels for radiant heating systems have been established.

Evaporation

Humid air is exhaled, and further transfer of moisture from the body takes place by evaporation from the skin and through clothing. Maintenance of a steady rate of moisture removal from the body is essential, and this is a mass transfer process depending on air humidity, temperature and speed as well as variables such as clothing and activity.

Ventilation

The quality of the air in a building depends upon the quantity, type and dispersal of atmospheric pollutants (Awbi, 1991, p. 27). Some of these, odorants, can be detected by the olfactory receptors in the nose. These are the odours, vapours and gases that ingress from the outdoor environment and are released from humans, animals, flora, furnishings and the structural components of the building. Solid particles of dust, pollen and other contaminants often have little or no smell. These might be seen in occasional shafts of sunlight, and become visible when they have settled. Cleaning fluids such as ammonia, cigarette smoke, hair spray, deodorants and perfumes can be most noticeable. The inflow of diesel exhaust fumes, road tar, paint vapours and creosote creates unpleasantly noticeable pollution, even when of short duration. The presence of harmful pollutants such as carbon monoxide and radon gases is not detectable by the occupant.

Professor Ole Fanger has introduced units of subjective assessment for odorants only. The olf quantifies the concentration of odorous pollutants. The decipol is the evaluation of the pollutant as determined by the recipient through the olfactory sensations from the nose. One olf is the emission rate of biological effluents from one standard person, or the equivalent from other sources. One decipol is the pollution caused by one standard person when ventilated with 10 l/s of unpolluted air. The number of olfs corresponding to different levels of human activity is shown in Table 1.1 (Fanger, 1988).

Office accommodation normally has one person for each 10 m^2 of floor area, so the biological effluent pollution load produced by normal occupancy is 0.1 olf/m^2. Smokers, building and furnishing materials and ventilation systems add to the pollution load. The average pollution in an existing building that has 40% of the occupants as smokers produces a load G of 0.7 olf/m^2. A low-pollution building with an absence of smoking has a load G of 0.2 olf/m^2. When there is complete mixing of the ventilation air with the air in the room, the rate of supply of outdoor air that is necessary to maintain the required standard of air quality is found from

$$Q = \frac{10 \times G}{C_i - C_o} \text{ l/s}$$

where C_i is the perceived air pollution within the enclosure (decipol), C_o is the perceived air pollution of outdoor air, usually 0.05 decipol but which may rise to 0.3 in a city with moderate pollution, and G is the concentration of pollution in the enclosure and the ventilation system (olf).

The perceived air pollution C_i within the enclosure is found from the percentage of the occupants who are dissatisfied with the conditions, PD, from

$$C_i = \frac{112}{(5.98 - \ln PD)^4} \text{ decipol}$$

Table 1.1 Olf values for human activities

Human activity	Number of olfs
Sedentary	1
Active	5
Highly active	11
Average for a smoker	6
During smoking	25

where *PD* is the percentage of the occupants who are dissatisfied (ASHRAE (1985) recommend 20%), and ln = logarithm to base *e* ($\log_e$)

EXAMPLE 1.1

Calculate the outdoor air ventilation rate, from the Fanger method, that is required to satisfy 90% of the occupants of an office where it is expected that none of them are smokers. The office building was constructed in 1980 and has only had routine maintenance since then. It is located in a rural town.

To satisfy 90% of the occupants, 10% are dissatisfied, so *PD* is 10%.

$$C_i = \frac{112}{(5.98 - \ln 10)^4} \text{ decipol}$$

$$= \frac{112}{(5.98 - 2.303)^4} \text{ decipol}$$

$$= \frac{12}{(3.677)^4} \text{ decipol}$$

For an existing building with no smokers, $G = 0.2$ olf/m^2. $C_o = 0.05$ decipol for clean city air. For 1 m^2 of office floor area

$$Q = \frac{10 \times G}{C_i - C_o} \text{ l/s m}^2$$

$$= \frac{10 \times 0.2}{0.612 - 0.05} \text{ l/s m}^2$$

$$= 3.56 \text{ l/s m}^2$$

EXAMPLE 1.2

A new aerobics gymnasium is being designed for the basement of a commercial building in the City of London. The room is to be 20 m long, 15 m wide and 3 m high. It has no exterior windows. There will be between 10 and 55 simultaneous users of the facility. There will not be any smoking and all the furnishings and building materials will have the low-pollution emission of 0.1 olf/m^2. At peak usage, all the occupants will be highly active. Calculate the outdoor air ventilation rate, from the Fanger method, that will be required so that 90% of the occupants will be satisfied. Recommend how the outdoor air ventilation system could be controlled for energy economy and comfort. Recommend an acceptable solution for the client.

(a) At full occupancy (55 people), to satisfy 90% of the occupants, 10% (6 people) are dissatisfied, so $PD = 10\%$.

$$C_i = \frac{112}{(5.98 - \ln 10)^4} \text{ decipol}$$

$$= 0.613 \text{ decipol}$$

The low-pollution building produces 0.1 olf/m^2.

At peak usage, there will be 55 people each producing 11 olf, from Table 1.1.

$$\text{gymnasium floor area} = 20 \text{ m} \times 15 \text{ m} = 300 \text{ m}^2$$

$$\text{pollution load } G = 0.1 \text{ olf/m}^2 + \frac{55 \text{ people} \times 11 \text{ olf/person}}{300 \text{ m}^2}$$

$$= 0.1 + 2.02 \text{ olf/m}^2$$

$$= 2.12 \text{ olf/m}^2$$

$C_o = 0.3$ decipol for vitiated city air that enters the basement from street level. For 1 m^2 of gymnasium floor area,

$$Q = \frac{10 \times G}{C_i - C_o} \text{ l/s m}^2$$

$$= \frac{10 \times 2.12}{0.613 - 0.3} \text{ l/s m}^2$$

$$= 67.7 \text{ l/s m}^2$$

For the whole gymnasium,

$$Q = 300 \text{ m}^2 \times 67.7 \text{ l/s m}^2$$

$$= 20\ 320 \text{ l/s}$$

$$\text{the room volume } V = 20 \text{ m} \times 15 \text{ m} \times 3 \text{ m} = 900 \text{ m}^3$$

$$\text{air change rate } N = 20\ 320 \frac{1}{\text{s}} \times \frac{1 \text{ m}^3}{10^3 \text{ l}} \times \frac{1 \text{ air change}}{900 \text{ m}^3} \times \frac{3600 \text{ s}}{1 \text{ h}}$$

$$= \frac{20\ 320 \times 3600}{10^3 \times 900} \text{ air change/h}$$

$$= 81.3 \text{ air change/h}$$

This is an unacceptably high air change rate and will result in large heating and cooling costs during winter and summer if the thermal conditions are to remain within acceptable ranges for the activity. The number of room occupants who can be satisfied with the air quality will need to be reduced. Air quality, here, refers to maintaining a low enough level of odours so that the occupants are satisfied. Alternative solutions to supplying 100% outdoor air into the basement are available.

(b) When the gymnasium has ten occupants with 90% satisfaction, only one person is dissatisfied, so *PD* is 10%, C_i is 0.613 decipol, *G* is 0.47 olf/m^2; when C_o is 0.3 decipol, *Q* becomes 15.0 l/s m^2, 4505 l/s for the whole room and *N* is 18.0 air change/h.

(c) The supply of outside air into the gymnasium is required to comply with the minimum of 10.4 l/s per person (Chartered Institution of Building Services Engineers (hereafter CIBSE), 1986).

$$\text{Minimum outside air supply } Q_o = 55 \text{ people} \times \frac{10.4 \text{ l}}{\text{s person}}$$

$$= 572 \text{ l/s}$$

$$\text{Air change rate } N = 572 \frac{\text{l}}{\text{s}} \times \frac{1 \text{ m}^3}{10^3 \text{ l}} \times \frac{1 \text{ air change}}{900 \text{ m}^3} \times \frac{3600 \text{ s}}{1 \text{ h}}$$

$$= \frac{572 \times 3600}{10^3 \times 900} \text{ air change/h}$$

$$= 2.3 \text{ air change/h}$$

(d) The air change rates of outdoor air that are, apparently, needed to satisfy the Fanger odour air quality criteria varies from zero, when the gymnasium is unoccupied, to 18.0 per hour when 10 people use the gymnasium, to 81.0 per hour when 55 people use the room. The minimum required outdoor air supply is 572 l/s; this corresponds to a room air change rate of 2.3 per hour when 55 people are using the gymnasium. Mechanical ventilation will be needed owing to the location of the room. The design engineer will recommend a minimum energy use by recirculating the maximum amount of room air and minimizing the use of outdoor air to match the room occupancy. All the incoming outdoor air can be passed through a flat-plate heat exchanger to recover heat that is being discharged to the atmosphere through the exhaust air duct, if this is possible for the duct installation. When the room is unoccupied, the outside air supply can be closed with a motorized damper. The number of occupants in the gymnasium can be detected from a carbon dioxide sensor in the extract air duct, which can be used to control the opening of the outside air motorized damper as well as the supply air and return air fan speeds. The varying air flow rates can be provided with a system of supply-and-extract air fans that have variable performance. Each fan can have a variable-speed electric motor which has its supply frequency controlled between 0 to 50 Hz from an inverter drive. The outdoor air that is supplied into the gymnasium can be heated to around 16 °C in winter and cooled to 20 °C in summer, for thermal comfort. When lower than the Fanger-recommended outside air intake quantities are used – that will normally be always – the odour air quality can be improved by passing the return air through an activated carbon filter or by injecting a deodorant spray into the supply air

Table 1.2 Comparison of ventilation rates for offices

Data source	Ventilation rate	
	(l/s m^2)	(l/s per person)
Fanger equation	5	50
ASHRAE	0.7	7
BS 5925	1.3	13
DIN 1946	1.9	19
CIBSE, open plan	1.3	13
CIBSE, office, heavy smoking	6	60

duct. Whichever alternative is selected, the energy-conscious design engineer will minimize the use of outside air to ventilate the room. Those using the gymnasium are fairly unlikely to find the presence of some body odour totally unacceptable. Significant energy savings will be achieved by the use of carbon dioxide sensing of the occupants in a gymnasium that is used intermittently each day. The outside air damper will remain closed whenever the room is unoccupied and it will only open in response to the actual occupancy level.

Awbi (1991, p. 31) shows a comparison between the current standards for ventilation with outdoor air. They are based upon one person for each 10 m^2 of office floor area, the standard person and activity plus some smoking. To these, the *CIBSE Guide* recommendations (CIBSE, 1986, Section A1) have been added, as shown in Table 1.2.

Other applications are taken on their merits: for example toilets 10, corridor 1.3 and commercial kitchen 20 l/s m^2 (CIBSE, Section A1). The designer needs to evaluate all the aspects of the need for outdoor air ventilation. These include the heating and cooling plant loads that will be generated, the potential for energy recovery between the incoming fresh air and outgoing air at room temperature, avoidance of draughts in the occupied rooms, the variation of load with the occupancy level and the ability to utilize outdoor air to provide free cooling to the building when the outdoor air is between 10 °C and 20 °C (Chadderton, 1997a, p. 5).

Comfort equation

The fundamental purpose of heating or cooling an occupied space is to maintain constant body temperature. Regulation of personal comfort is achieved by clothing choice, and a successful building can be said to be one where 5% or less of its occupants complain. Definitive work was done by Fanger (1972) to produce a complete mathematical statement of the heat transfer between the body and its surroundings, and then to conduct subjective tests and produce guidelines for building and engineering designers. Fanger's comfort equation is of the form

$$f(H, \text{clo}, v, t_r, t_a, p_s) = 0,$$

i.e. a balance, where f represents a mathematical function connecting all the variables contained in the brackets, H is the internal heat production in the human body, clo is the thermal insulation value of clothing, v is the air velocity, t_r is the mean radiant temperature, t_a is the air temperature and p_s is the atmospheric vapour pressure.

The energy released per unit time in the human body by oxidation processes is known as the metabolic rate M. Some of this can be converted into useful mechanical work W, so that

$$M = H + W$$

The mechanical efficiency of the body is given by

$$\frac{W}{M} \times 100\%$$

and this varies from zero while reading this book to a maximum of 20% during heavy physical work.

Internal heat production H varies from 35 W/m^2 (watts per square metre of body surface) while sleeping to 440 W/m^2 during maximum exertion. The thermal insulation value of clothing is expressed in clo units. One clo unit is equal to the total thermal resistance from the skin to the outer surface divided by 0.18. It is worth noting that 1 tog = 0.645 clo. Values of clo vary from zero when nude through 1 when wearing a normal business suit to 4 for a polar suit.

Moving-air velocity in normally occupied rooms will be between 0 and 2 m/s (metres per second), where the upper figure relates to a significantly uncomfortable hot or cold draught. Still air conditions are most unlikely to occur, as convection currents from people and warmed surfaces will promote some circulation. Room air movement patterns should be variable rather than monotonous, and ventilation of every part of the space is most important.

The mean radiant temperature is a measure of radiation heat transfers taking place between various surfaces and has an important bearing on thermal comfort. Heat transfer Q_r by radiation from a warm emitting surface to a receiving plane is given by

$$Q_r = \frac{5.67}{10^8} A_1 F_1 E_1 E_2 (T_1^4 - T_2^4) \text{ W}$$

where E_1 and E_2 are surface emissivities, which range from 0.04 for polished aluminium to 0.96 for matt black paint, and T_1 and T_2 are the absolute temperatures of the emitting and receiving surfaces in kelvins (°C + 273). F_1 is the configuration factor denoting the orientation of and distance between the two surfaces. Flat parallel surfaces are evaluated using

$$A_1 F_1 = \frac{A_1 A_2}{\pi l^2}$$

where A_1 and A_2 are the surface areas (m^2) and l is the distance between the surfaces (m). This only fits the simplest cases, such as radiation across the cavity in a wall. Other real problems become geometrically rigorous and complex.

EXAMPLE 1.3

A hot-water central heating system has a radiator of surface area 4.5 m^2 and a surface temperature of 70 °C. It faces a person 2 m away whose body surface area and temperature are 1 m^2 and 25 °C respectively. The surface emissivities of the radiator and the person are 0.85 and 0.90 respectively. Calculate the radiant heat transfer.

$$A_1F_1 = \frac{4.5\ \text{m}^2 \times 1\ \text{m}^2}{\pi(2\ \text{m}^2)^2} = 0.358\ \text{m}^2$$

$$T_1 = (273 + 70)\ \text{K} = 343\ \text{K}$$

$$T_2 = (273 + 25)\ \text{K} = 298\ \text{K}$$

$$Q_r = \frac{5.67}{10^8} \times 0.358 \times 0.85 \times 0.90 \times (343^4 - 298^4)\ \text{W}$$

$$= 92.5\ \text{W}$$

Because of the complexity of such calculations for a heated room, an approximation to the mean radiant temperature t_r is often made that estimates its value for the geometric centre of the room volume:

$$t_r \simeq t_m = \frac{A_1 t_{s_1} + A_2 t_{s_2} + \cdots + A_n t_{s_n}}{A_1 + A_2 + \cdots + A_n}$$

Thus t_m is seen to be the average room surface temperature weighted in proportion to the surface area, room geometry and point of measurement; n is the number of surfaces.

EXAMPLE 1.4

A room is heated with a hot-water panel radiator and the data shown in Table 1.3 were found during site measurements.

Table 1.3 Data for Example 1.4

Surface	Average surface temperature (°C)	Area (m^2)
Window	2	3
Radiator	65	4
Outside wall	15	20
Inner walls	19	30
Ceiling	22	20
Floor	14	20

Estimate the mean radiant temperature of the room at the centre of its volume.

Table 1.4 Calculations for Example 1.4

Surface	Area A(m^2)	t_s (°C)	At_s
Window	3	2	6
Radiator	4	65	260
Outside wall	20	15	300
Inner walls	30	19	570
Ceiling	20	22	440
Floor	20	14	280
	$\Sigma A = 97$		$\Sigma(At_s) = 1856$

ΣA and ΣAt_s are calculated as shown in Table 1.4. Then

$$t_m \simeq \frac{\Sigma(At_s)}{\Sigma A}$$

$$= \frac{1856}{97}$$

$$= 19.1\ °C$$

The air temperature t_a is that recorded by a mercury-in-glass dry-bulb thermometer freely exposed to the air stream. The thermometer is usually placed in a sling psychrometer.

The atmospheric vapour pressure p_s is that part of the barometric pressure produced by the water vapour in humid air. Standard atmospheric pressure at sea level is 1013.25 mb (millibars) and this comprises about 993.0 mb from the weight of dry gases and 20.25 mb from the water vapour, depending on the values of the barometric pressure, the air temperature and the humidity (Chapter 5).

The comfort equation balances to zero when heat transfers between the body and surroundings are stable. This is the thermally neutral condition when there are no feelings of discomfort. It is unlikely to be satisfied for a group of people, and a comfort zone is used to specify the range of acceptable levels for the majority.

Comfort measurement

Figure 1.1 shows the percentage of people dissatisfied (PPD) and predicted mean vote (PMV) that are used for the assessment of indoor thermal environments, from Fanger. The thermal comfort analogue computer that is used to make this assessment is shown in Fig 1.2. Its sensor has an ellipsoidal shape that has the same mathematical relationship between its heat loss by radiation and convection as the human body.

To obtain values of temperature, velocity and humidity in the atmosphere, each is measured separately as 0–10 V or 0–20 mA analogue signals, converted

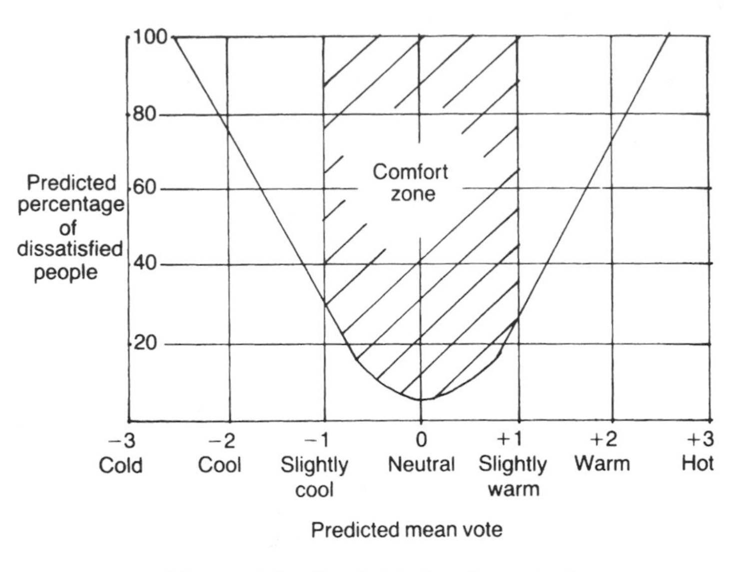

Figure 1.1 Comfy-Test meter output

Figure 1.2 Thermal comfort meter (reproduced by courtesy of Bruel & Kjaer (UK) Ltd)

into digital bytes with an analogue-to-digital converter and then stored in a data logger in random access memory. Numerous readings are easily retained in memory for later access through a desktop or portable computer. Lists of data numbers and their time of occurrence can be stored on hard and floppy disks for processing with dedicated software or any spreadsheet program. Graphical output can show the history of the recorded temperature, or other variable, during a period of time. Such graphs are termed trends. The software can be programmed to take readings at, say, intervals of 2 s or longer, or only whenever a significant change takes place. Figure 1.3 shows the arrangement of a data-logging system with a trend display of room air temperature.

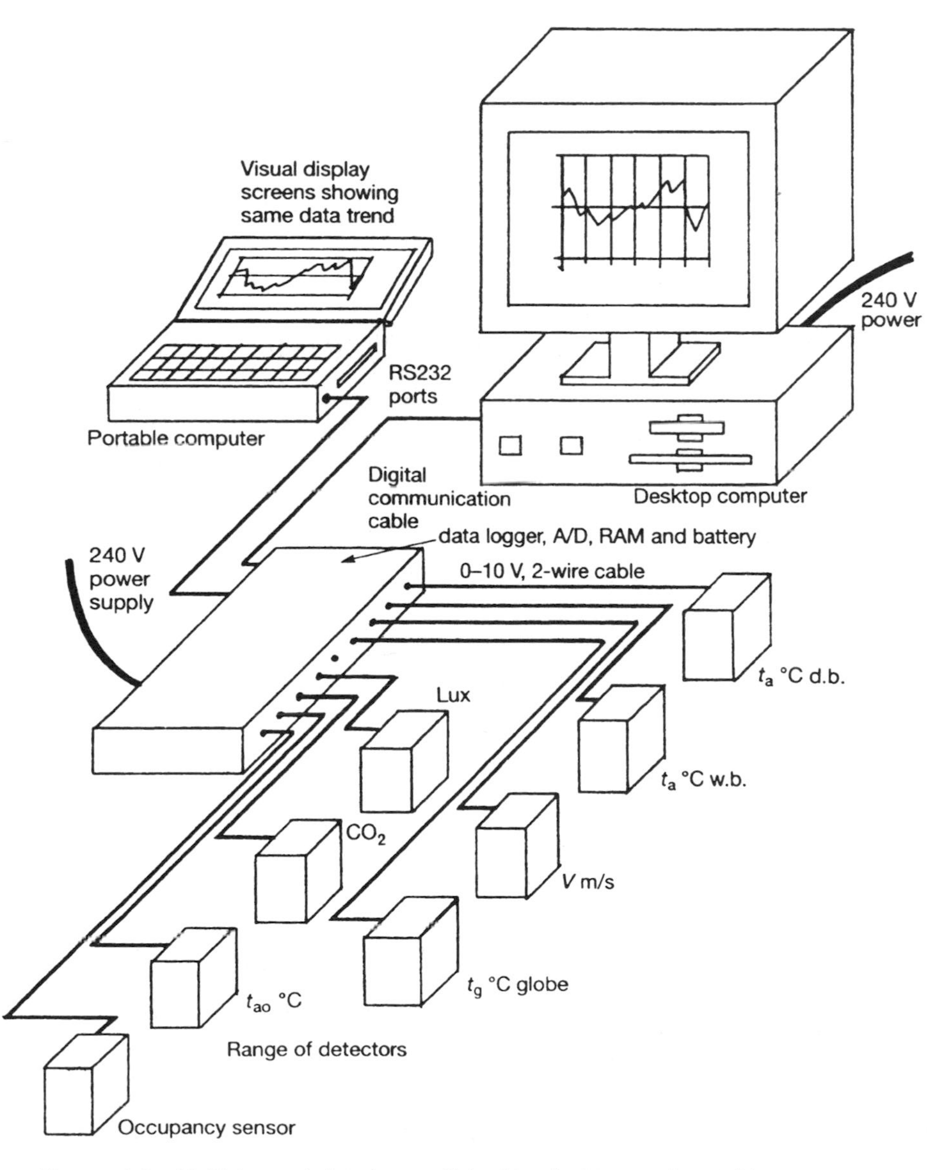

Figure 1.3 Multichannel data logger linked to desktop and portable computers

Research results indicate that nationality, geographic origin, location, age, sex and body build have no significant effect on basic comfort requirements. The achievement of acceptable comfort levels for the variety of activities that are conducted in offices, factories, schools, hospitals and homes can be very expensive in terms of building and services construction and energy costs. In practice, compromise solutions are made that satisfy most requirements for the majority of the occupants.

External environments

Extremes of external climate are mainly of concern to construction workers in severe environments. There are two main indices: wind chill index (WCI) and heat stress index (HSI). Wind chill index is given by

$$\mathrm{WCI} = (10.45 + 10\sqrt{v} - v)(33 - t_a)$$

and measures the cooling effect on the body of a moving airstream. The wind speed v is in metres per second, and the equation can be used for values of up to 22.0 m/s (79 km/h (kilometres per hour)). Frostbite should be avoided if the wind chill index is less than 1400 at an air temperature of −10.0 °C dry-bulb (d.b.) during a maximum exposure of 30 min. A convenient use of the wind chill index is to calculate the equivalent wind chill temperature (EWCT), which is the air temperature, calculated from measured data, that would provide the same chilling effect but at an air velocity of 1.78 m/s (6.4 km/h):

$$\mathrm{EWCT} = -(0.045\ \mathrm{WCI} + 33)\ ^\circ\mathrm{C}$$

EXAMPLE 1.5

In February on a site in Manchester the air velocity was measured as 5.5 m/s and the sling psychrometer air temperature showed −1 °C d.b. Find the wind chill index and the equivalent wind chill temperature.

$$v = 5.5\ \mathrm{m/s} \qquad t_a = -1\ ^\circ\mathrm{C}$$

$$\mathrm{WCI} = (10.45 + 10\sqrt{5.5} - 5.5)(33 - -1)$$

$$= 965.7$$

$$\mathrm{EWCT} = -(0.045 \times 965.7) + 34\ ^\circ\mathrm{C}$$

$$= -9.5\ ^\circ\mathrm{C}$$

Construction workers in hot climates are exposed to heat hazards, and severe cases will lead to an inevitable increase in body temperature. This is the zone of unacceptable body heating. In its attempt to dissipate the metabolic heat produced, which cannot be transferred to the high-temperature environment, the body raises its temperature to compensate. The symptoms are fatigue, headache, dizziness, vomiting, irritability, fainting and failure of normal blood circulation to cope with the problem. Heat exhaustion of this sort can normally be counteracted by removal to a cool place. Cramp may occur as a result of loss of some body salts. Salt tablets can be taken to redress the balance. Heatstroke occurs if the body temperature rises to 40.6 °C (normal body temperature is 36.9 °C), and in this condition sweating ceases, the body enters a comatose state, brain damage can occur and death is imminent.

The heat stress index is a measure of the maximum elevation of sweat rate, body temperature and heart rate that can be tolerated (Belding and Hatch, 1955):

$$\text{HSI} = \frac{E_{\text{req}}}{E_{\text{max}}} \times 100$$

where $E_{\text{req}} = M - Q_r - Q_c$ is the required evaporative cooling, $Q_r = 4.4(35 - t_r)$ is the heat lost from the clothed body by radiation, $Q_c = 4.6\,v^{0.60}(35 - t_a)$ is the heat lost from the clothed body by convection and $E_{\text{max}} = 7.6\,v^{0.60}(56 - p_a)$ is the maximum available evaporative cooling.

The numerical value of the heat stress index can be related to practical circumstances as shown in Table 1.5. The allowed exposure time (AET) is given by

$$\text{AET} = \frac{2440}{E_{\text{req}} - E_{\text{max}}} \text{ min}$$

When the heat stress index is less than 100.0, the exposure time is unrestricted. The heat stress index reduces and the allowed exposure time increases when a prevailing wind or fans produce extra cooling.

EXAMPLE 1.6

A construction worker is exposed to conditions of 36 °C d.b. and 20% saturation in the shade. His metabolic rate is 320 W/m^2 while planing timber. Local air movement is 0.30 m/s. Find the heat stress index and the allowed exposure time.

Assume that $t_r = t_a$. The atmospheric vapour pressure p_a is 12.48 mb.

Table 1.5 Evaluation of heat stress index

Heat stress index	Response to 8 h exposure
−20	Mild cold strain
0	Neutral
10–30	Mild strain
40–60	Unsuitable for mental effort
70–90	Selected personnel
100	Maximum possible for a fit acclimatized young man

$$E_{req} = 320 - 4.4(35 - 36) - 4.6 \times 0.30^{0.60}(35 - 36)$$
$$= 326.6\ \mathrm{W/m^2}$$
$$E_{max} = 7.6 \times 0.30^{0.60}(56 - 12.48)$$
$$= 160.6\ \mathrm{W/m^2}$$
$$\mathrm{HSI} = \frac{326.6}{160.6} \times 100$$
$$= 203.4$$
$$\mathrm{AET} = \frac{2440}{326.6 - 160.6}\ \mathrm{min}$$
$$= 14.7\ \mathrm{min}$$

By providing increased air movement around the carpenter, an elevation of v to 1 m/s would reduce the heat stress index to 100 and make exposure time unrestricted.

Environmental measurements

Measurement of air humidity is facilitated by using both dry- and wet-bulb mercury-in-glass thermometers in a sling psychrometer as shown in Fig. 1.4. Evaporation of water from the cotton wick cools the wet-bulb thermometer, and

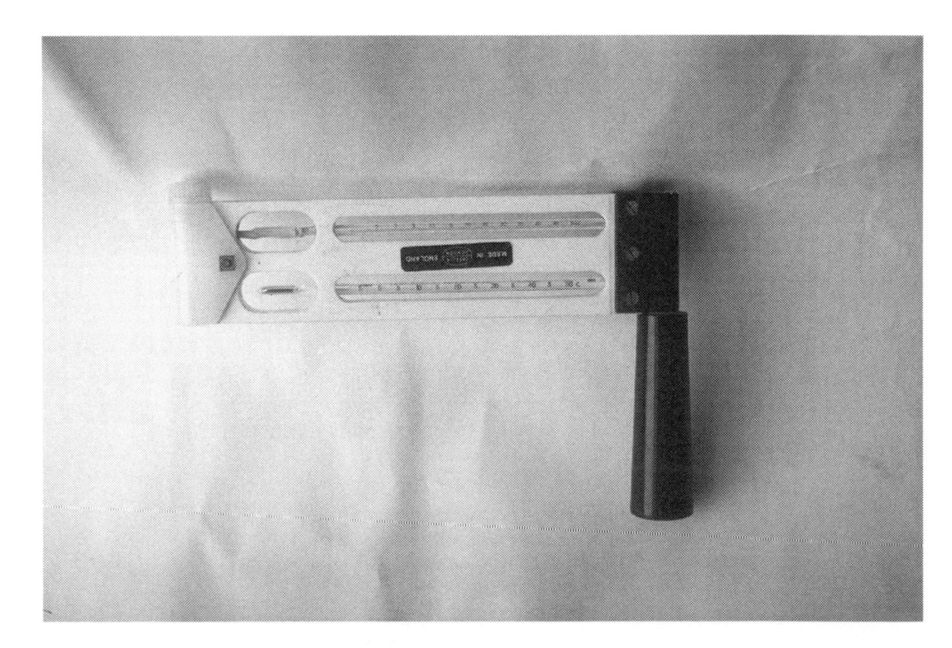

Figure 1.4 Sling psychrometer

its reading is known as the wet-bulb temperature t °C w.b. The difference between dry- and wet-bulb temperatures is the wet-bulb depression.

In order to find the mean radiant temperature t_r, the dry-bulb temperature t_a and the air velocity v m/s are measured, and the empirical relationship

$$t_r = t_g(1 + 2.35\sqrt{v}) - 2.35 t_a \sqrt{v}$$

where t_g is the globe temperature, is used. The globe temperature is measured at the centre of a blackened globe of diameter 150 mm suspended at the measurement location. Figure 1.5 shows the large and small globe thermometers used. The resultant temperature, found using a globe of diameter 100 mm, can be used, and t_r can be calculated by replacing 2.35 with 3.17 in the previous formula.

The Kata thermometer shown in Fig. 1.6 is used to measure the cooling power of room air movements and air velocity. The thermometer is heated by immersing the large bulb in hot water until the alcohol enters the upper reservoir. The bulb is dried, and the Kata thermometer is suspended at the location being investigated. The time taken for the meniscus to fall between the two marks on the stem is noted, the operation is repeated twice and the average time of cooling is found. A Kata factor, which is the amount of heat lost from the bulb surface during cooling (typically 475 mcal/cm^2 of bulb surface as the thermometer cools through 2.8 °C), is inscribed on the stem. The cooling power is defined as the Kata factor divided by the average cooling time in seconds. A cooling power of around 10 is suitable for sedentary occupation. Wet-bulb Kata readings can also be taken, and

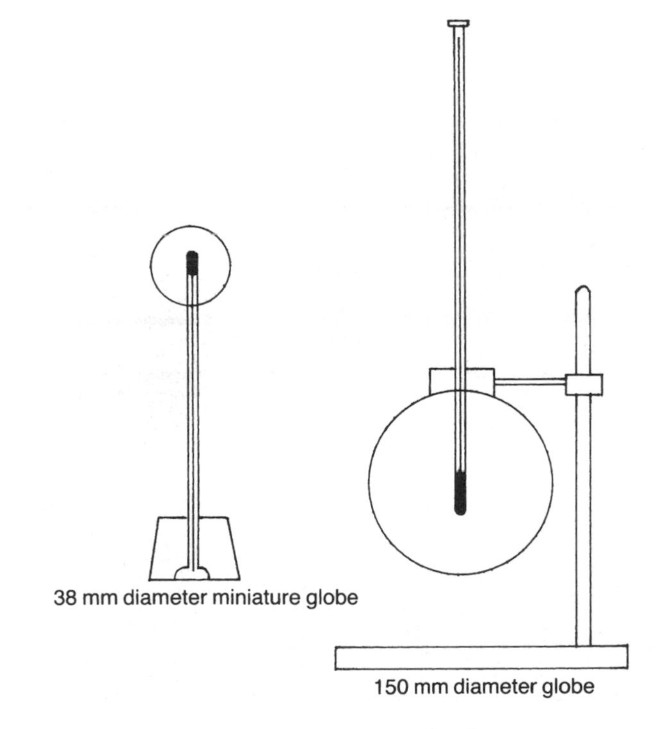

Figure 1.5 Large and small globe thermometers

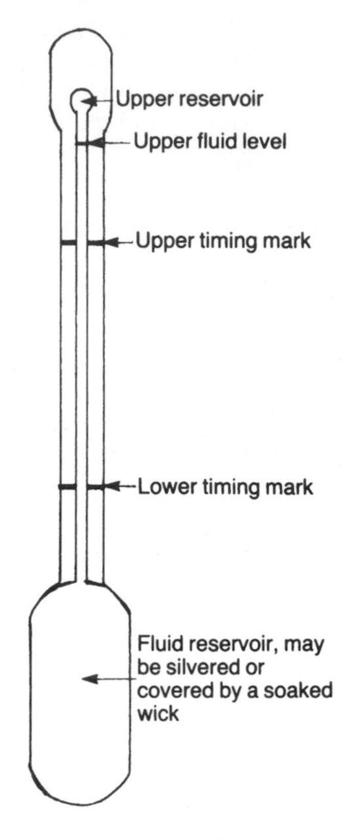

Figure 1.6 Kata thermometer

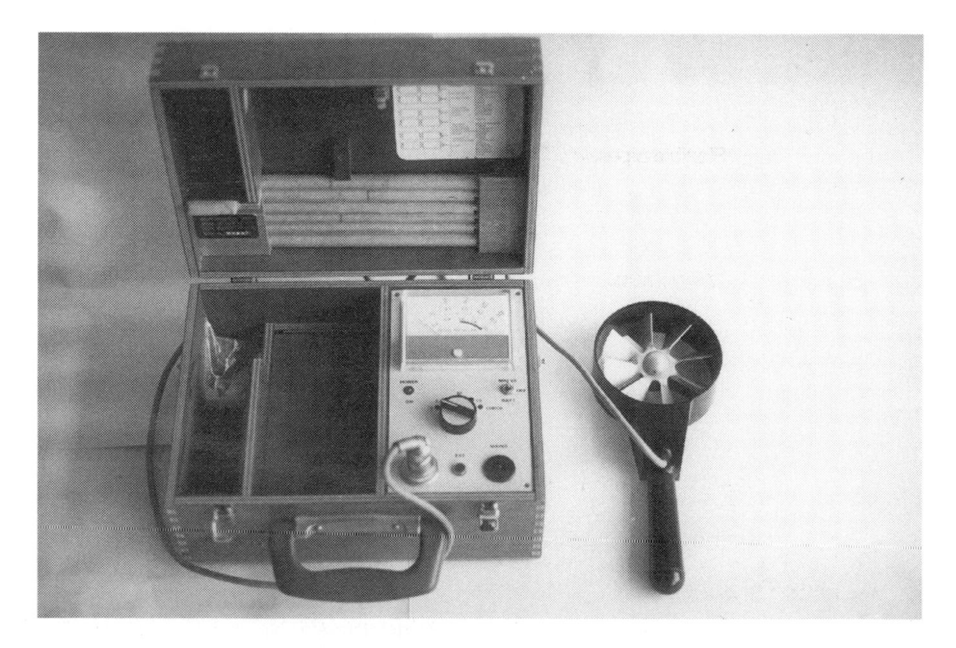

Figure 1.7 Vane anemometer

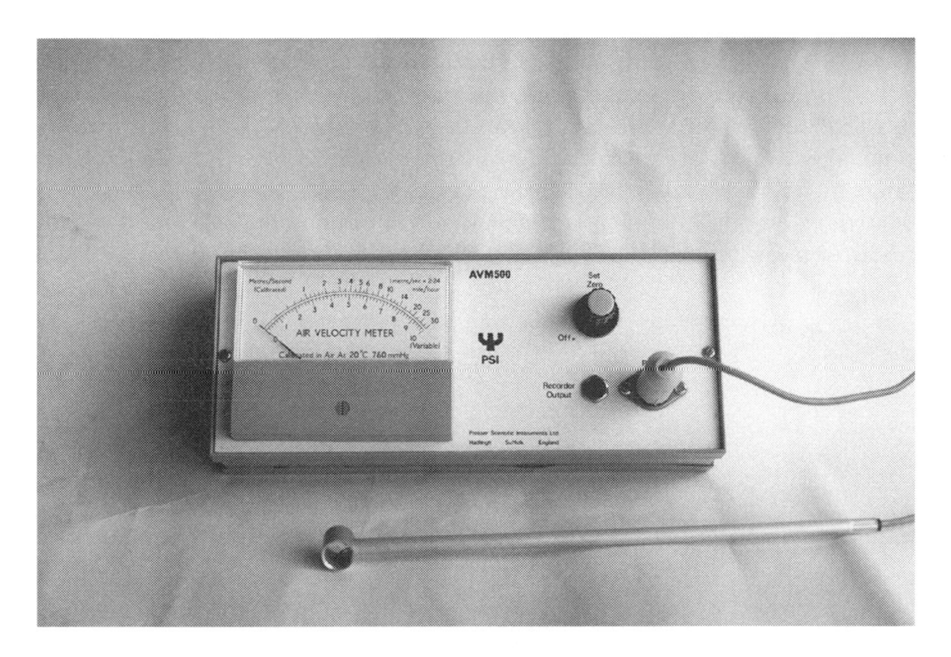

Figure 1.8 Thermistor anemometer

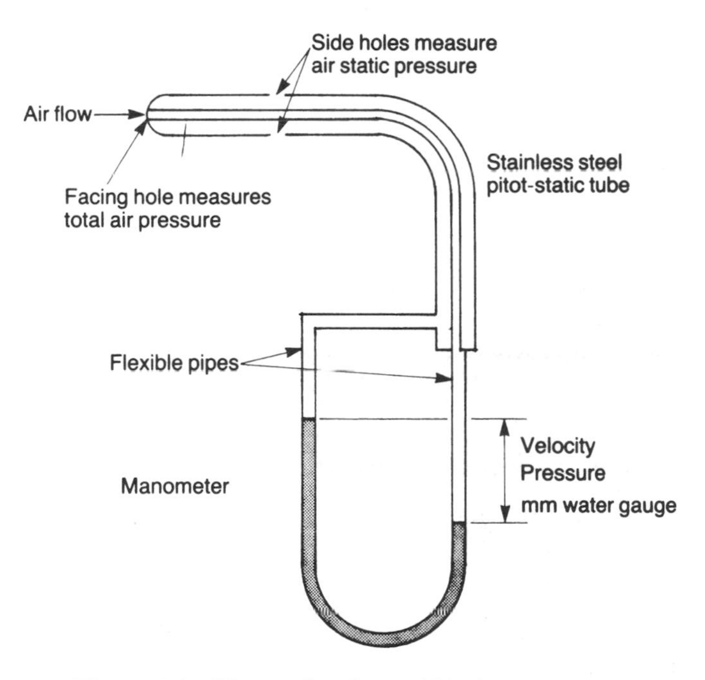

Figure 1.9 Pitot-static tube and U-tube manometer

these include humidity effects. A nomogram or formula can be used to find the air velocity at that location.

Direct-reading air velocity measuring instruments are shown in Figs 1.7–1.9. Vane anemometers are used to measure airstreams through ventilation grilles, where the rotational speed of the blades is magnetically counted. Thermistor and hot-wire anemometers utilize the airstream-cooling effect on the probe; the latter type is mainly used for research work. Duct air flow rates are found by inserting a pitot-static tube into the airway, taking up to 48 velocity readings at locations specified in BS 848: Part 1: 1980 and evaluating the air volume flow rate from the average air velocity found from all the readings.

The term for air humidity is percentage saturation, and the most reliable method of measurement is to take dry- and wet-bulb air temperature readings

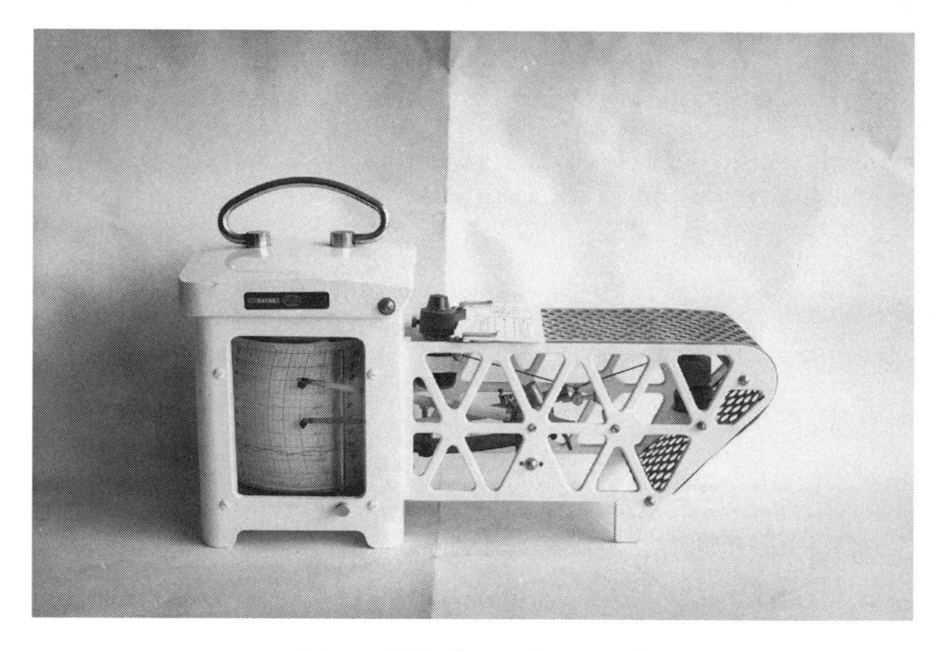

Figure 1.10 Thermohygrograph

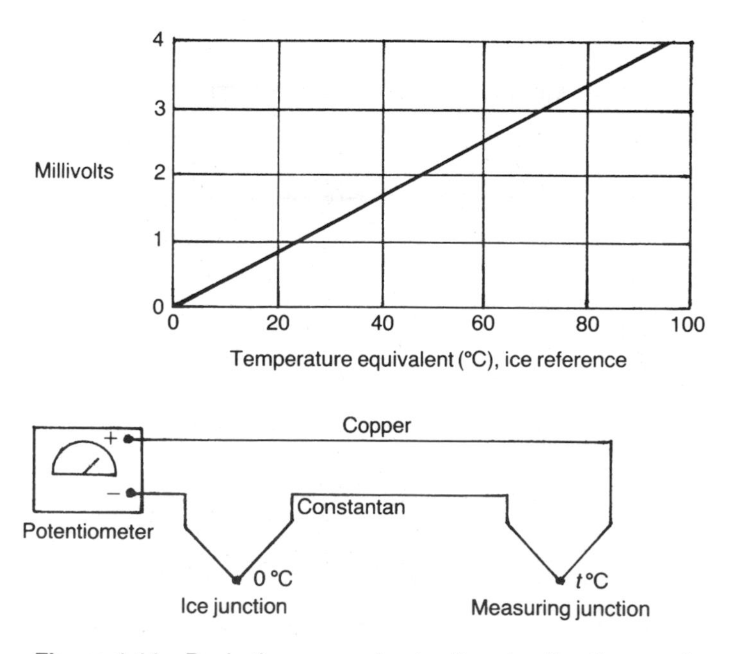

Figure 1.11 Basic thermocouple circuit and calibration graph

using a sling psychrometer and referring to a psychrometric chart (Chapter 5). Hair hygrometers can be used to display percentage saturation. A combined thermohygrograph can be used to monitor room conditions over a period of days, as shown in Fig. 1.10. Permanent monitoring and control requires an electronic sensor utilizing a hygroscopic salt covering a coil of wire. The output signal from this unit depends on the amount of moisture absorbed by the sensor.

Thermocouple temperature sensing enables supervision of large numbers of locations by manual or automatic scanning. Inaccessible areas can be reached by installing thermocouple wires during construction or by drilling. Figure 1.11 shows the basic construction of a copper–constantan thermocouple circuit. The small electrical voltages caused between the junctions of dissimilar metals at different temperatures produce an electron flow, and thus the potential difference between the wires can be measured and calibrated into temperature. Surface temperature measurements can be made using portable instruments to check the integrity of thermal insulation (Chapter 3) or in experimental investigations into the thermal transmittance of building components using multichannel recording systems. A two-channel thermocouple meter is shown in Fig 1.12 measuring the (negligible) temperature difference between two pipes.

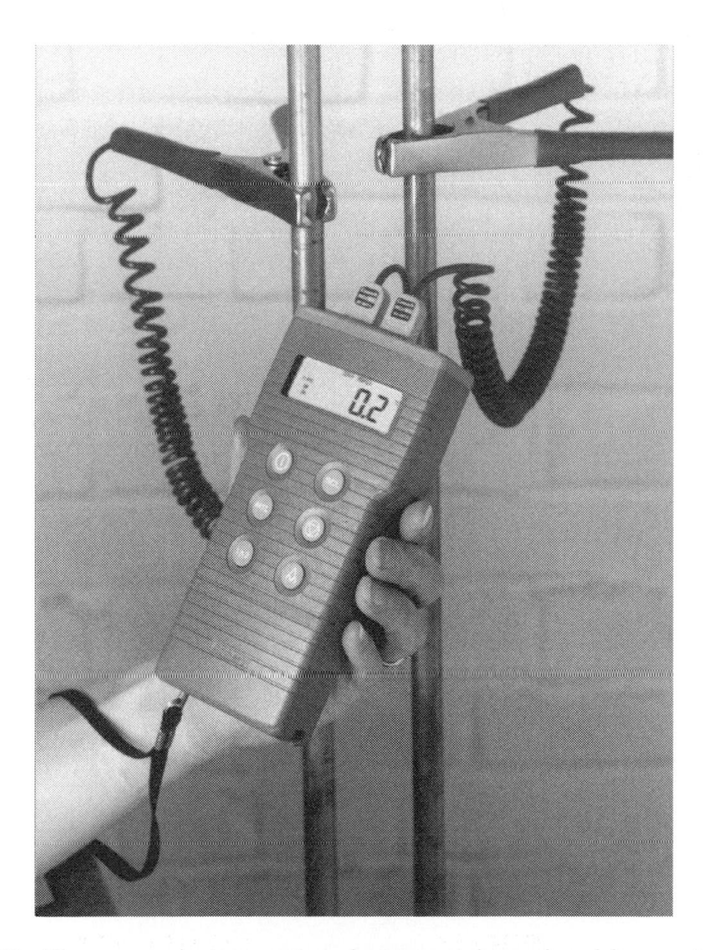

Figure 1.12 Thermocouple temperature instrument (reproduced by courtesy of KM Comark)

Figure 1.13 Hand-held infrared scanner

Non-touch temperature sensors receive infrared heat radiation emitted from all bodies that are above absolute zero, and they are calibrated to give the surface temperature. Small areas can be surveyed with portable instruments similar to that shown in Fig. 1.13, and large scans can be performed by mobile television camera equipment, which produces temperature contour maps. Ground-level or aerial surveys are used to detect energy waste from non-existent, inadequate or damaged thermal insulation in homes, in factories or where there are buried pipes in district heating systems.

Environmental temperature

Heat losses from buildings and the conditions required for thermal comfort both depend on the mean radiant temperature and the air temperature. The environmental temperature t_{ei} combines these two measures:

$$t_{ei} = 0.667\ t_r + 0.333\ t_{ai}$$

EXAMPLE 1.7

An hotel lounge is designed for sedentary occupation and is to have general air movement not exceeding 0.15 m/s and an air temperature of 24 °C d.b., in summer. It is expected that a globe temperature of 21 °C would be found at the centre of the room volume. What would be the mean radiant, resultant and environmental temperatures?

$$t_r = t_g(1 + 2.35\sqrt{v}) - 2.35t_a\sqrt{v}$$
$$= 21(1 + 2.35 \times \sqrt{0.15}) - 2.35 \times 24 \times \sqrt{0.15}$$
$$= 18.3\ °C$$

$$t_{res} = t_g(1 + 3.17\sqrt{v}) - 3.17t_a\sqrt{v}$$
$$= 21(1 + 3.17 \times \sqrt{0.15}) - 3.17 \times 24 \times \sqrt{0.15}$$
$$= 17.3\ °C$$

$$t_{ei} = 0.667t_r + 0.333t_{ai}$$
$$= 0.667 \times 18.3 + 0.333 \times 24$$
$$= 20.2\ °C$$

EXAMPLE 1.8

The room in Example 1.4 is now not to have a hot-water radiator but is to be heated by a ducted warm-air system. Calculate the new mean radiant temperature and the required air temperature if the environmental temperature is to remain at 20 °C.

In the example, 4 m^2 of outside wall at a surface temperature of 15 °C replaces the radiator. Thus

$$\sum A = 97\ m^2$$

$$\sum(At_s) = 1656 \text{ (with the warm-air system)}$$

$$t_r = \frac{1656}{97} = 17.1\ °C$$

$$t_{ei} = 0.667t_r + 0.333t_a$$

$$t_{ei} - 0.667t_r = 0.333t_a$$

Thus

$$t_a = \frac{1}{0.333}(t_{ei} - 0.667t_r)$$
$$= 3(20 - 0.667 \times 17.1)$$
$$= 25.8\ °C$$

A similar calculation for the room with the radiator shows that an air temperature of 21.8 °C would provide the required environmental temperature.

Dry resultant temperature

The dry resultant temperature t_{res} is the temperature recorded by a thermometer at the centre of a blackened globe of diameter 100 mm. This index is now recommended for use as the temperature to be specified for comfort conditions in the internal environment. Comfortable values range from 21 °C for a residential living room through 18 °C for a bedroom or lecture room to 16 °C in passageways. It is related to the mean radiant temperature, the air temperature and the air velocity v m/s by

$$t_{res} = \frac{t_r + t_{ai} \sqrt{(10v)}}{1 + \sqrt{(10v)}}$$

In normally occupied rooms the air temperature and the mean radiant temperature should be within a few degrees of each other. The air velocity within a habitable space should be barely discernible, in the region of 0.1 m/s. Under these conditions the dry resultant temperature at the centre of the room is

$$t_{res} = t_c = 0.5t_{ai} + 0.5t_m$$

There should be no significant effect upon comfort conditions when t_c is within 1.5 °C of its design value; this allows for some flexibility in the actual value of the mean radiant and air temperatures and the air velocity due to the occupant's changing position within the room or to weather variations.

Comfort criteria

The main comfort criteria for sedentary occupants in buildings in climates similar to that of the British Isles are as follows.

1. The dry resultant temperature should be in the range 19–23 °C depending on room use (CIBSE, 1986).
2. A feeling of freshness is produced when the mean radiant temperature is slightly above air temperature. A significant amount of radiant heating is needed in order to achieve this.
3. The air temperature and the mean radiant temperature should be approximately the same. Large differences cause either radiant overheating or excessive heat loss from the body to the environment, as would be experienced during occupation of a glasshouse through seasonal variations.
4. Percentage saturation should be in the range 40–70%.
5. Maximum air velocity at the neck should be 0.1 m/s for a moving-air temperature of 20 °C d.b. Both hot and cold draughts are to be avoided. Data are available for other temperature and velocity combinations.
6. Variable air velocity and direction are preferable to unchanging values of these variables. This is achieved by changes in natural ventilation from prevailing wind, movement of people around the building, on–off or high–low thermostatic operation of fan-assisted heaters or variable-volume air-conditioning systems.

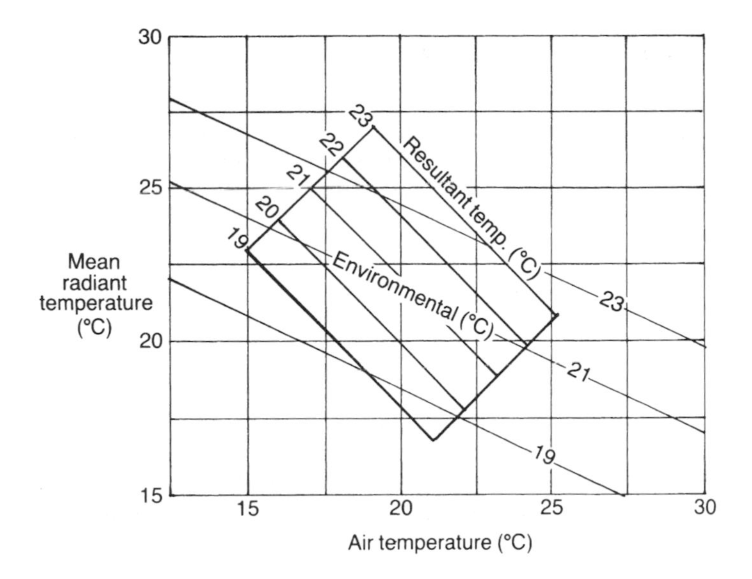

Figure 1.14 Sedentary comfort zone

7. The minimum quantity of fresh air for room use that will remove probable contamination from smoking, for example, is 10.4 l/s per person.
8. Mechanical ventilation systems should provide at least four air changes per hour to avoid stagnant pockets and ensure good circulation.
9. Incoming fresh air can be filtered to maintain a clean dust-free internal environment.
10. The difference between room air temperatures at head and foot levels should be no more than 1 °C.
11. Ventilation air quantity can be determined by some other controlling parameter: for example, removal of smoke, fumes or dust, solar or other heat gains and dilution of noxious fumes.

Figure 1.14 shows a comfort zone, and it can be seen that the data for the previous example produce an acceptable office environment.

Experimental work

Various measurements of the thermal environment can be made using relatively simple equipment, and some suggestions for practical projects are listed below. It is most important to take great care with observation of the conditions being measured. Record all the relevant details: temperatures, air speeds, solar radiation, number of people, clothing, activity level and, most important of all, where the measurements are being taken. Do not take all the readings near a draughty window and then claim that the results show average room conditions.

1. Effects of room air temperature gradients on comfort: measure floor-to-ceiling air temperature gradients with a multichannel thermocouple probe and discuss their effects on the comfort conditions encountered.

2. Room air velocity distribution: produce velocity contours for different parts of the room and compare with observed comfort conditions.
3. Measure the Kata cooling power and air velocity, the globe temperature and the dry-bulb air temperature. Compute the other temperature scales and compare with subjective assessments of the conditions.
4. Make and test a thermocouple temperature-measuring circuit with copper–constantan wire and an accurate potentiometer. Build junctions into walls and building structures, plot their temperature profiles and calculate their thermal transmittances.

Questions

Descriptive answers must be made in your own words based on a thorough understanding of the subject. Copied work displays a noticeable discontinuity of style between your own production and the reference material. Also, the work has not been fully comprehended if it can only be answered by copying. The ability to pass on reliable and concise information is a vitally important part of business and government work, and you should realize that as much practice and experience as possible is needed to become an effective communicator.

1. List and discuss the factors affecting thermal comfort.
2. Discuss the following statement: 'a building acts as an environmental filter'.
3. State how extremes of heat and cold affect the workers on a site, what environmental measurements can be taken, and the corrective actions possible to ensure safe and healthy working conditions.
4. Describe how an investigation into the thermal comfort provided in a building can be measured. State which instruments would be used and give reasons.
5. Describe with the aid of sketches how each of the following instruments functions: dry-bulb thermometer, wet-bulb thermometer, globe thermometer, Kata thermometer, vane anemometer, thermocouple, and infrared scanner.
6. Describe how heating, cooling and humidity control systems interact with the building to provide a comfortable environment.
7. A survey of a heated room revealed the information given in Table 1.6. The dry-bulb air temperature is 19 °C. Estimate the mean radiant and environmental temperatures and assess the thermal comfort conditions for sedentary occupation. State any remedial measures that may need to be taken.

Table 1.6

Surface	Average surface temperature (°C)	Area (m^2)
Window	0	7
Outside	10	20
Inner wall	17	28
Ceiling	23	30
Floor	12	30

8. Compare the wind chill index and the equivalent wind chill temperature on two sites operated by workers wearing similar clothes.

   Site A: air velocity, 5 m/s; air temperature, 2.0 °C d.b.

   Site B: air velocity, 0.6 m/s; air temperature, −10 °C d.b.

9. A scaffold erector works in air at −5 °C d.b. and 20% saturation with a metabolic rate of 300 W/m^2. The wind velocity gusts up to 2.5 m/s. Given that the atmospheric vapour pressure is 0.8 mb, calculate the heat stress index produced. Assume that $t_r = t_a$.

10. Site work is being conducted in bright sunshine when the air peaks at 35 °C d.b. and 20% saturation at 3 pm. A worker's metabolic rate is 280 W/m^2. Atmospheric vapour pressure is 12 mb and the air speed is 0.8 m/s. Calculate the heat stress index produced and comment on the conditions. Assume that $t_r = t_a$.

11. Measurements in an office showed that the general air movement amounted to 0.25 m/s at a temperature of 23 °C d.b. The globe temperature was 19 °C. Calculate the mean radiant and environmental temperatures, and discover whether room conditions are within the comfort zone.

12. State two uses for the Kata thermometer.

13. Where are thermocouples used? What are their advantages?

14. How can heat leakages due to inadequate thermal insulation and damaged pipes or cables be detected?

15. What is the function of environmental temperature?

16. A high-temperature gas-fired radiant heater is used to provide warmth for site workers. The heater has a red-hot area of 300 mm × 500 mm at a temperature of 700 °C.

    A worker of surface area 1 m^2 and temperature 12.0 °C is 2.5 m from the heater. Surface emissivities are 1.0 for the heater and 0.9 for the worker. Calculate the radiant heat transfer to the worker.

17. State the factors that are taken into account when designing for the provision of ventilation with outdoor air.

18. Write a technical report on the aspects of the provision of outdoor air for the ventilation of the following:

    (a) a commercial building that has offices, retail shops and a pedestrian atrium;
    (b) an underground high-security manufacturing and storage facility for nuclear materials;
    (c) a large open-plan metal-fabrication factory;
    (d) a college or university;
    (e) an hotel;
    (f) a sports centre that has swimming, weight training, aerobics, racquet courts and restaurant facilities.

19. Explain the meaning of the terms olf and decipol and state how they are used in the design of ventilation systems.

20. List the atmospheric pollutants that are likely to be present within normally occupied buildings. Identify those pollutants that are used for the design of the ventilation system, the filtration equipment, acoustic insulation and general maintenance during occupation.

21. A ten-storey office building is located in Birmingham city centre alongside a highway that has continual, heavy traffic density. The building has 30 m of road frontage, is 20 m deep and each floor is 3 m high. The occupancy averages one person for each 10 m^2 of floor area. Calculate the outdoor air ventilation rate, from the Fanger method, that is required to satisfy 85% of the occupants of an office where it is expected that none of them are smokers. State the recommended location for the fresh air intake and discharge louvres. The office occupants can be taken as mainly sedentary. The interior has a low-pollution load of around 0.20 olf/m^2.

22. An open-plan office is being designed for a city centre development. Each floor is 50 m long, 25 m wide and 3.5 m high and is to have 100 workstations for the occupants. Use the data from Table 1.2 to evaluate the outdoor ventilation air requirement from the different authorities for this application for one floor. What is the maximum room air change rate that could be produced?

23. A new lecture theatre is being designed for the Bournemouth City University. The theatre is to be 25 m long, 12 m wide and 5 m high. It has no exterior windows. The peak occupancy can be 125 and can be considered to be sedentary. There will not be any smoking, the building has a low pollution load of 0.1 olf/m^2 and each occupant creates a load of 1 olf. The external atmosphere is considered to have a low-pollution count. Calculate the outdoor air ventilation rate, from the Fanger method, that will be required so that 90% of the occupants will be satisfied. Recommend how the outdoor air ventilation system could be controlled for energy economy and comfort.

24. Discuss the difference between the Fanger method of assessing the provision of outdoor air with that based on the supply of a fixed quantity, such as

10.4 l/s. State the implications for the energy used in heating and cooling the supply of outdoor air for both systems.

25. Explain, with the aid of sketches, how the ventilation design can be arranged to minimize the use of outdoor air in the provision of acceptable indoor air quality and temperature control for human thermal comfort. The reader can make use of Chapter 5 in answering this question.

2 Energy economics

Learning objectives

Study of this chapter will enable the reader to:

1. understand the basis of energy auditing and design an energy audit;
2. be able to calculate energy costs for all applications and different types of measuring unit within SI (Système International);
3. use multiples and submultiples of SI units;
4. identify usefully employed energy;
5. evaluate energy costs in pounds per useful gigajoule;
6. understand degree days and their use;
7. compare the energy efficiency of different buildings;
8. calculate the economic thickness of thermal insulation for both hot surfaces and building structures;
9. analyse financial return on investment in energy-saving measures;
10. calculate energy-use targets for buildings.

Key terms and concepts

Introduction

Buildings are such major consumers of primary energy, i.e. coal, oil, gas, nuclear energy and renewable sources such as hydroelectric, wind, solar and wave energy, that accounting for its use and calculating the consequent financial implications are of paramount importance to all who are involved in building design and operation.

Logical methods of dealing with the calculations are introduced to enable the new user to cope with the complex conversion equations and calculation of energy costs per standard unit, the annual energy cost and the economic thickness of thermal insulation.

Degree days are explained and their use as an accounting tool is explored.

The financial implications of purchasing or leasing energy saving hardware equipment are investigated.

Energy demand targeting is outlined.

Energy audit

Management of the energy that is used for buildings has three major components:

1. initial design;
2. retrofitting energy-saving measures;
3. maintenance practices.

Design engineers should provide heating, ventilating and air-conditioning systems that consume the minimum amount of fossil fuel energy in satisfying the needs of the site. The initial installation is designed in conjunction with the architecture and the client's requirements, in accordance with statutory legislation and in compliance with the standards of good engineering practice. The design includes the means of controlling the use of energy. Control can mean anything from switching building services plant on and off manually, up to a fully automated computer-based system that gives audible and visual alarms when something goes wrong. The best efforts of the design engineer are limited by the initial construction cost of the new installation, which is usually minimized. The installation of energy-saving systems often increase this cost. Some buildings are designed to be low-energy users. Most building services engineering systems are designed to provide thermal comfort for the conditions that are found in the building. For example, perimeter heating to overcome the heat loss through large areas of glazing and thermostatically controlled ventilation louvres in a naturally ventilated building in a cool climate.

Energy-saving measures that are installed after the first few years of use of a new building, or during a major upgrade, can be justified for two primary reasons. Either the owner of the building has decided to refurbish it, and has found the capital funding that is needed for all the work, or the operating cost of the site is significantly greater than comparable facilities, and the owner or tenant is prepared to invest in measures that will reduce annual outgoings. The owner may be forced to provide lettable office space that has competitive energy and maintenance costs. A building that has a labour-intensive maintenance and supervision workload, from steam boilers, manual switching of mechanical plant and lighting systems, unreliable water chillers, poorly maintained closed-circuit water conditions and highly stressed belt drives on fans, is not attractive to a new user.

Many sites have maintenance practices that encourage the provision of breakdown repairs and replacements, rather than by preventing breakdowns through good-quality methods. The financial controller of the business may view the annual maintenance budget as expendable, through a lack of understanding about engineering equipment. This is understandable. The maintenance engineer has only to ask the finance director whether it is preferable for a company car to be taken for regular servicing or to wait for the car to break down on a motorway during inclement weather, because the engine has run out of oil, the engine cooling system has boiled dry and the brake pads have worn down to the metal! This is how the maintenance budget of some sites is managed, that is, breakdown maintenance only. Many building services are critical to the life safety of the users. These life safety systems are not just the emergency exit lighting, smoke spill ventilation fans, stairway air pressurization fans and electrical earth leakage circuit breakers, but also include the air conditioning to hospital operating theatres, lifts, outside air ventilation dampers, domestic hot water and cooling tower bacteria controls. Proficient maintenance practice helps to prevent breakdowns by:

1. monitor the condition of plant;
2. optimizing the maintenance activity to replace items only when they are needed;
3. keeping the maintenance team well motivated;
4. planning expenditure;
5. comprehensive maintenance record keeping;
6. enabling a quick response to problems, such as the failure of a fan motor, before the tenants complain of experiencing poor quality air conditions. (The building maintenance manager usually has about half an hour from when an air-conditioning fan ceases to function to when the tenants complain on the telephone. If the plant failure has been monitored through the building management system computer with audible and visual alarms, and an automatically sent message to the engineer's pager or mobile phone, the corrective response can be made within 5 minutes and the tenants provided with a briefing.)

The energy audit engineer assesses the practical and financial viability of energy-saving measures for each site, as is appropriate. The purpose of the energy-saving analysis is to identify suitable investments in capital equipment that will reduce the use of energy and labour, so that the savings will provide a payback on the investment in a reasonable period. This period will vary from 1 year, for those only interested in this year's profits, to 3 years for those who rely upon their bank for capital funding, to 5 years for those who can source capital funds from an equity performance contracting partner, to the longer terms of 10 to 25 years when the user is a government department and is to retain ownership of the public buildings indefinitely. The retrofit energy-saving measures that are usually considered include the following:

1. thermal insulation of the building;
2. solar shading;
3. changing the fuel source for heating and cooling;
4. heat pumps;
5. heat reclaim;
6. cogeneration of electricity with heating or cooling;
7. computer-based building management system;
8. digital control refrigerant circuit of the water chiller;
9. hot-water, chilled-water or ice thermal storage;
10. load shedding large electrical loads at critical times for short periods;
11. energy tariff change;
12. reducing the lighting system power usage;
13. variable speed drives of fan and pump motors;
14. reducing the usage of water by taps and in toilets;
15. economy air recycling ductwork and motorized damper controls;
16. air-to-air heat exchange between exhaust and incoming outside air ducts;

17. occupancy-sensing with infrared, acoustic or carbon dioxide sensing to control lighting and the supply of outside air;
18. air curtains at doorways;
19. oxygen sensing in the boiler flue gas to modulate the combustion air supplied to the burner;
20. replacement of old inefficient boilers and heating systems;
21. distribution of domestic hot water at 45 °C with a mixing valve and temperature control;
22. replacement of steam-to-water heat exchangers and calorifiers with local gas-fired heating and domestic hot-water systems;
23. thermal insulation of heating, cooling and steam pipework and heat exchangers;
24. recovery of the maximum quantity of condensate in a steam distribution system;
25. replacement and overhaul of steam traps and condensate pumping.

An energy audit of an existing building or a new development is carried out in a similar manner to a financial audit but it is not only money that is accounted. All energy use is monitored and regular statements are prepared showing final uses, costs and energy quantities consumed per unit of production or per square metre of floor area as appropriate. Weather data are used to assess the performance of heating systems. Monthly intervals between audits are most practical for building use, and in addition an annual statement can be incorporated into a company's accounts.

Payne (1978) and the booklets on fuel and energy efficiency published by the Department of Energy (1983, 1988) are useful further reading. An initial energy audit has certain basic aims. To:

1. establish total costs of energy purchased;
2. locate the principal energy-consuming areas;
3. notice any obvious losses or inefficient uses of heat, fuel and electricity;
4. take overall data to gain initial results quickly, which can be refined later and broken down into greater detail;
5. find where additional metering is needed;
6. take priority action to correct wastage;
7. survey buildings and plant use at night and weekends as well as during normal working hours;
8. initiate formal records monitored by the energy manager;
9. compare all energy used on a common basis (kilowatt-hours, therms or megajoules);
10. list energy inputs and outputs to particular buildings or departments.

A vital part of auditing is enlisting the cooperation of all employee groups, and explaining the problem not just in financial terms but also in quantities of energy. A joint effort by all staff is needed. Posters, stickers and prizes for ideas can be used to stimulate interest.

An overall energy audit will list each fuel, the annual quantity used and the cost for the year, including standing charges and maintenance; then a comparison is made with other fuels by converting to a common unit of measurement.

Energy use performance factors (EUPFs) enable comparisons to be made between similar buildings or items of equipment. These can be litres of heating oil per degree day, kilowatt-hours of electricity consumption per square metre of floor area, megajoules of energy per person per hour of building use, or other accounting ratios as appropriate. For example, car manufacturers may analyse energy used per car. As experience is gained in auditing a particular building, data can be refined to monthly energy use in conjunction with degree day figures for this period.

This detailed analysis can be made for each building or department of a large site, each large room or factory area, each type of heating, air-conditioning or power-using system, each industrial process and each item of plant. The most serious deficiency in the acquisition of data is likely to be the lack of sufficient metering stations. Electricity, gas and other fuels are metered by the supply authority at the point of entry to the building or site; further metering is the responsibility of the site user. Frequently, no further meters are installed and capital expenditure is needed to obtain data. A careful cost–benefit approach is required to assess the viability of this equipment (Moss, 1997).

Unity brackets

In order to deal with the numbers involved, a degree of familiarity with the units and conversion between the various types is needed. A handling technique known as the unity bracket helps to avoid errors being made when dealing with unfamiliar combinations of units. Suppose that we wish to convert 1260 mm into metres.

$$\text{length} = 1260 \text{ mm}$$

Now, $1000 \text{ mm} \equiv 1 \text{ m}$. Divide each side by 1 m: thus

$$\frac{1000 \text{ mm}}{1 \text{ m}} \equiv 1$$

The left-hand side is now a unity bracket exactly equal to 1 (or unity). Similar unity brackets can be formed for any suitable conversion problem. Now, multiply length by the unity bracket:

$$\text{length} = 1260 \text{ mm} \times \frac{1 \text{ m}}{1000 \text{ mm}}$$

$$= 1.26 \text{ m}$$

Notice that the unity bracket is arranged so that its denominator units cancel the original units. A long chain of conversions can easily be handled and the method avoids errors of logic that can occur if an attempt is made to cope with the problem using mental arithmetic.

EXAMPLE 2.1

British Gas sold gas by the therm up to 1992. Harmonization of the units of measurement in Europe caused the change to kWh units. If 1 therm is equal to 105.5 MJ, how many kilowatt-hours are there in 1 therm?

$$1 \text{ therm} = 105.5 \text{ MJ}$$

$$1 \text{ MJ} = 10^6 \text{ J}$$

$$1 \text{ h} = 3600 \text{ s}$$

$$1 \text{ watt} = 1 \text{ J/s}$$

$$1 \text{ kWh} = 1 \text{ kWh} \times \frac{3600 \text{ s}}{1 \text{ h}} \times \frac{10^3 \text{ W}}{1 \text{ kW}} \times \frac{1 \text{ J}}{1 \text{ Ws}} \times \frac{1 \text{ MJ}}{10^6 \text{ J}} \times \frac{1 \text{ therm}}{105.5 \text{ MJ}}$$

$$= \frac{3600 \times 10^3}{10^6 \times 105.5} \text{ therms}$$

$$= 0.0341 \text{ therms}$$

Therefore

$$1 \text{ therm} = \frac{1}{0.0341} \text{ kWh}$$

$$= 29.3056 \text{ kWh}$$

Gross calorific value of a fuel

The total heat energy content of a fuel is known as the gross calorific value (GCV) and is usually expressed in megajoules per kilogram (MJ/kg).

> **EXAMPLE 2.2**
>
> Domestic central heating oil of Redwood no. 1 viscosity has a specific gravity of 0.83 and a GCV of 45.5 MJ/kg. Find its heat content in kWh per litre.

$$\text{GCV} = 45.5 \frac{\text{MJ}}{\text{kg}} \times \frac{0.83 \text{ kg}}{1 \text{ l}} \times \frac{10^3 \text{ kJ}}{1 \text{ MJ}} \times \frac{1 \text{ kWs}}{1 \text{ kJ}} \times \frac{1 \text{ h}}{3600 \text{ s}}$$

$$= \frac{45.5 \times 0.83 \times 10^3}{3600} \text{ kWh/l}$$

$$= 10.4903 \text{ kWh/l}$$

EXAMPLE 2.3

A factory uses 36000 l of oil of GCV 43.0 MJ/kg and specific gravity 0.85 costing £4700 and 10^6 kWh of electricity costing 6.2 p per unit. The fixed charge for the electrical installation is £950 and the servicing cost for the central heating system is £1800. The period of use being considered is 1 year. Draw up an overall energy audit for the year.

$$\text{oil GCV} = 43.0\ \frac{\text{MJ}}{\text{kg}} \times \frac{0.85\ \text{kg}}{1\ \text{l}} \times \frac{10^3\ \text{kJ}}{1\ \text{MJ}} \times \frac{1\ \text{kWs}}{1\ \text{kJ}} \times \frac{1\ \text{h}}{3600\ \text{s}}$$

$$= 10.153\ \text{kWh/l}$$

$$\text{total cost of oil} = £4700 + £1800$$

$$= £6500$$

$$\text{kWh of oil} = 36\ 000\ \text{l} \times 10.153\ \frac{\text{kWh}}{\text{l}}$$

$$= 365\ 508\ \text{kWh}$$

$$\text{cost per kWh for oil} = \frac{£6500}{365\ 508} \times \frac{100\ \text{p}}{£1}$$

$$= 1.78\ \text{p/kWh}$$

$$1\ \text{kWh} = 1\ \text{unit of electricity}$$

$$\text{total cost of electricity} = 10^6\ \text{kWh} \times \frac{6.2\ \text{p}}{\text{kWh}} \times \frac{£1}{100\ \text{p}} + £950$$

$$= £62\ 950$$

The data are shown in Table 2.1

$$\text{Average cost of all energy used} = \frac{£69\ 450}{1\ 365\ 508\ \text{kWh}} \times \frac{100\ \text{p}}{£1}$$

$$= 5.09\ \text{p/kWh}$$

Table 2.1 Fuel cost data for Example 2.3

Fuel	Quantity	Total cost	kWh	Cost per kWh
Oil	36 000 l	£6 500	365 508	1.78
Electricity	10^6 kWh	£62 950	1 000 000	6.30
Total		£69 450	1 365 508	

Energy cost per useful gigajoule

The supplying authority or company quotes energy costs in the unitary system most convenient to their industry, and there is no obvious method of comparing the real cost of providing a specific amount of useful heat or power in the building. A decision can be made to reduce all costs to a common base unit and this may be the therm, the kilowatt-hour or the gigajoule, where

$$1\ \mathrm{GJ} = 10^9\ \mathrm{J} = 10^6\ \mathrm{kJ} = 10^3\ \mathrm{MJ}$$

If the overall efficiency of the energy conversion process is included in the unit cost, then the incurred cost of using that system of heat or power can be realistically assessed. The cost per useful kWh is the cost of providing 1 kWh of useful heat, or energy, at the place of use. The overall efficiency of a central heating system will include the following.

1. Combustion efficiency of the fuel. Regular maintenance is necessary with all fuel-burning appliances to ensure that the correct fuel-to-air ratio is maintained.
2. Heat transfer efficiency of the appliance. Both flue gas and water-side surfaces must be kept clean.
3. Heat losses from distribution pipework. All hot-water pipes and surfaces must be adequately insulated unless they provide a useful heating surface in rooms. It is not good practice to allow heating system pipes to be bare metal as the uncontrolled heat transfer will lead to high fuel costs.
4. Ability of the final heat emitter to transfer warmth to the occupants. A hot-water central heating radiator placed under a window-sill should counteract down-draughts and provide a reasonably adequate air temperature at the window and at the inner surface of the outside wall; this has the effect of increasing heat flows through the window and wall. Placing the radiator on a warm internal wall will improve its useful heat output but at the loss of some warm usable space in the room as the window region will be colder.
5. Thermal storage capacity of the heating system and building. Large amounts of heat are stored in the water in heating systems and the dense fabric of buildings. An insensitive automatic control system, or the lack of such a system, will lead to wild swings above and below the desired resultant temperature and cause excessive fuel use.

The estimated overall system efficiencies are listed in Table 2.2 from Uglow (1981).

The cost in use of a fuel or source of energy, UC, can be calculated from the basic price C and the overall efficiency η. For gas,

$$\mathrm{UC} = C\,\frac{\mathrm{p}}{\mathrm{kWh}} \times \frac{100}{\eta} + \frac{\text{annual standing charge}}{\text{annual kWh}}$$

Thus once the total kWh used during the year are assessed, the annual standing charge can be apportioned to each kWh. This is not necessary if the standing charge has already been incurred by another use, for example cooking and water heating, and UC is being evaluated for an additional heating system.

Table 2.2 Overall efficiencies of heating systems

Fuel	Appliances	Overall efficiency η (%)
Electricity	Individual appliances	100
	Storage radiators	90
	Storage warm air	90
	Storage underfloor	90
Gas	Individual appliances	55
	Boiler and hot-water radiators	65–70
	Ducted warm air	70–75
Solid fuel	Open grate fire	35
	Closed stove	60
	Boiler and hot-water radiators	60
Oil	Boiler and hot-water radiators	65–70
	Ducted warm air	70–75

The higher figures relate to intermittent heating system operation.

EXAMPLE 2.4

A gas-fired central heating and hot-water system is to be installed in a residential property. The gas tariff is 1.2 p/kWh plus a standing charge of 8 p per day. The estimated annual heat energy that will be used by the occupants is 120 000 kWh. The annual maintenance works cost £75. Find the total energy bill and the average cost per kWh, for the year.

From Table 2.2 the overall efficiency of the heating system, η, is 70%.

$$\text{energy usefully consumed} = 120\ 000\ \text{kWh/year}$$

$$\text{energy paid for at the gas meter} = 120\ 000 \times \frac{100}{70}\ \text{kWh/year}$$

$$= 171\ 400\ \text{kWh/year}$$

The nearest whole number of kWh is the significant number. Do not waste time with too many decimal places. Fractions of a millimetre, watt, pascal or kWh are of little or no significance to the reality of the calculation.

$$\text{total cost of energy} = 1.2\ \frac{\text{p}}{\text{kWh}} \times 171\ 400\ \frac{\text{kWh}}{\text{year}} \times \frac{£1}{100\ \text{p}}$$

$$= £2057$$

$$\text{annual standing charge} = 365\ \frac{\text{days}}{\text{year}} \times 8\ \frac{\text{p}}{\text{day}} \times \frac{£1}{100\ \text{p}}$$

$$= £29$$

$$\text{total annual energy bill} = £2057 + £29 + £75$$
$$= £2161$$

$$\text{overall average cost of energy} = \frac{£2161}{171\ 400\ \text{kWh}} \times \frac{100\ \text{p}}{£1}$$
$$= 1.26\ \text{p/kWh}$$

The cost in use, UC, for other fuel or energy sources is calculated from the basic price. These are pence per litre for oils, £ per tonne for solid fuels and £ per kg refill for liquefied petroleum gas. The specific gravity of liquid fuel, SG, is used to convert volume to mass measurements. When the specific gravity of an oil is 0.83, 1 l of the oil weighs 0.83 kg. One litre of water at 4 °C weighs 1 kg. Oil and paraffin is sold by the litre but its heat content, gross calorific value GCV, is usually listed as around 45 MJ/kg. For oil

$$\text{UC} = \text{C}\,\frac{\text{p}}{\text{l}} \times \frac{1\ \text{l}}{\text{SG kg}} \times \frac{\text{kg}}{\text{GCV MJ}} \times \frac{100}{\eta} \times \frac{1\ \text{MJ}}{10^3\ \text{kJ}} \times \frac{1\ \text{kJ}}{1\ \text{kWs}} \times \frac{3600\ \text{s}}{1\ \text{h}}$$

$$= \frac{\text{C} \times 100 \times 3600}{\text{SG} \times \text{GCV} \times \eta \times 10^3}\ \text{p/kWh}$$

For solid fuel

$$\text{UC} = \text{C}\,\frac{\text{p}}{\text{kg}} \times \frac{\text{kg}}{\text{GCV MJ}} \times \frac{100}{\eta} \times \frac{1\ \text{MJ}}{10^3\ \text{kJ}} \times \frac{1\ \text{kJ}}{1\ \text{kWs}} \times \frac{3600\ \text{s}}{1\ \text{h}}$$

$$= \frac{\text{C} \times 100 \times 3600}{\text{GCV} \times \eta \times 10^3}\ \text{p/kWh}$$

For electricity

$$\text{UC} = \text{C}\,\frac{\text{p}}{\text{kWh}} \times \frac{100}{\eta}$$

$$= \frac{\text{C} \times 100}{\eta}\ \text{p/kWh}$$

Liquefied petroleum gas (LPG) (butane or propane) can be used from refillable cylinders on sites that are remote from the mains gas distribution of methane. LPG is sold at a refill charge per number of kilograms, depending upon the size of the cylinder. Convert the refill cost into a price per kilogram of LPG by dividing by its weight. For example, a 25 kg refill might cost £18.50:

$$\text{C}\,\frac{\text{p}}{\text{kg}} = \frac{£18.50}{25\ \text{kg}} \times \frac{100\ \text{p}}{£1}$$
$$= 74\ \text{p/kg}$$

$$UC = C\,\frac{\text{p}}{\text{kg}} \times \frac{\text{kg}}{\text{GCV MJ}} \times \frac{100}{\eta} \times \frac{1\ \text{MJ}}{10^3\ \text{kJ}} \times \frac{1\ \text{kJ}}{1\ \text{kWs}} \times \frac{3600\ \text{s}}{1\ \text{h}}$$

$$= \frac{C \times 100 \times 3600}{\text{GCV} \times \eta \times 10^3}\ \text{p/kWh}$$

The calculation for paraffin is the same as that for oils:

$$UC = \frac{C \times 100 \times 3600}{\text{SG} \times \text{GCV} \times \eta \times 10^3}\ \text{p/kWh}$$

EXAMPLE 2.5

Construct a table of fuel cost per useful kWh from the data given below, assuming that standing charges have already been allocated to other services and need not be included here.

Gas: 1.3 p/kWh, $\eta = 75\%$, GCV = 39.0 MJ/m^3

Oil: 28 Redwood No. 1, SG = 0.83, GCV = 45.5 MJ/kg, $\eta = 70\%$, 13.0 p/l

Anthracite: GCV = 26.7 MJ/kg, $\eta = 60\%$, 6.0 p/kg

Electricity: daytime 8.0 p/kWh, $\eta = 100\%$; night-time 2.70 p/kWh, $\eta = 90\%$

LPG (propane): 14.6 p/kg, $\eta = 70\%$, GCV = 50 MJ/kg

Paraffin: 18.0 p/l, $\eta = 80\%$, GCV = 46.4 MJ/kg, SG = 0.79

$$\text{gas cost} = 1.30\ \frac{\text{p}}{\text{kWh}} \times \frac{100}{75}$$

$$= 1.73\ \text{p/kWh}$$

$$\text{oil cost} = \frac{13.0 \times 100 \times 3600}{70 \times 0.83 \times 45.5 \times 1000}\ \text{p/kWh}$$

$$= 1.77\ \text{p/kWh}$$

$$\text{anthracite cost} = \frac{6.0 \times 100 \times 3600}{60 \times 26.7 \times 1000}\ \text{p/kWh}$$

$$= 1.35\ \text{p/kWh}$$

$$\text{electricity (day)} = 8.0 \times \frac{100}{100}\ \text{p/kWh}$$

$$= 8.0\ \text{p/kWh}$$

$$\text{electricity (night)} = 2.70 \times \frac{100}{90} \text{ p/kWh}$$
$$= 3.0 \text{ p/kWh}$$
$$\text{LPG (propane) cost} = \frac{14.6 \times 100 \times 3600}{70 \times 50.0 \times 1000} \text{ p/kWh}$$
$$= 1.50 \text{ p/kWh}$$
$$\text{paraffin cost} = \frac{18.0 \times 100 \times 3600}{80 \times 0.79 \times 46.4 \times 1000} \text{ p/kWh}$$
$$= 2.20 \text{ p/kWh}$$

Notice that the cost of energy in £ per useful GJ is likely to be within the range from £2 to £25 for the foreseeable future. £3/GJ may apply to the lowest grades of solid and liquid fuels, which are used in large power-generating stations. £25/GJ will be the upper limit for electricity consumed during the day by household consumers.

Deregulation of the electricity in the UK and Australia has led to falling prices for peak electricity. Natural gas, where it is available, remains the most popular means of generating electrical power, heating and cooling, owing to it's cleanliness, convenience and lack of site storage requirement.

EXAMPLE 2.6

Evaluate the fuel cost in pounds per useful gigajoule and add them to Table 2.3.

$$\text{gas cost} = 1.30 \frac{\text{p}}{\text{kWh}} \times \frac{100}{70} \times \frac{1 \text{ kWs}}{1 \text{ kJ}} \times \frac{1 \text{ h}}{3600 \text{ s}} \times \frac{10^6 \text{ kJ}}{1 \text{ GJ}} \times \frac{£1}{100 \text{ p}}$$
$$= 1.30 \times \frac{100}{70} \times \frac{10^6}{3600 \times 100} £/\text{GJ}$$
$$= 1.86 \times 2.778 \ £/\text{GJ}$$
$$= £5.16 \text{ per GJ}$$

Table 2.3 Summary of fuel costs

Fuel	Cost per useful unit	
	kWh (pence)	GJ (£)
Gas	1.73	5.16
Electricity (night)	3.0	8.33
Paraffin	2.2	6.14
Oil	1.77	4.90
Anthracite	1.35	3.75
LPG (propane)	1.5	4.17
Electricity (day)	8.0	22.2

The 2.778 is the constant which, when multiplied by the fuel cost in pence per useful kWh, will give the equivalent in £/GJ.

$$\text{oil cost} = \frac{13.0 \times 100 \times 10^3}{70 \times 0.83 \times 45.5 \times 100} \text{ £/GJ}$$

$$= \text{£}4.90 \text{ per GJ}$$

$$\text{anthracite cost} = \frac{6.0 \times 100 \times 10^3}{60 \times 26.7 \times 100} \text{ £/GJ}$$

$$= \text{£}3.75 \text{ per GJ}$$

$$\text{electricity (day)} = \frac{8.0 \times 100 \times 10^6}{100 \times 3600 \times 100} \text{ £/GJ}$$

$$= \text{£}22.2 \text{ per GJ}$$

$$\text{electricity (night)} = \frac{2.7 \times 100 \times 10^6}{90 \times 3600 \times 100} \text{ £/GJ}$$

$$= \text{£}8.33 \text{ per GJ}$$

$$\text{LPG (propane) cost} = \frac{14.6 \times 100 \times 10^3}{70 \times 50.0 \times 100} \text{ £/GJ}$$

$$= \text{£}4.17 \text{ per GJ}$$

$$\text{paraffin cost} = \frac{18.0 \times 100 \times 10^3}{80 \times 0.79 \times 46.4 \times 100} \text{ £/GJ}$$

$$= \text{£}6.14 \text{ per GJ}$$

Annual fuel costs

Annual fuel costs can be estimated in advance of their occurrence from a knowledge of the following:

1. energy cost per useful unit;
2. length of heating season;
3. operational hours of the system;
4. mean internal building temperature;
5. design external temperature;
6. degree days for the locality.

The design steady-state building heat loss is known as the design external air temperature (Chapter 3) and ranges from −1 °C to −5 °C. Throughout the

Table 2.4 Degree day data showing 20-year averages

	Month											
Region	Sep	Oct	Nov	Dec	Jan	Feb	Mar	Apr	May	Jun	Jul	Aug
Thames	56	129	252	333	349	306	281	200	113	49	25	27
NE Scotland	125	196	321	382	399	362	340	274	203	112	86	86

heating season, the heat loss will fluctuate with the cyclic variations in ambient temperature. Fortuitous heat gains will reduce fuel consumption provided that the automatic controls can reduce heating system performance sufficiently and avoid overshooting the desired room temperatures. Weather variations are evaluated by using the degree day data issued monthly by the Department of Energy. Table 2.4 shows degree days for 2 of the 17 geographical regions covering the UK (Moss, 1997).

Degree days are recorded temperature data that facilitate the production of a climatic correction or load factor for calculation of heating system operational costs and efficiency over months or yearly time intervals. They are applied to normally occupied buildings where the heat loss from the warm interior is balanced by gains of heat from the sun, occupants, lighting, cooking and hot-water usage at an external air temperature of 15.5 °C; this is known as the base temperature t_b. The value taken for the base temperature is an estimate of the conditions under which there will be no net heat loss from a traditionally constructed residence; thus no fuel will be consumed at this and higher outside temperatures.

Calculation of the actual base temperature for a particular building may reveal another value; consequently, care is needed in the application of degree day data, and correction factors may be included for other than traditionally constructed dwellings: for example, highly insulated or commercial structures and where internal heat gains from electrical equipment are high (CIBSE, 1986).

The standard method of use is to assess the daily difference between the base temperature and the mean value of the external air temperature during each 24 h period. A modified calculation is made when the base temperature is below the external mean temperature, as this would indicate a net heat gain to the building. Degree day data are not used for air-conditioning cooling-load calculations as they are not appropriate. Figure 2.1 shows a typical fluctuation in external air temperature relative to base temperature. As the maximum and minimum air temperatures are 10 °C and 6 °C respectively, the 24 h mean is 8 °C. Therefore, as there is a difference of 7.5 °C per day, 7.5 degree days are added to the accumulated total for that month.

The maximum possible number of degree days for a particular location and period of heating system operation can be found as shown in the following example.

EXAMPLE 2.7

A house is continuously occupied during a 30-week heating season. The design external air temperature is −1 °C. Find the maximum possible number of degree days.

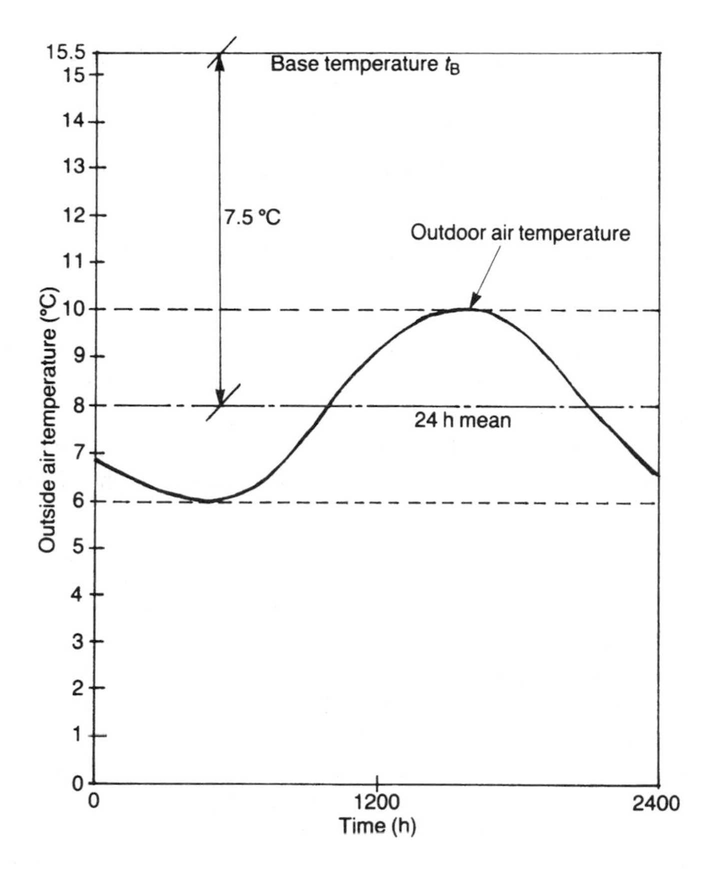

Figure 2.1 Method of calculating degree days during a 24 h period

$$\text{days} = 30\ \text{weeks} \times \frac{7\ \text{days}}{1\ \text{week}}$$

$$= 210\ \text{days}$$

$$\text{maximum temperature difference} = [15.5 - (-1.0)]\ ^\circ\text{C}$$

$$= 16.5\ ^\circ\text{C}$$

$$\text{maximum degree days} = 210\ \text{days} \times 16.5\ ^\circ\text{C}$$

$$= 3465\ \text{degree days}$$

The load factor L is the ratio of actual to maximum degree days and is used to find the average rate of heat loss from a building over the heating season:

$$L = \frac{\text{degree days for locality}}{\text{maximum possible degree days}}$$

EXAMPLE 2.8

Find the average rate of boiler power used during the heating season when there were 2300 degree days, and steady-state heat losses were calculated to be 18 kW at an outside air temperature of −1 °C.

$$L = \frac{2300}{3465}$$

$$= 0.66$$

$$\text{seasonal average rate of heat loss} = \text{design heat loss} \times \text{load factor}$$

$$= 18 \text{ kW} \times 0.66$$

$$= 11.88 \text{ kW}$$

The boiler will have an average heat output of 11.9 kW over the heating season, i.e. in addition to the hot-water service requirement and heat losses from pipework.

Degree days can be used to monitor fuel consumption and check that it is not being used wastefully. Incorrectly serviced fuel-burning appliances would show an increasing use of energy per degree day rather than a constant rate. Deterioration of the performance of an automatic control system or lack of proper manual regulation of ventilation openings would also result in a departure from expected ratios. A graph of energy consumption against degree days should be linear for a building, and any major divergence will show that corrective action is needed. Figure 2.2 shows an example.

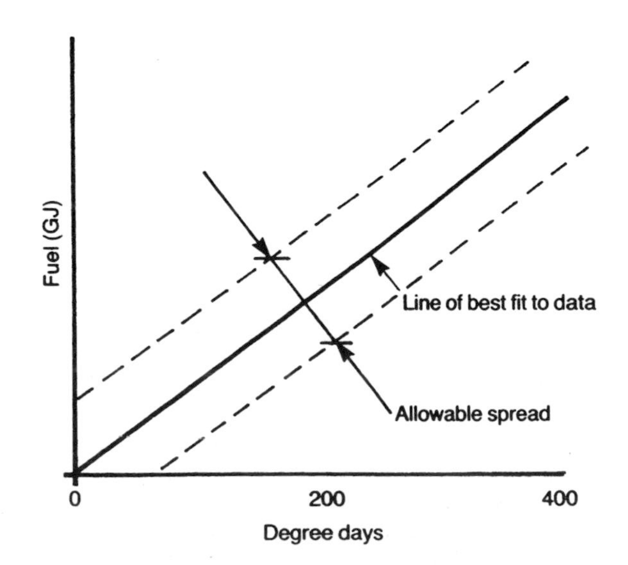

Figure 2.2 Relationship of fuel consumption to degree days

The calculation of expected annual fuel costs is now a matter of finding the number of gigajoules or therms consumed in the building and then the cost of providing this useful amount of heat.

EXAMPLE 2.9

A hospital is heated for 24 h per day, 7 days per week for 30 weeks a year. Its steady-state heat loss at −1 °C outside is 3000 kW, and a gas-fired boiler with hot-water radiator central heating system is used. Estimate the annual fuel cost for heating the building if there are likely to be 2300 degree days in that locality. Maximum degree days are 3465.

From previous calculations, gas costs £5.16 per useful gigajoule.

$$\text{load factor} = \frac{2300}{3465} = 0.664$$

$$\begin{aligned}\text{annual energy} &= \text{heat loss} \times \text{load factor} \times \text{operating time}\\ &= 3000\text{ kW} \times 0.664 \times \frac{24\text{ h}}{\text{day}} \times \frac{210\text{ days}}{\text{year}} \times \frac{3600\text{ s}}{1\text{ h}}\\ &\quad \times \frac{1\text{ kJ}}{1\text{ kW s}} \times \frac{1\text{ GJ}}{10^6\text{ kJ}}\\ &= 3000 \times 0.664 \times 24 \times 210 \times 3600 \times 10^{-6}\text{ GJ}\\ &= 36\ 143\text{ GJ/year}\end{aligned}$$

$$\begin{aligned}\text{annual cost} &= \text{useful energy required}\ \frac{\text{GJ}}{\text{year}} \times \frac{\text{cost£}}{\text{useful GJ}}\\ &= 36\ 143\ \frac{\text{GJ}}{\text{year}} \times \frac{£5.16}{\text{GJ}}\\ &= £186\ 498\text{ per annum}\end{aligned}$$

EXAMPLE 2.10

An initial energy audit of a hospital revealed the data shown in Table 2.5. The data were for a month that had 300 degree days, and the energy manager required energy use performance factors of total cost per square metre of floor area, heating system energy used per degree day and electrical energy used per person per hour. All gas consumed was for the heating system and cost 1.3 p/kWh plus £200 per month standing charge and £250 per month for maintenance.

Electricity cost was 5 p/kWh and maintenance costs amounted to £900 per month.

Table 2.5 Energy audit data for Example 2.10

Location	Electricity (kWh)	Gas (kWh)	Floor (m^2)	Usage (h)	Occupants
Medical	10 000 000	15 000 000	40 000	720	2000
Administration	1 000 000	1 500 000	4 000	360	250
Engineering	200 000	300 000	800	400	25
Totals	11 200 000	16 800 000	44 800	1480	2750

Energy use performance factor of total cost per square metre of floor area for the month, $EUPF_1$:

$$\text{electricity cost} = 11\ 200\ 000\ \text{kWh} \times 5\ \frac{\text{p}}{\text{kWh}} \times \frac{£1}{100\ \text{p}} + £900$$

$$= £560\ 900$$

$$\text{gas cost} = 16\ 800\ 000\ \text{kWh} \times 1.3\ \frac{\text{p}}{\text{kWh}} \times \frac{£1}{100\ \text{p}} + £200 + £250$$

$$= £218\ 850$$

$$EUPF_1 = \frac{£560\ 900 + £218\ 850}{44\ 800\ \text{m}^2}$$

$$= £17.4\ \text{per m}^2\ \text{floor area}$$

$EUPF_2$, gas heating system energy used per degree day:

$$EUPF_2 = \frac{16\ 800\ 000\ \text{kWh}}{300\ \text{degree days}}$$

$$= 56\ 000\ \text{kWh per degree day}$$

$EUPF_3$, electrical energy used per person per hour:

$$\text{total occupation} = \text{sum of (occupants} \times \text{usage hours)}$$

$$= (2000 \times 720) + (250 \times 360) + (25 \times 400)$$

$$= 1\ 540\ 000\ \text{person-hours}$$

$$EUPF_3 = \frac{11\ 200\ 000\ \text{kWh}}{1\ 540\ 000\ \text{person-hours}}$$

$$= 7.27\ \text{kWh per person per hour}$$

(You may wish to evaluate the energy use performance factors for various locations for comparison.)

Economic thickness of thermal insulation

A balance needs to be made between the capital cost of thermal insulation of buildings or hot surfaces and the potential reduction in fuel costs in order to obtain the lowest total cost combination of these two cash flows. Capital cost is normally expected to be recovered from fuel cost savings during the first 2–3 years of use; however, longer periods than this are needed for major structural items, such as cavity fill and double glazing, and there will be additional benefits, such as improved thermal storage capacity, reduced external noise transmission, fewer draughts and added value to the property, that do not fit easily into a financial treatment of their worth.

For a flat surface, the cost of heat loss per square metre through the structure can be represented as follows:

$$\text{Fuel cost} = U\,\frac{\text{W}}{\text{m}^2\,\text{K}} \times (t_{\text{ai}} - t_{\text{ao}})\,\text{K} \times L \times \text{S}\,\frac{\text{h}}{\text{year}} \times \frac{3600\text{ s}}{1\text{ h}} \times \frac{1\text{ J}}{1\text{ W s}} \times \frac{1\text{ GJ}}{10^9\text{ J}} \times \frac{\pounds C}{\text{GJ}}$$

$$= \pounds[U(t_{\text{ai}} - t_{\text{ao}})LS \times 3.6C \times 10^{-6}] \text{ per m}^2 \text{ year}$$

The cost of fuel usage for a range of thermal transmittances U can be calculated for a particular structure. This is usually a decreasing curve for increasing insulation thickness as each additional layer reduces the thermal transmittance by progressively smaller amounts.

If the cost $\pounds I/\text{m}^3$ of the thermal insulation as installed is known, then the cost for each thickness per square metre of surface area can be found from

$$\text{insulation cost} = \frac{\pounds I}{\text{m}^3} \times \frac{\text{thickness m}}{\text{repayment time years}}$$

Data from these equations can be drawn on a graph. Addition of the two curves produces a total cost curve. The lowest point on this curve gives the optimum insulation thickness; if its lower part is fairly flat, then any one of a number of commercially available thicknesses will be economic.

EXAMPLE 2.11

Expanded polystyrene board is to be added to the internal face of a wall having an initial thermal transmittance of 3.30 W/m^2 K in thicknesses of 25, 50, 75, 100, 125 and 150 mm. Insulated wall thermal transmittances will be 0.96, 0.56, 0.40, 0.31, 0.25 and 0.21 W/m^2 K. The insulation costs £48 per cubic metre fitted and the capital recovery period is to be 3 years. Fuel costs £8.93 per useful gigajoule. Internal and external design temperatures are 21 °C and −1 °C respectively, the load factor is 0.608 and the building is to be heated for 3000 h per year. Use the information provided to find the economic thickness of insulation.

$$\text{fuel cost} = U[21 - (-1)] \times 0.608 \times 3000 \times 3.6 \times 8.93 \times 10^{-6}\ £/\text{m}^2\ \text{year}$$

$$= £1.29U \text{ per m}^2 \text{ year}$$

$$\text{insultation cost} = \frac{£48}{\text{m}^3} \times \frac{\text{thickness } l \text{ m}}{3 \text{ years}}$$

$$= £16\ l \text{ per m}^2 \text{ year}$$

The results are shown in Table 2.6. Figure 2.3 shows that the total cost curve can be drawn by adding the fuel and insulation cost curves for each insulation thickness. The economic thickness is 50 mm.

Table 2.6 Cost data for Example 2.11

Thickness l (m)	0.025	0.050	0.075	0.100	0.125	0.150
U (W/m^2 K)	0.96	0.56	0.40	0.31	0.25	0.21
Insulation cost (£)	0.40	0.80	1.20	1.60	2.00	2.40
Fuel cost (£)	1.24	0.72	0.52	0.40	0.32	0.27
Total cost (£)	1.64	1.52	1.72	2.00	2.32	2.67

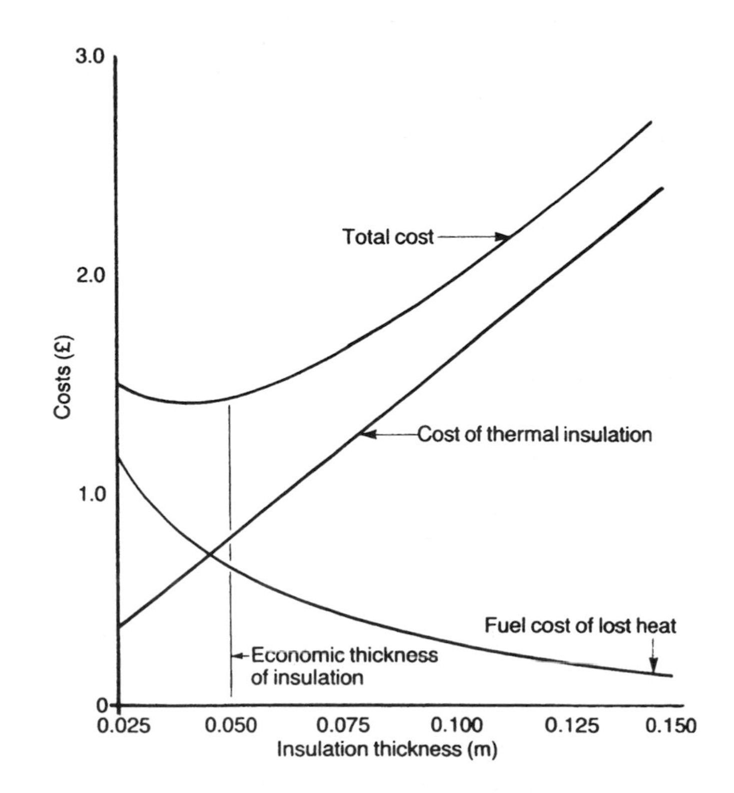

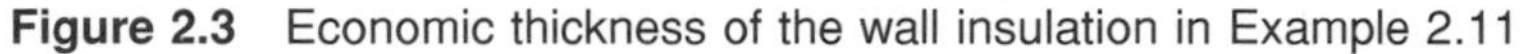

Figure 2.3 Economic thickness of the wall insulation in Example 2.11

The economic thermal transmittance of a structure with a flat surface can be evaluated from the following equation (Diamant, 1977):

$$U_e = \left[\frac{\lambda \alpha I}{8.64C(Y + S\Delta t)}\right]^{1/2}$$

where U_e is the economic U value (W/m^2 K)), λ is the thermal conductivity (W/m K), α is the depreciation and interest charges (%), I is the total cost of thermal insulation fitted into the building (£/m^3), C is the cost of useful heat (£/MJ), Y is the annual degree days for a base temperature of 18 °C, S is the length of the heating season (days) and Δt is the difference between the average internal air temperature and 18 °C (°C). Once U_e has been found, the thermal insulation thickness can be found from

$$l = \lambda \left(\frac{1}{U_e} - \frac{1}{U}\right) \text{ m}$$

where U is the uninsulated thermal transmittance (W/m^2 K).

EXAMPLE 2.12

The roof of a factory in Scotland is insulated with expanded polystyrene (EPS) slab. The locality has 3000 degree days for a base temperature of 18 °C and a heating season of 240 days. The average internal air temperature is 21 °C. The EPS costs £43 per cubic metre fitted, interest is charged at 8% and the life expectancy of the insulation is 50 years. Heat costs £5.20 per gigajoule. The thermal conductivity of EPS is 0.035 W/m K. Find the economic insulation thickness.

The original thermal transmittance of the single sheet roofing is 6.5 W/m^2 K.

$$\alpha = 8\% \text{ interest per annum} + 2\% \text{ depreciation per annum}$$
$$= 10\%$$

$$C = \frac{£5.20}{\text{GJ}} \times \frac{1 \text{ GJ}}{10^3 \text{ MJ}}$$
$$= £5.20 \times 10^{-3} \text{ per MJ}$$

$$U_e = \left[\frac{0.035 \times 10 \times 43}{8.64 \times 5.2 \times 10^{-3} \times (3000 + 240 \times (21 - 18))}\right]^{1/2} \text{ W/m}^2\text{ K}$$
$$= 0.3 \text{ W/m}^2\text{ K}$$

$$l = 0.035 \left(\frac{1}{0.3} - \frac{1}{U}\right)$$

$U = 6.5$ W/m^2 K for the uninsultated roof.

Therefore

$$l = 0.035\left(\frac{1}{0.3} - \frac{1}{6.5}\right)$$

$$= 0.111 \text{ mm}$$

Thus the economic thickness of insulation for this roof is 111 mm. As EPS slab is manufactured in 100 mm thickness, the insulated U value will be 0.35 W/m^2 K.

Accounting for energy-economizing systems

Once the capital cost and fuel cost savings have been assessed for thermal insulation, fuel-saving equipment or automatic controls, the capital repayment period or return on capital investment can be calculated in simple terms:

$$\text{capital repayment period} = \frac{\text{capital cost}}{\text{energy savings per year}}$$

$$\text{percentage return on investment} = \frac{\text{energy savings per year}}{\text{capital cost}} \times 100$$

Further refinements such as discounted cash flow, loan interest charges, tax allowances and grants can be included to improve accuracy.

Cash flow statements for limited companies are handled differently from those for home-owners, as allowances for capital expenditure and taxation on increased profitability due to energy economies can markedly improve estimates of payback times. A purchase costing £500 000, which would save £200 000 in the first year's energy bill, would appear to take 2.5 years for capital recovery, but the cash flow projection may be as shown in Table 2.7. Figures in parentheses are outward cash flows from the business. Energy costs are indicated to increase by 5% per year, so savings increase by the same amount.

This investment commences cash generation during the second year of equipment use and would start to provide funds for further investment. Certain

Table 2.7 Cash flow forecast for the purchase of an energy-economizing system

		Year 1	Year 2	Year 3
A	Cash balance brought forward	0	(315 000)	(53 000)
B	Capital purchase	(500 000)	0	0
C	Energy saving	200 000	210 000	220 500
D	Capital allowance, 1 year in arrears 25% × B	0	125 000	0
E	Cash balance (A − B + C + D)	(300 000)	20 000	167 500
F	Interest, say 10% × E × 0.5	(15 000)	1 000	8 375
G	Tax, 1 year in arrears, 40% × (C + F)	0	(74 000)	0
H	Net cash flow (C + D + F − B − G)	(315 000)	262 000	228 875
I	Cash balance (A + H)	(315 000)	(53 000)	175 875

Table 2.8 Cash flow forecast for the leasing of an energy-economizing equipment

		Year 1	Year 2	Year 3
A	Cash balance brought forward	0	157 500	270 375
B	Leasing payment, 10% × cost	(50 000)	(50 000)	(50 000)
C	Energy saving	200 000	210 000	220 500
D	Capital allowance	0	0	0
E	Cash balance (A − B + C + D)	150 000	317 500	440 875
F	Interest, say 10% × E × 0.5	7 500	15 875	22 044
G	Tax, 40% × (C + F − B), 1 year in arrears	0	(63 000)	(70 350)
H	Net cash flow (C + D + F − B − G)	157 500	112 875	122 194
I	Cash balance (A + H)	157 500	270 375	392 569

items of equipment can be leased rather than purchased, and this releases cash earlier but calls for continuous payments to the leasing company. Table 2.8 shows a sample cash flow forecast. Cash flow is always positive to the company, but leasing payments are made for 10 years and then at a reduced rate after that period. Self-contained items of plant such as heat pumps or electricity generators may be leased.

Low-energy buildings

Low-energy buildings are those that utilize energy efficiently to maintain a comfortable thermal environment suitable for the purpose of the building. Energy design targets may be proposed that include uses for heating, ventilation, hot-water services, lighting and electrical power. The total demand target T for a building is assessed by adding the thermal demand target T_T to the electrical demand target T_E in the CIBSE *Building Energy Code* (CIBSE, 1981).

In heated and naturally ventilated buildings, the rate of heat loss is related to the floor area by the dimensionless building envelope number B:

$$B = \frac{A_W}{A_f} + \frac{K_1}{n_f} + K_2 H$$

where A_W is the gross external walling surface area (m^2), A_f is the total floor area (m^2), n_f is the number of storeys and H is the floor-to-ceiling height (m). When hot-water services are included

$$T_T = C_1 B + C_2 \text{ W/m}^2$$

where K_1, K_2, C_1 and C_2 are constants given in Table 2.9, C_3 is the mean electrical power requirement for the lighting system (W/m^2) and

$$T_E = C_3 + 0.10 T_T \text{ W/m}^2$$

which shows that the electrical power consumption associated with the heating services is expected to be 10% of the thermal target.

Table 2.9 Values of constants for demand target

Building type	K_1	K_2	C_1	C_2	C_3
Office, 5 days/week	0.5	0.1	13	−5	24
Shop, 6 days/week	0.5	0.1	16	−6	27
Factory, 5 days/week, single shift	1.1	0.2	6	−3	8
Hotel	1.0	0.1	15	−3	15
Warehouse	1.1	0.2	6	−2	6
Hospital	1.0	0.1	17	+12	15
Institutional residence	1.0	0.1	15	+4	15
Educational	0.5	0.2	16	−4	13

EXAMPLE 2.13

Find the total demand target for a proposed ten-storey hospital medical building 50 m long, 30 m wide and 3 m floor-to-ceiling height, and compare it with an alternative design having the same floor area but of single-storey design, 172 m long and 87 m wide. Comment on the relative energy use of these alternative configurations.

For the ten-storey block

$$A_f = 50 \times 30 \times 10 \text{ m}^2$$

$$= 15\,000 \text{ m}^2$$

$$A_W = 10 \times 3 \times 2 \times (50 + 30) \text{ m}^2$$

$$= 4800 \text{ m}^2$$

$$n_f = 10$$

$$h_c = 3 \text{ m}$$

From Table 2.9 $K_1 = 1$ and $K_2 = 0.1$. Then

$$B = \frac{4800}{15\,000} + \frac{1}{10} + (0.1 \times 3)$$

$$= 0.72$$

Also from Table 2.9 $C_1 = 17$, $C_2 = +12$ and $C_3 = 15$ W/m^2. Then

$$T_T = 17 \times 0.72 + 12 \text{ W/m}^2$$

$$= 24.24 \text{ W/m}^2$$

$$T_E = 15 + (0.1 \times 24.24) \text{ W/m}^2$$

$$= 17.4 \text{ W/m}^2$$

Thus

$$T = T_T + T_E$$

$$= 24.24 + 17.4\ \text{W/m}^2$$

$$= 41.6\ \text{W/m}^2$$

Similarly, for the single-building

$$B = 1.4$$

$$T_T = 35.8\ \text{W/m}^2$$

$$T_E = 18.6\ \text{W/m}^2$$

$$T = 54.4\ \text{W/m}^2$$

The single-story building has a better chance of being ventilated by assisted natural systems, rather than the mechanical plant needed in multi-story designs. The walls and perimeter glazing are more easily shaded from solar heat gains; floor usage is more efficient as vertical service shafts, lifts and stairways are unnecessary; wind exposure is reduced; and maintenance of the external surfaces of the building is less costly.

Both buildings have the same floor area; the lower fatter configuration has less external wall surface area and consequently a higher total demand target.

Compliance with the demand target is achieved when the total demand of the proposed building does not exceed its target figure; the thermal demand may be up to 10% greater than its target value, but the total demand target must not be surpassed.

The effect on gas consumption of thermal insulation in houses

When the design steady-state heat loss from a dwelling is reduced by the addition of thermal insulation and draught-proofing, increased standards of thermal comfort are provided. However, the full potential saving due to the extra insulation may not be reflected in the fuel bills as expected. Field measurements (British Gas, 1980) have shown a correlation between domestic gas consumption, design heat loss, occupancy and degree days for the locality:

$$\text{annual therms} = 61 + \frac{70YQ}{2222} + 59\ N$$

where Y denotes annual degree days, N is the number of persons in the household and Q is the design heat loss in kilowatts. This relationship permits an assessment of anticipated gas consumption for heating and hot-water services in housing and quantification of thermal insulation savings.

Therms are no longer used, so they are converted into kWh by multiplying by 29.3056:

$$\text{annual gas consumption} = 29.3056 \times \left(61 + \frac{70YQ}{2222} + 59N\right) \text{ kWh}$$

> **EXAMPLE 2.14**
>
> A house in the Thames region has a design heat loss of 36 kW and six occupants. Added thermal insulation reduces the design heat loss to 32 kW. Estimate the probable cost savings for the gas-fired central heating and hot-water system.
>
> Natural gas costs 2 p/kWh.

From Table 2.4 the total degree days for the year are 2120. Then gas consumption before insulation

$$= 29.3056 \times \left(61 + \frac{70 \times 2120 \times 36}{2222} + 59 \times 6\right) \text{ kWh/year}$$

$$= 82\ 622 \text{ kWh/year}$$

gas consumption after insulation = 74 793 kWh/year

$$\text{gas cost saving} = (82\ 622 - 74\ 793) \text{ kWh} \times \frac{2}{\text{kWh}} \times \frac{\pounds 1}{100 \text{ p}}$$

$$= \pounds 156.58 \text{ during the first year}$$

Questions

1. State the function of an energy audit. What data are collected? How are the data presented? What is likely to be the most serious barrier to data collection?
2. Explain the uses of energy use performance factors.
3. How can the costs of different fuels be compared with each other?
4. Ascertain current energy prices and update Table 2.3.
5. Explain the term 'degree day' and state its use.
6. How is the load factor calculated and how is it used?
7. A factory uses 20 000 l of oil for its heating and hot-water systems, 160 000 kWh of electrical power and 300 000 kWh of gas for furnaces in a year. Fixed charges are £800 for the oil, £700 for the electrical equipment and £1200 for gas equipment. Use the data provided in this chapter and current energy prices to produce an overall energy audit based on the gigajoule unit and find the average cost of all the energy used.
8. Calculate the annual cost of a gas-fired heating system in a house with a design heat loss of 30 kW at −2 °C for 16 h per day, 7 days per week for 30 weeks in the year. Use the data provided in this chapter and the current fuel price.

9. Find the total annual cost of running a gas-fired heating and hot-water system in a house with four occupants if its design heat loss is 32 kW. Maintenance charges amount to £160 per year.

10. Determine the following energy use performance factors for offices A and B:

    (a) total energy cost per square metre of floor area;
    (b) heating system energy used per degree day;
    (c) total energy used per person per occupation hour.

Table 2.10

Building	Electricity (kWh)	Gas (kWh)	Floor (m^2)	Usage (h)	Occupants
Office A	115 000	720 000	2400	2600	160
Office B	100 000	850 000	2600	2300	210

The data required are given in Table 2.10. They refer to a year having 2150 degree days for office A and 2310 degree days for office B. Maintenance costs are £800 for the gas system and £750 for the electrical installation in each building. All the gas is used for heating systems. Compare the performance of the two office buildings.

11. Calculate the economic thickness of Rockwool thermal insulation which is to be applied to a wall with an uninsulated thermal transmittance of 1.6 W/m^2 K.. The thermal conductivity of Rockwool is 0.04 W/m K. The locality has 2900 degree days for a base temperature of 18 °C and a heating season of 240 days. The average internal temperature is 20 °C. Rockwool costs £50 per cubic metre fitted. Interest is charged at 9% and depreciation at 10%. Fuel oil for heating costs £5.00 per useful gigajoule.

12. Find the total demand target for an eight-storey office 22 m long, 15 m wide, 3 m floor-to-ceiling height. How will compliance with this target be achieved?

3 Heat loss calculations

Key terms and concepts

Learning objectives

Study of this chapter will enable the reader to:

1. identify and use the thermal conductivity and resistivity of building materials;
2. calculate the thermal resistance of a composite structure;
3. use building exposure categories;
4. use surface and cavity thermal resistance values;
5. identify high- and low-emissivity building materials;
6. calculate, or find, the thermal transmittance of walls, flat and pitched roofs, floors and windows;
7. use the proportional area method to calculate the average *U* value of a thermally bridged wall or other structure;
8. calculate the fabric and ventilation building heat loss components for a building;
9. calculate air and environmental temperatures produced in a room from a specified resultant temperature;
10. identify the use of admittance *Y* values;
11. calculate hot-water storage boiler power requirements;
12. calculate total boiler power.

Introduction

The terms and techniques for handling thermal properties of building materials are introduced to enable calculation of the thermal resistance *R*, thermal transmittance *U* and use of admittance *Y* of composite building elements. Chartered Institution of Building Services Engineers (CIBSE) data are used throughout and representative values are given.

Calculation of building heat loss allows load estimation for heating equipment.

Thermal resistance of materials

The thermal resistance of a slab of homogeneous material is calculated by dividing its thickness by its thermal conductivity:

$$R = l/\lambda$$

where R is the thermal resistance (m^2 K/W), l is the thickness of the slab (m) and λ is the thermal conductivity (W/mK). Resistance to heat flow by a material depends on its thickness, density, water content and temperature. The latter two parameters result from the material's location within the structure. Insulating materials are usually protected from moisture and the possibility of physical damage as they are of low density and strength. The thermal conductivity of masonry can be found from the bulk dry density and the moisture content, which depends on whether it is exposed to the climate or is in a protected position. Table 3.1 shows data for building materials taken from the *CIBSE Guide* (CIBSE, 1986 [1980, Section A3]), to which further reference can be made.

EXAMPLE 3.1

Find the thermal resistance of a 150 mm thickness of expanded polystyrene (EPS) slab.

From Table 3.1, $\lambda = 0.035$ W/m K and $l = 0.15$ m. Therefore

$$R = \frac{0.15}{0.035} \text{ m}^2\text{K/W}$$

$$= 4.286 \text{ m}^2\text{K/W}$$

EXAMPLE 3.2

An architect wishes to replace an aerated concrete slab 120 mm thick in the design of a roof with phenolic foam having the same thermal resistance. What thickness of insulation could be used?

The values of λ are 0.16 W/m K for the aerated concrete slab and 0.04 W/m K for the phenolic foam slab.

$$R \text{ (concrete)} = \frac{0.12}{0.16}$$

$$R \text{ (phenolic)} = \frac{l}{0.04}$$

Therefore for the same resistance

$$\frac{0.12}{0.16} = \frac{l}{0.04}$$

Hence

$$l = 0.04 \times \frac{0.12}{0.16}\,\text{m} \times \frac{10^3\ \text{mm}}{1\ \text{m}}$$

$$= 30\ \text{mm}$$

Table 3.1 Thermal conductivities of materials

Material	Density (kg/m^3)	Thermal conductivity λ (W/m K)	Specific heat capacity (J/kg K)
Walls (external and internal)			
Asbestos cement sheet	700	0.36	1050
Asbestos cement decking	1500	0.36	1050
Brickwork (outer leaf)	1700	0.84	800
Brickwork (inner leaf)	1700	0.62	800
Cast concrete (dense)	2100	1.40	840
Cast concrete (lightweight)	1200	0.38	1000
Concrete block (heavyweight)	2300	1.63	1000
Concrete block (mediumweight)	1400	0.51	1000
Concrete block (lightweight)	600	0.19	1000
Fibreboard	300	0.06	1000
Plasterboard	950	0.16	840
Tile hanging	1900	0.84	800
Surface finishes			
External rendering	1300	0.50	1000
Plaster (dense)	1300	0.50	1000
Plaster (lightweight)	600	0.16	1000
Roofs			
Aerated concrete slab	500	0.16	840
Asphalt	1700	0.50	1000
Felt–bitumen layers	1700	0.50	1000
Screed	1200	0.41	840
Stone chippings	1800	0.96	1000
Tile	1900	0.84	800
Wood wool slab	500	0.10	1000
Floors			
Cast concrete	2000	1.13	1000
Metal tray	7800	50.00	480
Screed	1200	0.41	840
Timber flooring	650	0.14	1200
Wood blocks	650	0.14	1200
Insulation			
Expanded polystyrene (EPS) slab	25	0.035	1400
Glass fibre quilt	12	0.040	840
Glass fibre slab	25	0.035	1000
Mineral fibre slab	30	0.035	1000
Phenolic foam	30	0.040	1400
Polyurethane board	30	0.025	1400
Urea formaldehyde (UF) foam	10	0.040	1400

Thermal transmittance (*U* value)

Thermal transmittance is found by adding the thermal resistances of adjacent material layers, boundary layers of air and air cavities, and then taking the reciprocal. Boundary layer or surface film thermal resistances result from the near-stationary air layer surrounding each part of a building, with an allowance for the radiant heat transfer at the surface. Heat transmission across cavities depends upon their width, ventilation and surface emissivities. The external surface resistance depends upon the building's exposure.

Sheltered: up to the third floor of buildings in city centres
Normal: most suburban and rural buildings; fourth to eighth floors of buildings in city centres
Severe: buildings on coastal or hill sites; floors above the fifth in suburban or rural districts; floors above the ninth in city centres.

Surface resistances are shown in Tables 3.2–3.5.

Table 3.2 Inside surface resistance R_{si}

Building element	Heat flow	R_{si} (m^2 K/W)
Wall	Horizontal	0.12
Ceiling, floor	Upward	0.10
Ceiling, floor	Downward	0.14

These values are for the high emissivity surfaces ($E = 0.90$) common to most building components.

Table 3.3 Outside surface resistance R_{so}

Building element	Surface emissivity	R_{so}(m^2 K/W)		
		Sheltered	Normal	Severe
Wall	High	0.08	0.06	0.03
Wall	Low	0.11	0.07	0.03
Roof	High	0.07	0.04	0.02
Roof	Low	0.09	0.05	0.02

Table 3.4 Thermal resistance R_a of ventilated air spaces

Air space of 25 mm or more	R_a (m^2 K/W)
Loft space between flat plaster ceiling and pitched roof with tiles on felt	0.18
Air space behind tiles on tile hung wall	0.12
Air space in cavity wall	0.18
Air space between high- and low-emissivity surfaces	0.30

Table 3.5 Thermal resistances R_a for unventilated air spaces

Air space thickness (mm)	Surface emissivity	R_a (m^2 K/W)		
		Horizontal	Upward	Downward
5	High	0.10	0.10	0.10
5	Low	0.18	0.18	0.18
25 or more	High	0.18	0.17	0.22
25 or more	Low	0.35	0.35	1.06

EXAMPLE 3.3

An external wall consisting of 105 mm brick, 50 mm unventilated cavity, 105 mm brick and 13 mm dense plaster has a severe exposure. Find its U value.

The calculation of ΣR is shown in Table 3.6. The thermal transmittance U is calculated as follows:

$$U = \frac{1}{\Sigma R}$$

$$= \frac{1}{0.606} \text{ W/m}^2 \text{ K}$$

$$= 1.65 \text{ W/m}^2 \text{ K}$$

EXAMPLE 3.4

Calculate the thermal transmittance of the wall in Example 3.3 if the cavity is filled with urea formaldehyde (UF).

Table 3.6

Element	Length l (m)	λ (W/m K)	R (m^2 K/W)
R_{so}			0.030
Brick	0.105	0.84	0.125
R_a			0.180
Brick	0.105	0.84	0.125
Plaster	0.013	0.5	0.026
R_{si}			0.120
			$\Sigma R = 0.606$

Table 3.7

Element	Length l (m)	λ (W/m K)	R (m^2 K/W)
Previous ΣR			0.606
Less R_a			0.180
			0.426
UF foam	0.050	0.040	1.250
			$\Sigma R = 1.676$

The calculation of ΣR is shown in Table 3.7. Then the new value of U is 0.6 Wm^2 K.

Elements of buildings that are bridged by a material of noticeably different thermal conductivity, such as a dense concrete or steel lintel in a lightweight concrete wall, can be handled by combining the U values of the two constructions using the proportional area method. If U_1 and P_1 are the thermal transmittance and the unbridged proportion respectively of the gross wall area, and U_2 and P_2 are the same parameters for the bridging material, the overall U value is given by

$$U = P_1U_1 + P_2U_2$$

EXAMPLE 3.5

In a concrete-framed industrial building the external walling of brick and concrete has a U value of 1.2 W/m^2 K. The building has a gross perimeter of 200 m and is 4 m high. Forty dense concrete pillars 200 mm wide penetrate the walling from inside to outside. The exposure is normal and the wall thickness is 300 mm.

Find the overall U value.

For the concrete pillar

$$U_2 = \frac{1}{R_{si} + l/\lambda + R_{so}}$$

$$= \frac{1}{0.12 + 0.30/1.40 + 0.06} \text{ W/m}^2 \text{ K}$$

$$= 2.54 \text{ W/m}^2 \text{ K}$$

$$\text{surface area of pillars} = 40 \times 0.2 \times 4 \text{ m}^2$$

$$= 32 \text{ m}^2$$

$$\text{gross wall area} = 200 \times 4 \text{ m}^2$$

$$= 800 \text{ m}^2$$

$$P_2 \text{ (pillars)} = \frac{32}{800} = 0.04$$

$$P_1 \text{ (walling)} = \frac{800 - 32}{800} = 0.96$$

Thus the overall value of U is given by

$$U = (0.96 \times 1.2) + (0.04 \times 2.54) \text{ W/m}^2 \text{ K}$$

$$= 1.25 \text{ W/m}^2 \text{ K}$$

Where only one leaf of a structure containing a cavity is bridged, the resistance of each leaf is calculated separately using the proportional areas as appropriate and then the resistances can be added. The centre line of the cavity can be chosen as the dividing line between the two leaves and half the air space resistance added into each side of the structure. Heat bridges are thermal routes having a lower resistance than the surrounding material that cause distortions to the otherwise uniform temperature gradients. Precise calculation of overall thermal transmittance may require the use of a finite-element computer program that investigates the two- or three-dimensional heat conduction process taking place.

The thermal transmittance of a flat roof is calculated in the manner outlined for walls, but attention must be paid to tapered components. For a pitched roof

$$U = \frac{1}{R_A \cos\beta + R_R + R_B}$$

where R_A is the combined resistance of the materials in the pitched part of the roof including the outside surface resistance (m^2 K/W), β is the pitch angle of the roof (degrees), R_R is the resistance of the roof void (m^2 K/W) and R_B is the combined resistance of the materials in the flat part of the ceiling including the inside surface resistance (m^2 K/W).

Heat flow through solid ground floors in contact with the earth depends on the thermal resistance of the floor slab and the ground which, in turn, is largely determined by its moisture content. The thermal conductivity of the earth can

Table 3.8 *U* values for solid and suspended floors

		U (W/m^2 K)		
Length (m)	Breadth (m)	Four exposed edges	Two perpendicular exposed edges	Suspended floor
100	100	0.10	0.05	0.11
40	40	0.21	0.12	0.22
20	20	0.36	0.21	0.37
10	10	0.62	0.36	0.59
10	4	0.90	0.54	0.83
4	4	1.22	0.73	0.96

vary from 0.70 to 2.10 W/m K depending upon the moisture content. Table 3.8 was evaluated for $\lambda = 1.40$ W/m K, which is about the same as for a concrete slab, and so the *U* values given can be used for floors of any thickness. Dense floor-finishing materials will not influence the quoted *U* values.

The thermal resistance of insulation placed under the screed of a solid floor, or on netting between the joists of a suspended timber floor, can be added to the reciprocal of the *U* value of the uninsulated floor and the new thermal transmittance can be calculated. An insulation material placed vertically around the edge of a concrete floor slab which has a thermal resistance of at least 0.25 m^2 K/W and a depth of 1 m, for example a 10 mm thickness of expanded polystyrene, will reduce the *U* value of the floor by the percentages shown in Table 3.9.

The thermal transmittance of windows depends on glazing and frame types and exposure. If a low-emissivity reflective metallic film is applied to the inside surface of the glass, then the internal surface resistance value can be significantly increased, resulting in a lower *U* value and reduced heat and light transmission from outside. Glass and metal window frames, in themselves, offer negligible resistance to heat flow, but when resistive materials are used the overall *U* value can be found using the proportional area method. Table 3.10 shows window *U* values assuming that the frame takes up 10% of the gross opening in the wall.

Table 3.9 Corrections to *U* values of solid floors with edge insulation

Length (m)	Breadth (m)	Reduction in *U* value (%)
100	100	16
40	40	18
20	20	19
10	10	22
10	4	25
4	4	28

Table 3.10 *U* values for typical windows

	U (W/m^2 K)		
Windows	Sheltered	Normal	Severe
Single frame			
Wood frame	4.7	5.3	6.3
Aluminium frame	5.3	6.0	7.1
Aluminium frame with thermal break	5.1	5.7	6.7
Double glazing			
Wood frame	2.8	3.0	3.2
Aluminium frame	3.3	3.6	4.1
Aluminium frame with thermal break	3.1	3.3	3.7

The Building Regulations may specify a maximum permitted thermal transmittance for external surfaces and if, for walling, it includes openings for windows and doors, then this has the effect of controlling the areas of glazing.

We now make the following definitions:

A_1 area of single glazing (m^2);
A_2 area of double glazing (m^2);
A_3 gross perimeter surface area of the wall and openings (m^2);
A_W walling area (m^2);
F_G fraction of gross wall area taken up by openings having the same U value;
U_1 thermal transmittance of single glazing (W/m^2 K);
U_2 thermal transmittance of double glazing (W/m^2 K);
U_3 thermal transmittance of mean external surface (W/m^2 K);
U_W thermal transmittance of wall (W/m^2 K).

The heat balance for a 1 °C difference in temperature between inside and outside is given by

$$\Sigma(UA) = U_3 A_3$$

Therefore

$$U_3 = \frac{\Sigma(UA)}{A_3}$$

$$= \frac{U_1 A_1 + U_2 A_2 + U_W(A_3 - A_1 - A_2)}{A_3} \ \mathrm{W/m^2\,K}$$

If the maximum allowable area of openings is to be calculated for particular thermal transmittances, then

$$U_3 = \frac{U_W A_W + U_1 A_1}{A_W + A_1}$$

$$= \frac{A_W}{A_W + A_1} U_W + \frac{A_1}{A_W + A_1} U_1$$

Now

$$F_G = \frac{A_1}{A_W + A_1}$$

and

$$\frac{A_W}{A_W + A_1} = 1 - F_G$$

so that

$$U_3 = U_W(1 - F_G) + U_1 F_G$$

$$= U_W - U_W F_G + U_1 F_G$$

$$= U_W + F_G(U_1 - U_W)$$

and

$$F_G = \frac{U_3 - U_W}{U_1 - U_W}$$

Heat loss from buildings

Heat loss occurs by convection and radiation from the outside of the building, and by infiltration of outdoor air. Heating equipment is sized on the basis of steady-state heat flows through the building fabric, with an estimation of the effect of non-steady influences relating to the thermal storage capacity of the structure, adventitious heat gains from people, lighting and machines, and the intermittency of heating system operation.

The steady-state heat loss Q_u through the building fabric is

$$Q_u = \Sigma(AU)(t_{ei} - t_{ao}) \text{ W}$$

where $\Sigma(AU)$ is the sum of the products of the area and thermal transmittance of each room surface. Heat flows to adjacent rooms that are warmer than the outdoor air are found by using the appropriate temperature difference between them.

The ventilation heat Q_v required to warm the natural infiltration of outdoor air is

$$Q_v = 0.33\, NV\, (t_{ai} - t_{ao}) \text{ W}$$

The total heat requirement for each room is

$$Q_p = Q_u + Q_v$$

The values of environmental and air temperature used in the calculations depend upon the type of heating system employed, and the following temperature ratios are used:

$$F_1 = \frac{t_{ei} - t_{ao}}{t_c - t_{ao}}$$

$$F_2 = \frac{t_{ai} - t_{ao}}{t_c - t_{ao}}$$

These two ratios are substituted into the equations for heat requirements Q_u and Q_v. The total heat requirement Q_p then becomes

$$Q_p = [F_1\Sigma(AU) + 0.33F_2NV](t_c - t_{ao}) \text{ W}$$

For buildings with average external U values in the range 0.60–3.0 W/m^2 K, including openings, which covers the majority of habitable structures, the temperature ratios have the following values (with an accuracy to 5.0%):

$$F_1 = 1.00 \qquad F_2 = 1.10$$

for panel radiator heating systems;

$$F_1 = 0.92 \qquad F_2 = 1.23$$

for forced warm air heating systems. Further values are tabulated in CIBSE (1986).

To check the comfort conditions produced by the heating system in a room we use

$$t_{ai} = F_2(t_c - t_{ao}) + t_{ao}$$

where t_c is the dry resultant temperature specified for the centre of the room from consideration of the application and t_{ao} has been specified for the location. The environmental temperature produced in the room is given by

$$t_{ei} = F_1(t_c - t_{ao}) + t_{ao}$$

EXAMPLE 3.6

A community building 24 m long by 12 m wide and 4 m high is to have a hot-water panel radiator heating system that will maintain a dry resultant temperature of 20 °C at the centre of the room at an external air temperature of −3 °C. There are 12 windows each of area 3 m^2 and two doors each of area 6 m^2. The roof can be taken as being flat. Infiltration of outside air amounts to 0.5 air changes per hour. Thermal transmittances are as follows: windows, 5.7 W/m^2 K; walls, 0.4 W/m^2 K; doors, 2.9 W/m^2 K; roof, 0.4 W/m^2 K; floor, 0.5 W/m^2 K. Find the total rate of heat loss from the building under steady-state conditions, the room air temperature and the environmental temperature.

The calculation of $\Sigma(AU)$ is given in Table 3.11. Using $F_1 = 1.0$, $F_2 = 1.10$, $N = 0.50$, $V = 1152$ m^3, $t_c = 20$ °C and $t_{ao} = -3$ °C

$$Q_p = (1 \times 600 + 0.33 \times 1.1 \times 0.5 \times 1152)(20 + 3)$$

$$= 19\,418 \text{ W}$$

$$= 19.418 \text{ kW}$$

$$t_{ai} = 1.1[20 - (-3)] - 3 \text{ °C}$$

$$= 23.4$$

$$t_{ei} = 1[20 - (-3)] - 3 \text{ °C}$$

$$= 20 \text{ °C}$$

Table 3.11

Surface	Area A (m^2)	U (W/m^2 K)	AU (W/K)
Windows	36	5.7	205.2
Doors	12	2.9	34.8
Walls	252	0.4	100.8
Roof	288	0.4	115.2
Floor	288	0.5	144.0
			$\Sigma(AU) = 600.0$

EXAMPLE 3.7

The air-conditioning system in a computer room breaks down, and it is thought that there would be a risk of condensation forming from the moisture in the air if the room air temperature were to fall below 12 °C. Assess the likely room air temperature from the following information: room dimensions, 10 m × 8 m × 3 m high; dimensions of window in one long exterior wall, 3 m × 2 m; ventilation rate, 0.25 air changes per hour. The surrounding rooms are all at $t_{ei} = 18$ °C. The outside air temperature is −3 °C and the U values of the external wall, the window and the internal surfaces are 0.5 W/m^2 K, 2.8 W/m^2 K and 1.7 W/m^2 K respectively.

The surrounding rooms will steadily transfer heat into the computer room and then this heat will escape through the one external wall and by natural ventilation. The air temperature of the computer room should stabilize at some value t_1 °C. A balance of heat flows into and out of the room can be made:

$$\text{heat flow in} = \text{heat flow out}$$

$$\Sigma(UA\,\Delta t) = \Sigma(UA\,\Delta t) + Q_v$$

We can assume that, initially, the computer room environmental temperature is the same as its air temperature. The internal partition surface area is 238 m^2, the window area is 6 m^2 and the external wall area is 24 m^2. Then

$$\text{heat flow in} = 1.7\,\frac{\text{W}}{\text{m}^2\,\text{K}} \times 238\ \text{m}^2 \times (18 - t_1)\ \text{K}$$

$$= 404.6(18 - t_1)\ \text{W}$$

$$\text{heat flow out} = 2.8\,\frac{\text{W}}{\text{m}^2\,\text{K}} \times 6\ \text{m}^2 \times (t_1 + 3)\ \text{K}$$

$$+ 0.5\,\frac{\text{W}}{\text{m}^2\,\text{K}} \times 24\ \text{m}^2 \times (t_1 + 3)\ \text{K}$$

$$+ 0.33\,\frac{\text{W}}{\text{m}^3\,\text{K}} \times 0.25 \times 10 \times 8 \times 3\ \text{m}^3 \times (t_1 + 3)\ \text{K}$$

$$= 16.8(t_1 + 3) + 12(t_1 + 3) + 19.8(t_1 + 3)\ \text{W}$$

$$= 48.6(t_1 + 3)\ \text{W}$$

Therefore

$$404.6(18 - t_1) = 48.6(t_1 + 3)$$

$$8.33(18 - t_1) = t_1 + 3$$

$$149.9 - 8.33t_1 = t_1 + 3$$

$$146.9 = 9.33t_1$$

Table 3.12 Thermal transmittance and admittance factors for complete structural components with normal exposure

Construction	U (W/m^2 K)	Y (W/m^2 K)
Walls		
1. 220 mm brick, 13 mm light plaster	1.90	3.6
2. 220 mm brick, 25 mm cavity, 10 mm plasterboard on dabs	1.50	2.5
3. 220 mm brick, 25 mm cavity, 10 mm foil-backed plasterboard on dabs	1.20	1.8
4. 220 mm brick, 20 mm glass fibre quilt, 10 mm plasterboard	1.00	1.4
5. 220 mm brick, 25 mm polyurethane slab, 10 mm plasterboard	0.66	1.0
6. 19 mm render, 40 mm expanded polystyrene slab, 200 mm lightweight concrete block, 13 mm light plaster	0.40	2.2
7. 10 mm tile hanging, 25 mm air gap, 100 mm glass fibre quilt, 10 mm plasterboard	0.36	0.67
8. 105 mm brick, 25 mm cavity, 105 mm brick, 13 mm dense plaster	1.50	4.4
9. 105 mm brick, 50 mm UF foam, 105 mm brick, 13 mm light plaster	0.55	3.6
10. 105 mm brick, 25 mm cavity, 100 mm lightweight concrete block, 13 mm light plaster	0.92	2.2
11. 100 mm lightweight concrete block, 75 mm glass fibre, 100 mm lightweight concrete block, 13 mm lightweight plaster	0.29	2.4
Roof		
12. 5 mm asbestos cement sheet	6.5	6.5
13. 10 mm tile, loft-space, 10 mm plasterboard	2.6	2.6
14. 10 mm tile, loft-space, 100 mm glass fibre quilt, 10 mm plasterboard	0.34	0.66
15. 19 mm asphalt, 25 mm stone chippings, 150 mm heavyweight concrete block	2.3	5.2
16. 19 mm asphalt, 13 mm fibreboard, 25 mm air gap, 75 mm glass fibre quilt, 10 mm plasterboard	0.40	0.69
Floor		
17. Concrete	—	6.0
18. Concrete, carpet or woodblock	—	3.0
19. Suspended timber and carpet	—	1.5
Partitions		
20. Heavyweight partition walls	—	3.0
Windows		
21. Single-glazed	—	5.6
22. Double-glazed	—	3.2

Hence

$$t_1 = 15.7\ ^\circ\mathrm{C}$$

Therefore condensation is unlikely.

Where a building is occupied only occasionally, for example a traditional heavyweight stone church or a brick-built assembly hall, the heating system is used intermittently and steady-state heat loss calculations are inappropriate. Admittance factors are used to evaluate the heat flow into the thermal storage of the structure, rather than through it. The heat output required for the heating system is

$$Q_p = [F_1 \Sigma (AY) + 0.33 F_2 NV] (t_c - t_{ao}) \text{ W}$$

The Y values given in Table 3.12 are for a 12 h on, 12 h off heating cycle. To obtain other cycle times, multiply the Y values by $(12/\text{cycle hours})^{0.5}$. This gives higher heat input rates for shorter periods.

Boiler power

The boiler power required for a building is found from the sum of the following:

1. peak heat input rate $Q_f + Q_v$ W to the heating system;
2. heat loss from the distribution pipe or duct system, which can initially be taken as 10% of $Q_f + Q_v$ and refined later when pipe sizes and lengths are known;
3. rate of energy supply Q_{HWS} W to the hot-water services system where this is supplied from the same boiler plant.

Then

$$Q_{HWS} = \frac{\text{mass of stored water kg}}{\text{heating period}} \times 4.186 \frac{\text{kJ}}{\text{kg K}} \times (t_{HWS} - 10) \text{ K}$$

The specific heat capacity of water is 4.186 kJ/kg K and the temperature of mains cold water is normally about 10 °C. The mass of stored hot water at 65 °C either will be 135 l for a small domestic residence or can be found from the number of occupants N and the expected daily hot-water usage per person, which will vary from 4 l per person for an office or shop to 70 l per person for a hotel. The time period to raise the storage cylinder contents to the desired temperature can be varied to suit the site conditions; 3 h is acceptable for housing.

EXAMPLE 3.8

Calculate the boiler power required for a hotel having a peak heat loss of 550 kW, a low-pressure hot-water radiator heating system, 500 occupants and a 3 h heating period for the hot-water storage cylinder. Water is to be stored at 65 °C and daily consumption is expected to be 70 l per person. 1 l of water weighs 1 kg.

$$Q_{HWS} = \frac{500 \text{ people}}{3 \text{ h}} \times 70 \frac{1}{\text{person}} \times \frac{1 \text{ kg}}{1 \text{ l}} \times 4.186 \frac{\text{kJ}}{\text{kg K}} \times (65 - 10) \text{ K}$$

$$\times \frac{1 \text{ h}}{3600 \text{ s}} \times \frac{1 \text{ kWs}}{1 \text{ kJ}}$$

$$= 746 \text{ kW}$$

The boiler power is obtained as follows:

$$\text{building heat loss } Q_f + Q_v = 550 \text{ kW}$$

$$10\% \text{ distribution losses} = 55 \text{ kW}$$

$$Q_{HWS} = 746 \text{ kW}$$

$$\text{boiler power} = 1351 \text{ kW}$$

Thermal transmittance measurement

The current stock of buildings creates the need for the energy engineer to be able to discover the value of the thermal transmittance of existing structures. The building has design *U* values that were calculated in accordance with standard practice and regulations, but what is the reality of the designer's intentions? Does the design *U* value exist in the components that have been constructed? When the components of the building, such as walls, windows, corners of walls, and interfaces between walls and floors, are taken together as an integrated package, are the *U* values achieved? Has the process of construction destroyed the designer's work? Such possibilities have a lasting influence upon the energy consumption of the building.

A large proportion of the energy consumed in the UK is used to keep the inside of buildings warm. Monitoring the quality of *in-situ* thermal insulation and for the retrofitting of additional insulation to existing structures is an important part of energy management. The built thermal transmittance (*U* value) of a structure can be calculated from measurements of air and surface temperatures from a thermocouple temperature instrument or multichannel data logger (Figs 1.3, 1.11 and 1.12). It is not necessary to know the constructional details of the wall, roof, floor, door or window in order to discover its *U* value. The visiting surveyor will not wish to drill holes through brick, concrete and timber to measure the thickness of each material. Even if this is done, the quality of the materials remains largely unknown and assumptions about the water content and the integrity of each layer would have to be made. The constructional detail is unknown. There may be air spaces, vapour barriers and layers of thermal insulation in place, but these are hidden from view.

Figure 3.1 represents a cross-section through the unknown structure. It could be an external wall, internal wall, roof, floor, glazing or door. All that can be realistically assessed are the temperatures on either side, at nodes 1 and 4, and on the surfaces at nodes 2 and 3. A shielded surface-contact thermocouple probe can be used to measure each surface temperature. An exposed thermocouple junction or a sling psychrometer can be used to find the air temperatures. The values for the inside and outside surface film resistances, R_{si} and R_{so} m^2 K/W, are assumed to be their normal, tabulated values for the appropriate applications. The heat transfer equations (Chapter 10) that describe the heat flow through the structure, Q W, are as follows.

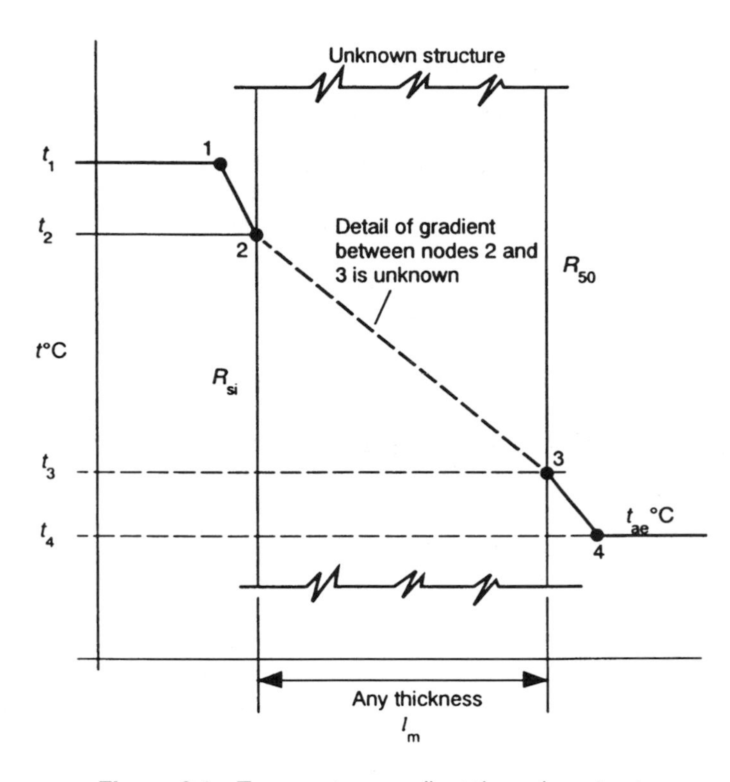

Figure 3.1 Temperature gradient through a structure

For the whole structure

$$Q = U \frac{\mathrm{W}}{\mathrm{m}^2\,\mathrm{K}} \times A\,\mathrm{m}^2 \times (t_1 - t_4)\,\mathrm{K}$$

For the interior film

$$Q = \frac{1}{R_{\mathrm{si}}} \frac{\mathrm{W}}{\mathrm{m}^2\,\mathrm{K}} \times A\,\mathrm{m}^2 \times (t_1 - t_2)\,\mathrm{K}$$

For the unknown structure

$$Q = \frac{1}{R} \frac{\mathrm{W}}{\mathrm{m}^2\,\mathrm{K}} \times A\,\mathrm{m}^2 \times (t_2 - t_3)\,\mathrm{K}$$

where $R\mathrm{m}^2$ K/W is the resistance of the unknown parts of the construction. For the exterior film

$$Q = \frac{1}{R_{\mathrm{so}}} \frac{\mathrm{W}}{\mathrm{m}^2\,\mathrm{K}} \times A\,\mathrm{m}^2 \times (t_3 - t_4)\,\mathrm{K}$$

The heat flow is considered to be under steady-state conditions: that is, it remains at a stable rate over several hours and certainly while the measurements are taken. This is the same assumption that is made for calculating thermal transmittances. It is also true that the daily cyclic variation in outdoor air temperature and solar heat gains, plus the intermittent cooling effects of wind and rain, cause unsteadiness in the flow of heat from the building. The analysis of such heat transfers requires dedicated software, weather data and a computer model of the whole building. An awareness of the overall problem is helpful, however.

There needs to be as large a temperature difference between indoors and outdoors as reasonably practical on the day of test. This is to minimize the effect of any errors in the measurement of the temperatures. When the indoor and external air temperatures are 20 °C and 10 °C, an error of 0.5 °C in one of the temperatures will be $100 \times 0.5/(20 - 10)\%$, 5% of the difference. If overall inaccuracies can be kept within 5%, a reasonably reliable outcome can be obtained. The heating system and weather should also be functioning under steady conditions during the test period. Take the values of R_{si} and R_{so} to be 0.12 m^2 K/W and 0.06 m^2 K/W, as they would be for walls with normal exposure. Use other values if necessary. If, on a test, the temperatures t_1, t_2, t_3 and t_4 are 21 °C, 17 °C, 0 °C and −2 °C, the rate of heat flow Q, thermal transmittance U and resistance R of the structure can be calculated by using the R_{si} or R_{so} equations:

$$Q = \frac{1}{R_{si}} \frac{\text{W}}{\text{m}^2\ \text{K}} \times A\ \text{m}^2 \times (t_1 - t_2)\ \text{K}$$

The surface area $A\text{m}^2$ is taken as 1 m^2:

$$Q = \frac{1}{0.12} \times (21 - 17)\ \text{W}$$

$$= 33.33\ \text{W}$$

The same answer results from the use of R_{so}:

$$Q = \frac{1}{R_{so}} \frac{\text{W}}{\text{m}^2\ \text{K}} \times A\ \text{m}^2 \times (t_3 - t_4)\ \text{K}$$

$$Q = \frac{1}{0.06} \times (0 - (-2))\ \text{W}$$

$$= 33.33\ \text{W}$$

Find the U value from

$$Q = U \frac{\text{W}}{\text{m}^2\ \text{K}} \times A\ \text{m}^2 \times (t_1 - t_4)\ \text{K}$$

$$33.33 = U \frac{\text{W}}{\text{m}^2\ \text{K}} \times 1\ \text{m}^2 \times (21 - (-2))\ \text{K}$$

$$U = \frac{33.33}{(21 - (-2))}\ \text{W/m}^2\ \text{K}$$

$$= 1.45\ \text{W/m}^2\ \text{K}$$

This is an elderly wall, which has a higher thermal transmittance than for modern standards. Consideration can be given as to how much additional thermal insulation is possible. The thermal resistance of the existing structure, without the surface film resistances, can be found from

$$Q = \frac{1}{R}\,\frac{\mathrm{W}}{\mathrm{m^2\,K}} \times A\,\mathrm{m}^2 \times (t_2 - t_2)\,\mathrm{K}$$

$$33.33 = \frac{1}{R}\,\frac{\mathrm{W}}{\mathrm{m^2\,K}} \times 1\,\mathrm{m}^2 \times (17 - 0)\,\mathrm{K}$$

$$R = \frac{(17-0)}{33.33}\,\mathrm{m^2\,K/W}$$

$$= 0.51\,\mathrm{m^2\,K/W}$$

When the thermal transmittance is known from design calculations or *in-situ* measurements, the thickness of additional thermal insulation that is needed to reduce heat loss can be calculated. This may be desirable in order to align the building with current regulations and improve its energy-using efficiency. Outdated building designs will be less attractive to potential users than new or recently refurbished, low-energy consumption residential, commercial and industrial alternative sites.

The wall U value that was considered here could be lowered from 1.45 $\mathrm{W/m^2\,K}$ to, say, 0.4 $\mathrm{W/m^2\,K}$ by the addition of thermal insulation. If the insulation can be injected into the wall cavity no further constructional measures are needed. Where there is no cavity, or if rainwater penetration could result, then an additional internal or exterior layer of material is required. Thermal insulation may not be structurally rigid and it often does not provide a hard-wearing or weatherproof surface. Adding layers to either side of a wall necessitates architectural changes, particularly to fenestration and doorways. If polyurethane board and an internal surface finish of 10 mm plasterboard can be fitted to the interior surfaces, the necessary thickness of insulation can be calculated as follows.

From Table 3.1, the thermal conductivities are:

$$\text{plasterboard: } \lambda = 0.16\,\mathrm{W/m\,K}$$

$$\text{polyurethane board: } \lambda = 0.025\,\mathrm{W/m\,K}$$

$$\text{new thermal resistance of whole structure } R_n = \frac{1}{U_n}\,\frac{\mathrm{m^2\,K}}{\mathrm{W}}$$

$$= \frac{1}{0.4}\,\frac{\mathrm{m^2\,K}}{\mathrm{W}}$$

$$= 2.5\,\mathrm{m^2\,K/W}$$

$$\text{resistance of plasterboard} = \frac{0.01 \text{ m}}{0.16} \frac{\text{m K}}{\text{W}}$$

$$= 0.0625 \text{ m}^2 \text{ K/W}$$

$$\text{resistance of existing wall} = \frac{1}{1.45} \frac{\text{m}^2 \text{ K}}{\text{W}}$$

$$= 0.69 \text{ m}^2 \text{ K/W}$$

The additional thermal insulation that is needed is found by subtracting the existing thermal resistance, and that for the new surface finish, from the target thermal resistance:

$$\text{additional resistance needed} = (2.5 - 0.69 - 0.0625) \text{ m}^2 \text{ K/W}$$

$$= 1.748 \text{ m}^2 \text{ K/W}$$

$$\text{insulation resistance} = \frac{l \text{ mm}}{\lambda} \frac{\text{m K}}{\text{W}} \times \frac{1 \text{ m}}{10^3 \text{ mm}}$$

$$1.748 = \frac{l \text{ mm}}{0.025} \frac{\text{m K}}{\text{W}} \times \frac{1 \text{ m}}{10^3 \text{ mm}}$$

$$\text{insulation thickness } l \text{ mm} = 1.748 \frac{\text{m}^2 \text{ K}}{\text{W}} \times 0.025 \frac{\text{W}}{\text{m K}} \times \frac{10^3 \text{ mm}}{1 \text{ m}}$$

$$= 43.7 \text{ mm}$$

Materials are available in standard dimensions. The thickness to be used will be the next larger size, 50 mm. Check that the additional insulation calculations have been correctly made and find the real new U value:

$$R_n = \frac{1}{1.45} + \frac{0.01}{0.16} + \frac{0.05}{0.025} \frac{\text{m}^2 \text{ K}}{\text{W}}$$

$$= 0.69 + 0.0625 + 2.0 \text{ m}^2 \text{ K/W}$$

$$= 2.752 \text{ m}^2 \text{ K/W}$$

$$U_n = \frac{1}{R_n} \frac{\text{W}}{\text{m}^2 \text{ K}}$$

$$= \frac{1}{2.752} \frac{\text{W}}{\text{m}^2 \text{ K}}$$

$$= 0.36 \text{ W/m}^2 \text{ K}$$

The new thermal transmittance does not exceed the desired value of 0.4 $W/m^2 K$ and is suitable. If the wall has an air cavity between the inner and outer surfaces, it may be possible to inject urea formaldehyde or phenolic foam, or blown rock wool, and achieve the desired result without architectural effects. Another possibility is the addition of insulation and a protective layer to the exterior surface.

EXAMPLE 3.9

A 20-year-old 150 mm thick ribbed concrete flat roof over an office is supported on a structural steel frame. A typical cross-section through the roof is shown in Fig. 3.2. The concrete is waterproofed with 19 mm asphalt that is topped with 25 mm of white stone chippings. Beneath the concrete, there is a 400 mm deep unventilated air space for service cables and pipes. The ceiling tiles are 12 mm thick fibreboard supported on a lightweight galvanized steel frame. The ceiling tile frame is suspended from the structural steel by galvanized wires and self-tapping screws. All the lighting, electrical and other services that are within the ceiling space are supported by hangers from the roof structural steel frame. The roof has normal exposure and its thermal transmittance is to be reduced to 0.25 $W/m^2 K$. Thermocouple temperature sensors were used to assess the average thermal transmittance of the roof structure. On the day of test, the temperatures t_1, t_2 and t_5 were 19 °C, 17.6 °C and 5 °C. The temperature at node 4 could not be measured owing to the roughness of the surface. The temperature at node 3 could be measured but it is not needed. Describe the features of two methods that could be used to insulate the roof. Decide which materials would be suitable and find the correct thickness for the insulation.

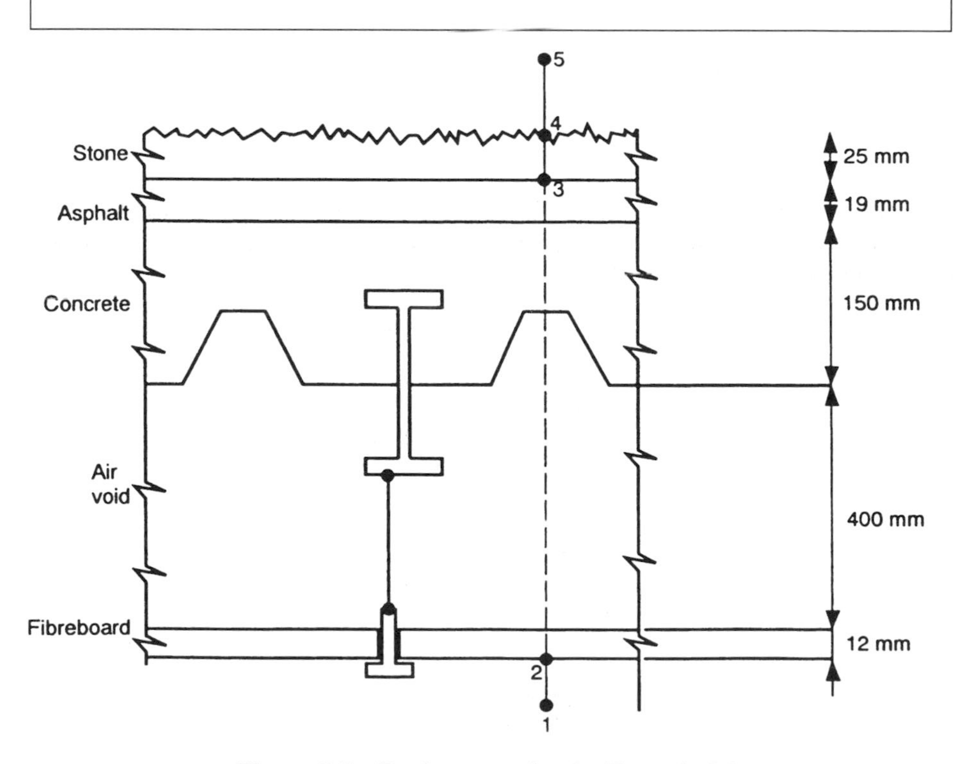

Figure 3.2 Roof construction for Example 3.9

From Tables 3.2, 3.3 and 3.5, $R_{si} = 0.1\ m^2\ K/W$, $R_{so} = 0.05\ m^2\ K/W$ (low emissivity), $R_a = 0.17\ m^2\ K/W$ (high emissivity).

From Table 3.1, the thermal conductivities of the materials are:

cast concrete, lightweight	$\lambda = 0.38$ W/m K
asphalt	$\lambda = 0.5$ W/m K
stone chippings	$\lambda = 0.96$ W/m K
fibreboard	$\lambda = 0.06$ W/m K
glass fibre quilt	$\lambda = 0.04$ W/m K
polyurethane board	$\lambda = 0.025$ W/m K

The options to be tried are as follows:

(a) Remove some ceiling tiles and lay a lightweight blanket, such as glass fibre, on top of the tiles. This depends on whether the fibreboard tiles, support wires and screws are able to hold the additional weight. Extra support rods may be needed. There will be considerable disturbance to the room usage. This may preclude installation work during normal working hours. Removing the tiles will disturb dust and debris from the void and necessitate a cleaning operation in the room. Indoor scaffolding will be needed. Care must be taken not to lay the insulation on top of luminaires and electrical cables, to avoid the overheating of lamps and wiring.

$$Q = \frac{1}{R_{si}} \frac{W}{m^2\ K} \times A\ m^2 \times (t_1 - t_2)\ K$$

$$= \frac{1}{0.1} \times (19 - 17.6)\ W$$

$$= 14\ W$$

$$14\ W = U \frac{W}{m^2\ K} \times 1\ m^2 \times (19 - 5)\ K$$

$$U = \frac{14}{19 - 5}\ W/m^2\ K$$

$$= 1\ W/m^2\ K$$

$$R_n = \frac{1}{0.25} \frac{m^2\ K}{W}$$

$$= 4\ m^2\ K/W$$

$$\text{resistance of the existing roof} = \frac{1}{1} \frac{m^2\ K}{W}$$

$$= 1\ m^2\ K/W$$

$$\text{insulation resistance} = (4 - 1)\ \mathrm{m^2\,K/W}$$

$$= 3\ \mathrm{m^2\,K/W}$$

$$\text{glass fibre thickness } l \text{ mm} = 3\ \frac{\mathrm{m^2\,K}}{\mathrm{W}} \times 0.04\ \frac{\mathrm{W}}{\mathrm{m\,K}} \times \frac{10^3\ \mathrm{mm}}{1\ \mathrm{mm}}$$

$$= 120\ \mathrm{mm}$$

Glass fibre thickness to be used will be 150 mm.

$$R_\mathrm{n} = \frac{1}{1} + \frac{0.15}{0.04}\ \frac{\mathrm{m^2\,K}}{\mathrm{W}}$$

$$= 4.75\ \mathrm{m^2\,K/W}$$

$$U_\mathrm{n} = \frac{1}{4.75}\ \frac{\mathrm{W}}{\mathrm{m^2\,K}}$$

$$= 0.21\ \mathrm{W/m^2\,K}$$

The new thermal transmittance is below the desired value of 0.25 $\mathrm{W/m^2\,K}$ and is suitable.

Check that the new U value is correct:

$$R_\mathrm{n} = 0.1 + \frac{0.012}{0.06} + \frac{0.15}{0.04} + 0.17 + \frac{0.15}{0.38} + \frac{0.019}{0.5} + \frac{0.025}{0.96} + 0.05\ \frac{\mathrm{m^2\,K}}{\mathrm{W}}$$

$$= 4.73\ \mathrm{m^2\,K/W}$$

$$U_\mathrm{n} = \frac{1}{4.73}\ \frac{\mathrm{W}}{\mathrm{m^2\,K}}$$

$$= 0.21\ \mathrm{W/m^2\,K}$$

(b) Remove the stone chippings from the roof and lay sheets of rigid polyurethane or phenolic foam. An adhesive can be used to hold the sheets in place. The stone chippings are then placed on top of the foam. The foam is water-repellent and rot-resistant. Installation on the outer surface of the roof will not cause disturbance indoors. The roof will become a warm deck type (Chapter 10) and will gain the benefit of improved thermal storage capacity: that is, the building will remain warmed for longer periods. In summer, the concrete will be insulated from the solar heat gains and hot outdoor air, and will remain relatively cool:

$$\text{polyurethane thickness } l \text{ mm} = 3\ \frac{\mathrm{m^2\,K}}{\mathrm{W}} \times 0.025\ \frac{\mathrm{W}}{\mathrm{m\,K}} \times \frac{10^3\ \mathrm{mm}}{1\ \mathrm{m}}$$

$$= 75\ \mathrm{mm}$$

Polyurethane thickness to be used will be 100 mm and the new U value will be 0.2 W/m^2 K.

The installation can be validated by measuring the three temperatures when steady-state conditions have been re-established. Calculate the ceiling surface temperature with the glass fibre insulation in place on a day when the indoor and outdoor air temperatures are 21 °C and 3 °C:

$$Q = 0.21 \frac{W}{m^2 K} \times 1 \; m^2 \times (21 - 3) \; K$$

$$= 3.78 \; W$$

$$Q = \frac{1}{R_{si}} \frac{W}{m^2 K} \times A \; m^2 \times (t_1 - t_2) \; K$$

$$3.78 = \frac{1}{0.1} \times (21 - t_2) \; W$$

$$t_2 = 21 - 3.78 \times 0.1 \; °C$$

$$= 20.6 \; °C$$

Questions

1. State what is meant by the following terms:

   (a) thermal resistance;
   (b) thermal conductivity;
   (c) thermal resistivity;
   (d) specific heat capacity;
   (e) thermal transmittance;
   (f) orientation and exposure;
   (g) surface resistance;
   (h) cavity resistance;
   (i) emissivity;
   (j) admittance factor;
   (k) heavyweight and lightweight structures.

2. The following materials are being considered for the internal skin of a cavity wall:

   (a) 105 mm brickwork;
   (b) 200 mm heavyweight concrete block;
   (c) 150 mm lightweight concrete block;
   (d) 75 mm expanded polystyrene slab;
   (e) 100 mm mineral fibre slab and 15 mm plasterboard;
   (f) 40 mm glass fibre slab, 150 mm lightweight concrete block and 15 mm lightweight plaster.

   Compare their thermal resistances and comment upon their suitability for a residence.

3. Calculate the thermal transmittances of the following:

   (a) single-glazed window, severe exposure;
   (b) double-glazed window, sheltered exposure;
   (c) 220 mm brick wall and 13 mm lightweight plaster;
   (d) 220 mm brick wall, 50 mm glass fibre quilt and 10 mm plasterboard;
   (e) 105 mm brick wall, 10 mm air space, 40 mm glass fibre slab and 100 mm lightweight concrete block;
   (f) 40 pitched roof, 10 mm tile, roofing felt and 10 mm flat plaster ceiling with 100 mm glass fibre quilt laid between the joists;
   (g) 19 mm asphalt flat roof, 13 mm fibreboard, 25 mm air space, 100 mm mineral wool quilt and 10 mm plasterboard.

   All exposures are normal unless otherwise specified.

4. A lounge 7 m long × 4 m wide × 2.8 m high is maintained at a resultant temperature of 21 °C and

has 1.5 air changes per hour of outside air at −2 °C. There are two double-glazed wood-framed windows of dimensions 2 m × 1.5 m and an aluminium framed double-glazed door of dimensions 1 m × 2 m. Exposure is normal. One long and one short wall are external and constructed of 105 mm brick, 10 mm air space, 40 mm polyurethane board, 150 mm lightweight concrete block and 13 mm lightweight plaster. The internal walls are of 100 mm lightweight concrete block and are plastered. There is a solid ground floor with edge insulation. The roof has a thermal transmittance of 0.34 W/m^2 K. Adjacent rooms are at a resultant temperature of 18 °C. Calculate the steady-state heat loss from the room for a convective heating system.

5. A single-storey community building of dimensions 20 m × 15 m × 3 m high has low-temperature hot water radiant panel heaters. There are ten windows of dimensions 2.5 m × 2 m. Natural infiltration amounts to one air change per hour. Internal and external design temperatures are 20 °C and −1 °C. Thermal transmittances are as follows: walls, 0.6 W/m^2 K; windows, 5.3 W/m^2 K; floor, 0.5 W/m^2 K; roof, 0.4 W/m^2 K. Calculate the steady-state heat loss.

6. Calculate the environmental temperature produced in an unheated room within an occupied building, using the following information: room dimensions, 5 m × 4 m × 2.6 m high; a window of dimensions 2.5 m × 1.25 m in one long external wall; 0.5 air changes per hour; external air temperature, 3 °C; solid ground floor; surrounding rooms are all at an environmental temperature of 20 °C. The thermal transmittances are as follows external wall, 1 W/m^2 K, window, 5.3 W/m^2 K; floor, 0.7 W/m^2 K, internal walls, 1.2 W/m^2 K; ceiling, 1 W/m^2 K. Assume that the room environmental temperature is equal to the air temperature.

7. A single-storey factory is allowed to have 35% of its wall area as single glazing and 20% of its roof area as single-glazed roof-lights. Wall and roof *U* values are not to exceed 0.6 W/m^2 K. An architect proposes a building of dimensions 50 m × 30 m × 4 m high with a wall *U* value of 0.4 W/m^2 K, a roof *U* value of 0.32 W/m^2 K, 20 double-glazed windows each of area 16 m^2 having a *U* value of 3.3 W/m^2 K and 35 roof-lights each of area 10 m^2 having a *U* value of 5.3 W/m^2 K. Does the proposal meet the allowed heat loss limit?

8. Calculate the boiler power required for a building with a heat loss of 50 kW and an indirect hot-water storage system for 20 people, each using 50 l of hot-water at 65 °C per day. The cylinder is to be heated from 10 °C in 2.5 h. Add 10% for pipe and cylinder heat losses.

9. A single-storey building has dimensions 40 m × 20 m × 4 m high with windows of area 80 m^2 and a door of area 9 m^2. It is to be maintained at a resultant temperature of 20 °C when the outside is at −1 °C and natural ventilation amounts to one air change per hour. Thermal transmittances are as follows: walls, 1.8 W/m^2 K; windows, 5.3 W/m^2 K; door, 5 W/m^2 K; floor, 0.6 W/m^2 K; roof, 1.8 W/m^2 K. A convective heating system is used. It is proposed to reduce the *U* values of the walls and roof to 0.4 W/m^2 K and 0.3 W/m^2 K respectively. Calculate the percentage reduction in heater power that would be produced.

10 List the ways in which existing residential, commercial and industrial buildings can have their thermal insulation improved. Discuss the practical measures that are needed to protect the insulation from deterioration.

11. Review the published journals and find examples of buildings where the existing thermal insulation has been upgraded. Prepare an illustrated presentation or article on a comparison of the outcomes from the cases found.

12. Write a technical report on the argument in favour of adding thermal insulation to existing buildings. Support your case by referring to government encouragement, global energy resources, atmospheric pollution, legislation, cost to the building user and the profitability of the user's company.

13. A flat roof over a bedroom causes intermittent condensation during sub-zero outdoor air temperatures. The roof has normal exposure. The owners want to eliminate the condensation and reduce the thermal transmittance to 0.15 W/m^2 K. Thermocouple temperature sensors were used to assess the average thermal transmittance of the roof structure. On the day of test, the indoor air, ceiling surface and outdoor air temperatures were 16 °C, 11 °C and −2 °C. Calculate the existing thermal transmittance of the roof and the thickness of expanded polystyrene slab that would be needed.

14. An external solid brick wall is to be insulated with phenolic foam slabs held on to the exterior brickwork with UPVC hangers. Expanded metal is to be fixed onto the outside of the foam and then cement rendered to a thickness of 12 mm. The wall has a sheltered exposure. The intention is to reduce the thermal transmittance to 0.3 $W/m^2 K$. Thermocouple temperature sensors were used to assess the average thermal transmittance of the wall prior to the design work. On the day of test, the indoor air, interior wall surface and outdoor air temperatures were 15 °C, 12.7 °C and 6 °C. Calculate the existing thermal transmittance of the wall and the thickness of phenolic foam that would be needed. If the foam is only available in multiple thicknesses of 10 mm, state the thermal transmittance that will be achieved for the wall. Calculate the internal surface temperature that should be found on the wall for a day when the indoor and outdoor air temperatures are 18 °C and 0 °C.

15. The roof over a car-manufacturing area consists of 4 mm profiled aluminium sheets on steel trusses. Wood wool slabs, 25 mm, are fitted below the roof sheets. The roof trusses remain uninsulated as they protrude through the wood wool. The trusses cause condensation to precipitate onto the vehicle bodies during cold weather. The roof is to be insulated with polyurethane board, which will be secured to the underside of the roof trusses. The roof has a normal exposure. The intention is to reduce the thermal transmittance to 0.25 $W/m^2 K$. Thermocouple temperature sensors were used to assess the average thermal transmittance of the roof prior to the insulation. On the day of test, the indoor air under the roof was 13 °C, internal roof surface temperature was 11 °C and the outdoor air temperature was 2 °C. Calculate the existing thermal transmittance of the roof and the thickness of polyurethane that would be needed. The insulation is only available in multiple thicknesses of 10 mm. State the thermal transmittance that will be achieved for the roof. Calculate the internal surface temperature that should be found on the newly insulated roof for a day when the indoor and outdoor air temperatures are 16 °C and −5 °C.

4 Heating

Learning objectives

Study of this chapter will enable the reader to:

1. state the applications for hot-water radiators, natural and fan convectors, embedded pipe radiant panel systems and overhead radiant panels;
2. discuss the use of centralized and decentralized forms of heating system;
3. state the applications for electrical heaters such as radiators, convectors and thermal storage radiators;
4. demonstrate the use of underfloor and ceiling heating systems utilizing electrical energy;
5. apply appropriate heat emitters to the user's needs;
6. explain the use of warm-air heating methods;
7. understand the low-, medium- and high-pressure classifications and applications for water heating systems;
8. understand schematic pipe layouts and pump positioning;
9. design hot-water pipe systems;
10. understand the requirements of oil-firing equipment;
11. have an understanding of combustion;
12. describe flues for oil boilers;
13. calculate the room air temperature to be found during a commissioning test on a heating system;
14. understand the principles of electrical power generation and the use of combined heat and power plant;
15. describe the uses of district heating systems;
16. understand the uses of a BEMS and its terminology;
17. differentiate between local control and overall supervision;
18. understand how a computer system enhances the engineer's work.

Key terms and concepts

Introduction

Terminal heat emitters, such as radiators, convectors and warm-air methods, pipework layouts, and pipe and pump sizing are discussed. Oil-firing equipment is described and the combustion process is analysed. Basic flue arrangements are shown.

Heating systems only operate at their design heating duty when the outside air temperature coincides with that used for heat loss calculations; the commissioning engineer needs to relate heating performance on the day of test to the design figures. Such calculations are shown.

Electricity is generated at the expense of usable energy discharged to the atmosphere or the sea. The plant needed to convert this surplus into saleable heat for district heating is outlined. Interest in this subject will develop for various reasons, and the UK lags behind other European countries in the employment of combined heat and power stations.

The control and operational monitoring of heating, air conditioning and other building services has been enhanced by the use of computer-based techniques known as building energy management systems (BEMSs). These are explained and clear links with other services are shown.

Heating equipment

A wide variety of heating equipment is available that can heat the occupied space either directly by combustion of a fuel or indirectly by utilizing air, water or steam as a heat transfer fluid. The cost of electricity reflects the complexity of its production and distribution, but from the user's point of view it is a refined source of energy, which can be converted with 100% efficiency. Electrical energy purchased at night can be used to heat water, concrete or cast iron in insulated containers. This stored heat is released when needed.

An economic balance is sought between capital and running costs for each application, bearing in mind the building's use. Automatic controls can monitor water and air temperatures, operational times and weather conditions to minimize fuel and electricity consumption. In order to take maximum advantage of a building's thermal storage capacity, optimum-start controllers are used to vary start and stop times for systems that are used intermittently. Computer control is employed in large buildings, where the capital cost can be offset by reduced energy consumption and personnel savings.

Heat emitters can be classified as follows.

Radiators

Heat emitters providing radiation come into this group. A steel single-panel radiator emits about 15% of its total heat output by radiation and the remainder by convection. Radiant output from multiple panel and column types may be a lower percentage of the total. Electric, gas and coal appliances produce large amounts of convection and are partly convectors.

Types of radiator are as follows.

Hot water:	single, double or triple panel column skirting heaters recessed panels banks of pipes
Electricity:	off-peak storage heaters radiant appliances convectors radiant ceiling systems
Gas, coal and oil:	radiant appliances

The main characteristics of these appliances are as follows.

Steel single panels:	neat appearance, high heat output per square metre of surface area, easy to clean, narrow
Steel double panels:	greater heat output per square metre of wall area used, difficult to clean, protrude into the room, more costly
Cast iron panels:	heavy and more obtrusive, low heat output, very long service period
Steel and cast iron columns:	high heat output per square metre of wall area used, bulky, heavy, often mounted on feet, difficult to clean except the hospital pattern which are smooth finished
Radiant panels:	flat cast iron or steel plates with water pipes bonded to their back. They are often mounted at high level in industrial workshops and require a large surface area
Banks of pipes:	bare steel or copper pipes fitted at skirting level in rooms or storage areas to provide an inexpensive heating surface
Off-peak storage:	thermal storage heaters taking electricity at night during less expensive charging periods. The heat is stored at high temperature in cast iron or refractory bricks in an insulated casing. Heat is released continuously into the building unless the heater is fitted with a thermostatically controlled fan and a time switch that determines its operating period. The only other control is over the length of the charge period; this requires estimating the following day's weather pattern. Heaters are bulky and their weight requires attention to the floor structure to ensure sufficient strength

Figure 4.1 demonstrates how the dimensions and the type of radiator are selected so that they can complement the interior decor. Note that curtains across the front of a radiator will reduce its effectiveness at keeping the occupants warm.

Figure 4.1 Panel radiator locations (reproduced by courtesy of Caradon Heating Ltd)

Convectors

There are two types of convector, natural and fan.

Natural convectors

Natural convectors rely on gravity convection currents produced by the heater. Skirting heaters have a finned pipe inside a sheet metal casing as shown in Fig. 4.2. Their heat emission is about 480 W per metre run, they are light and easily handled and they are less obtrusive than taller equipment. Long lengths of

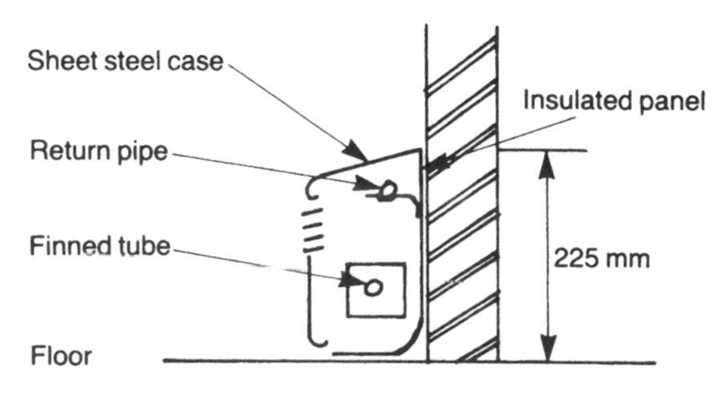

Figure 4.2 Skirting convector heater

unobstructed wall space are needed. Where they run behind furniture, the finned element is omitted and a plain pipe is installed to reduce heat output. They are always fitted onto two-pipe systems and the return pipe can be fitted inside the casing. Valves and air vents are enclosed in accessible boxes at the ends of continuous lengths. Natural convectors produce a uniformly rising current of warm air around the perimeter of the room and this is effective in producing a comfortable environment. There is negligible radiant heating.

Other natural convectors are either 1 m high or extend up to room height, as shown in Fig. 4.3. They create strong convection currents with little radiation and are particularly suitable for locations where elderly, very young or disabled people are being cared for as there are no hot surfaces that may cause skin burns or start fires.

Natural convectors have high heat outputs and can be built into walls, cupboards or adjacent rooms to improve their appearance. Electricity or low- or medium-temperature hot water can be used as the heating medium. The heating elements need periodic cleaning. Such heaters are used in locations where quiet operation and the lack of draughts or intense radiation are important design considerations, such as libraries, art galleries and antique furniture stores.

Fan convectors

Fan convectors have a similar construction to natural convectors with the addition of one or more centrifugal fans and an air filter. Heat output can be very high and fans may be operated at various fixed speeds or from variable-speed motors. Figure 4.4 shows a typical arrangement. Fan operation is controlled from built-in thermostats or remote temperature sensors.

Installation can be at low or high level and the heated airstream is directed away from sedentary occupants. Fan convectors can be usefully sited at doorways to oppose incoming cold air and rapidly reheat entrance areas.

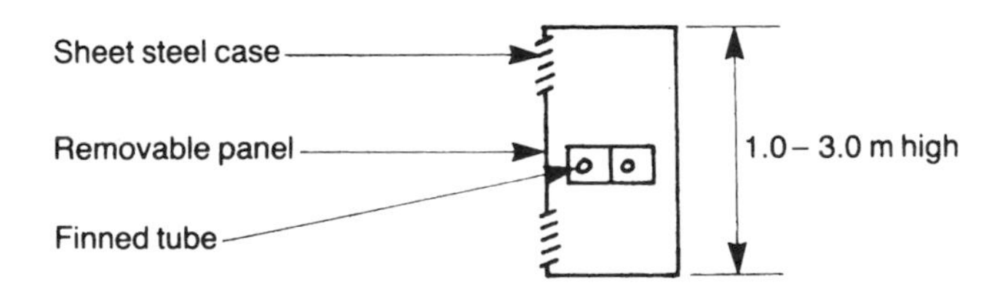

Figure 4.3 Natural convector heater

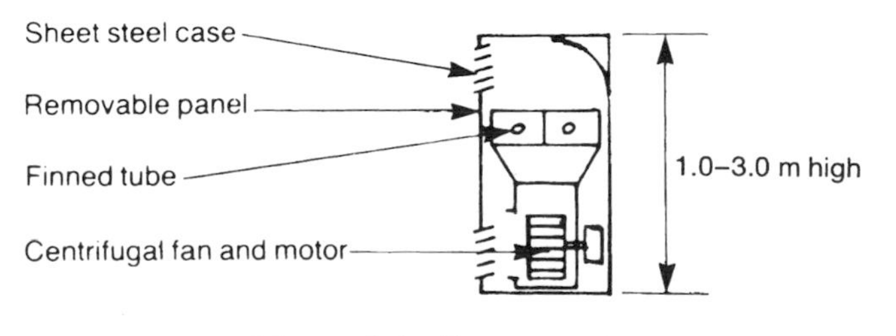

Figure 4.4 Fan convector

A two-pipe circuit must be used, and fan convectors are installed on separate circuits from hot-water radiators as their control characteristics are different. Constant-temperature hot-water is supplied to them, whereas radiators may have variable water temperatures to reduce heat output in mild weather.

Figure 4.5 shows a typical electric off-peak storage heater, which may be a natural or fan convector.

Embedded pipes and cables

Low-temperature hot-water heating pipes or electric heating cables are buried in concrete walls, floors or ceilings to provide a large low-temperature surface that is maintained at a few degrees above room air temperature. Floor-to-ceiling air temperature gradients tend to be less than those obtained with more concentrated forms of heat emission and a uniform distribution of comfort is produced. An example is shown in Fig. 4.6.

Soft copper pipes are laid in position on the concrete floor slab and held by clips, and the ends are connected to header pipes in service ducts. Joints are avoided for the underfloor sections. Steel or plastic pipes may be used in some

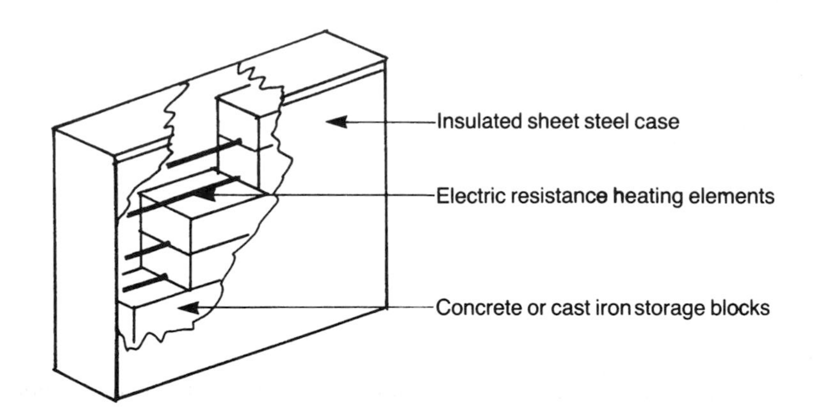

Figure 4.5 Off-peak electric storage heater

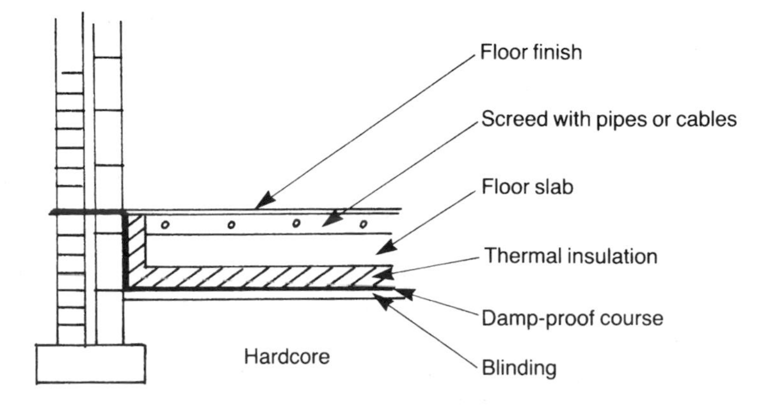

Figure 4.6 Embedded panel heating

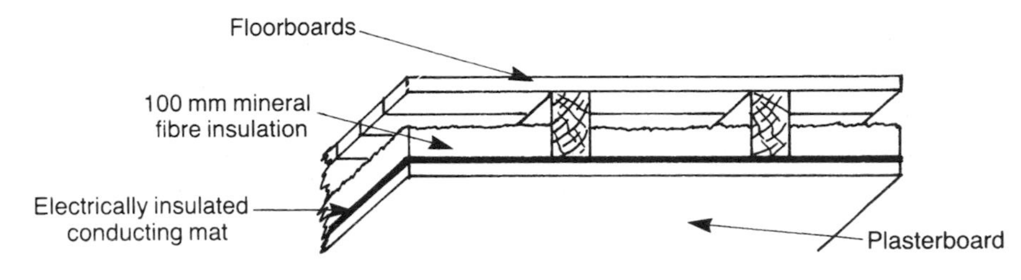

Figure 4.7 Radiant ceiling heating

situations. Thermal expansion and contraction of the pipework must be accommodated and the floor surface temperature is limited to avoid damage to the structure, surface finishes or occupants. This is done by enclosing the pipe in a hard asbestos sleeve on water pipes operating at 85 °C or by controlling water temperature to 45 °C with a mixing valve system. Pipes are buried in the floor screed. Heating elements are evenly distributed to provide uniform radiation and convection to the occupants.

Electric ceiling heating can consist of buried cables or a flexible conducting mat fixed between the ceiling joists and plasterboard, as shown in Fig. 4.7. The mat is electrically insulated from the structure and connected to 240 V 50 Hz supplies to rise to a surface temperature of 40 °C.

Radiant panels

Radiant panel systems employ either a high- or a low-temperature surface to transmit heat by radiation directly to the occupants, and to other unheated surfaces, producing an elevated mean radiant temperature. Comfort conditions can be maintained with lower air temperatures than with convective systems. This should result in economical running costs.

Convection heat output from the hot radiant source is minimized by placing thermal insulation over the reflector. High-temperature radiation is generated using gas combustion close to ceramic reflectors, which emit some heat in the visible part of the infrared region and consequently are seen to be contributing to a feeling of warmth. Domestic gas fires and industrial heaters are in this category. Covered pedestrian areas of shopping precincts can be warmed from recessed units in canopies.

The effect of using high-temperature panels is to produce a series of localized 'sun spots' over a small floor area. Careful siting is necessary to avoid overheating of people or objects.

Low-temperature systems utilize hot water, air or flue gas to heat a metal sheet or pipe, which emits long-wave infrared radiation outside the visible band. They can be installed in factory or office environments and produce a uniform overall warmth, assisted by re-radiation and convection from surfaces heated by the radiant source. Unlike convective systems, they are not adversely affected by room height. Complete systems can be suspended from the ceiling, leaving floors uncluttered.

An evacuated tube system is shown in Fig. 4.8. Flue products from the gas burners are drawn along steel pipes by a vacuum pump and discharged to the atmosphere.

Warm air

Recirculated room air is heated either directly or indirectly by the energy source. Direct firing of combustion gases into the air is permissible only in large well-ventilated factory premises. All other applications require a fuel-to-air heat

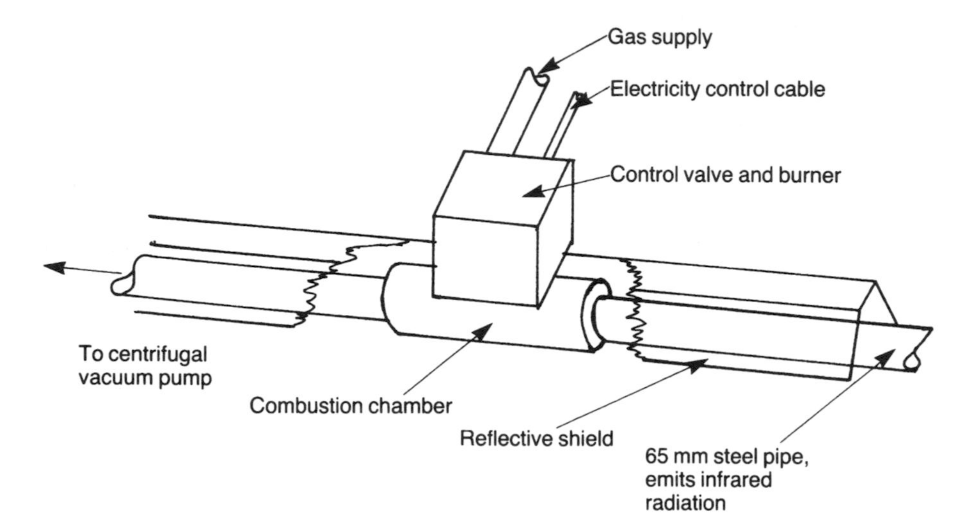

Figure 4.8 Nor Ray Vac industrial overhead radiant tube heating system

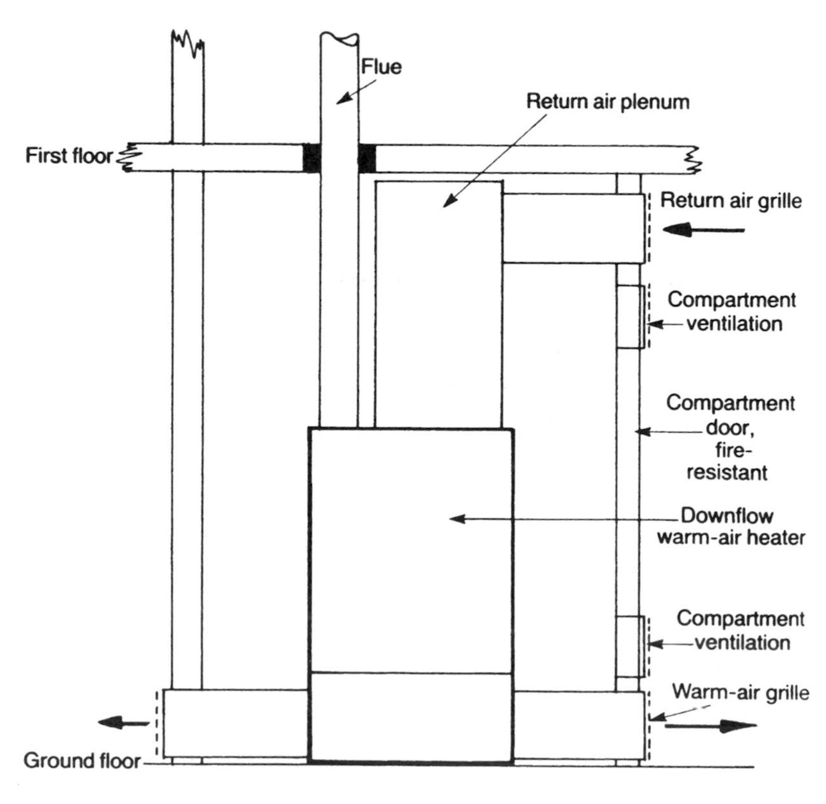

Figure 4.9 Warm-air heater installation

exchanger where the combustion products are enclosed in a sheet metal passageway. Room air is passed over the outside of this heating surface.

Heated air is passed through ducts to the occupied space. It is diffused into the room through a grille, which mixes it with room air convection currents and avoids draughts. Each grille has a damper to regulate the air flow. Extract grilles and ductwork return the air to the heater. Care is needed not to extract air directly from kitchens and bathrooms, as this would lead to odours and condensation in living areas.

The main advantage of warm-air systems is quick heating up and response to thermostatic control. A source of radiant heat is needed in the sitting room to complement the otherwise purely convective heating. A typical domestic installation is shown in Fig. 4.9, where the heater is fitted in a cupboard which is centrally located with respect to all the rooms. Stub ducts are used to connect the heater to the supply and recirculation grilles.

Hot-water heating

The basic arrangements of the various hot-water heating systems are shown in Figs 4.10–4.13. Hot-water heating systems are classified by the temperature and pressure at which they operate (Table 4.1).

The pump position relative to the cold feed and vent pipe connections is important in systems with an open expansion tank. The water pressure rise across the pump can be considerable, and if the arrangement is incorrect water can be pumped up the vent pipe and discharged into the open tank. The connection of the cold feed pipe to the circulation system is known as the neutral point (Fig. 4.15). It is here that the water pressure is always equal to the static height of water above it, with the pump exerting no additional pressure.

A satisfactory arrangement is shown in Fig. 4.15. The hydraulic gradient shows the variation of total water pressure throughout the circulation, i.e. the sum of

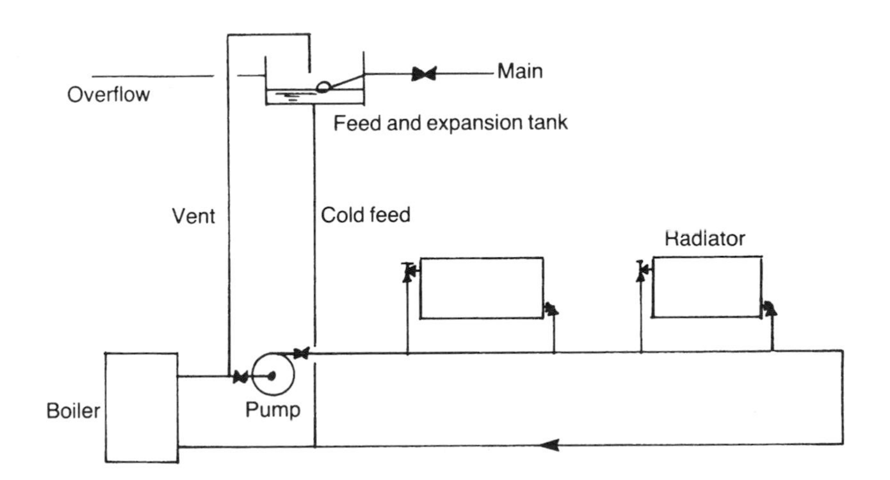

Figure 4.10 Low-temperature hot-water one-pipe heating system

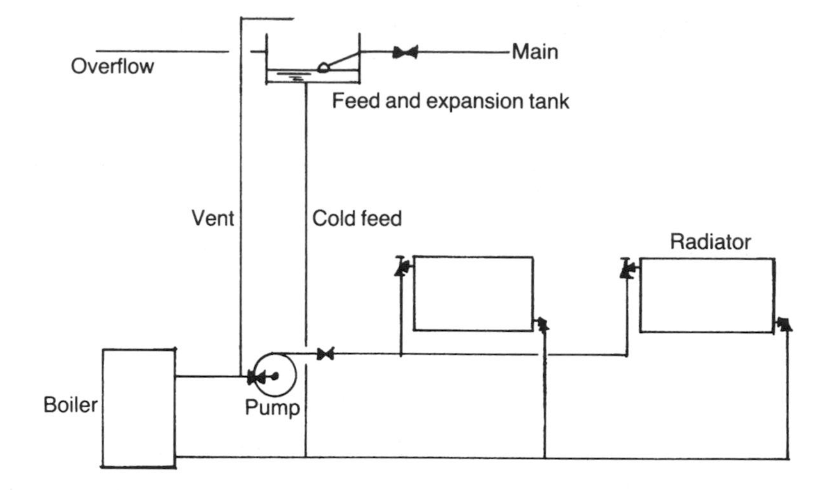

Figure 4.11 Low-temperature hot-water two-pipe heating system

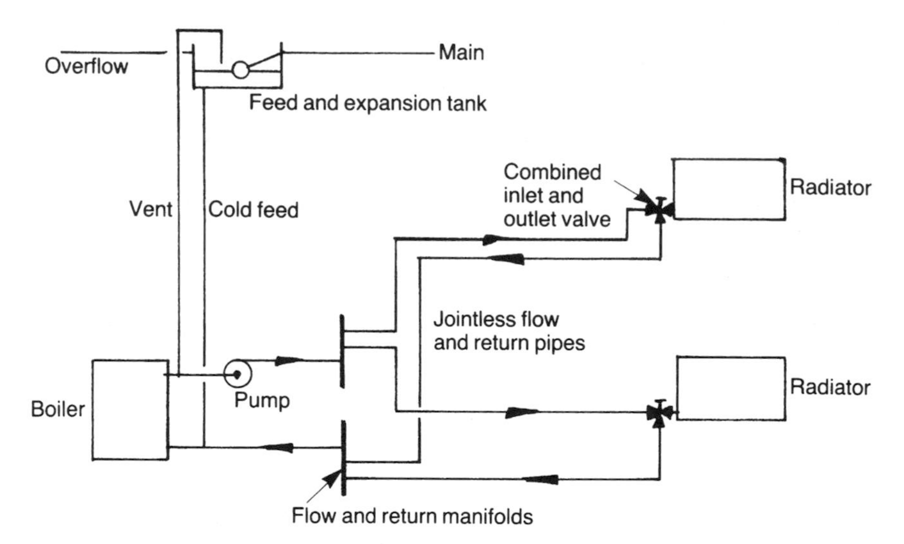

Figure 4.12 Low-temperature hot-water microbore heating system

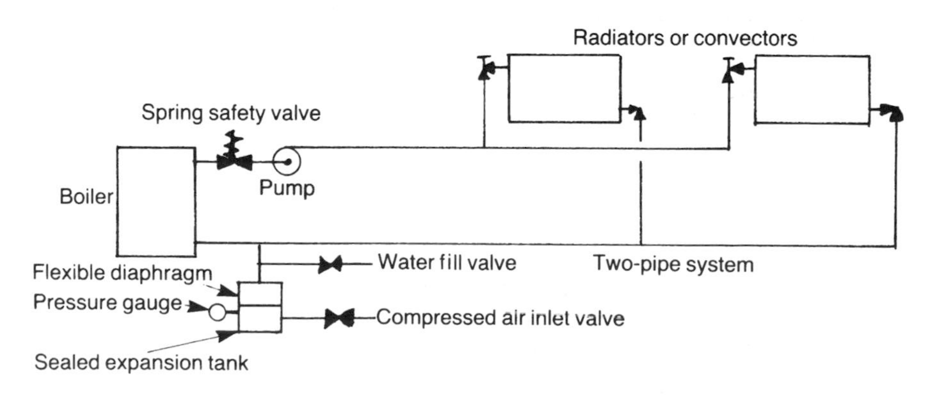

Figure 4.13 Low- or medium-temperature hot-water heating system using a sealed expansion tank

Table 4.1 Classification of hot-water heating systems

Type	Flow temperature (°C)	Pressure
LTHW	80	open to atmosphere, 1 bar
MTHW	120	7 m head above atmosphere, 2 bar
HTHW	over 120	in excess of 7 m head, 3 bar

LTHW, low-temperature hot-water; MTHW, medium-temperature hot-water (refer to Fig. 4.14); HTHW, high-temperature hot-water.

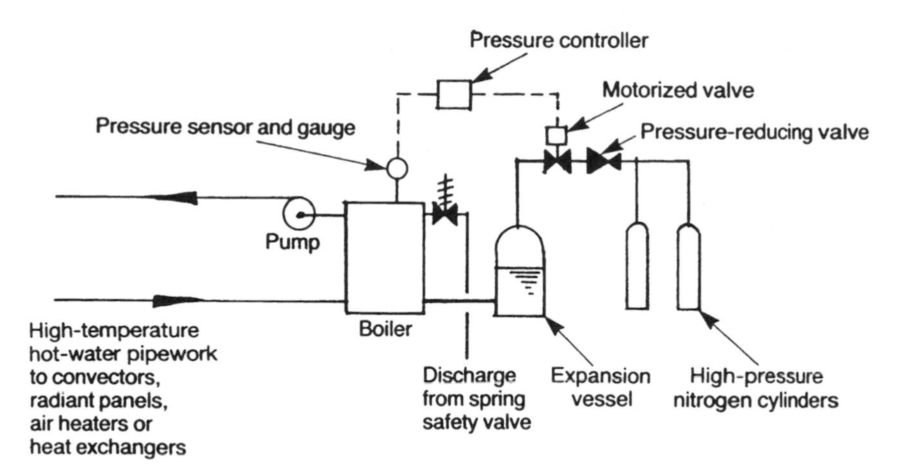

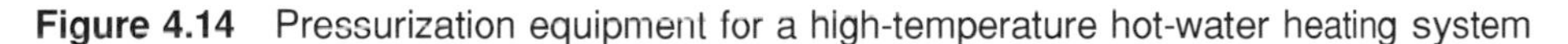
Figure 4.14 Pressurization equipment for a high-temperature hot-water heating system

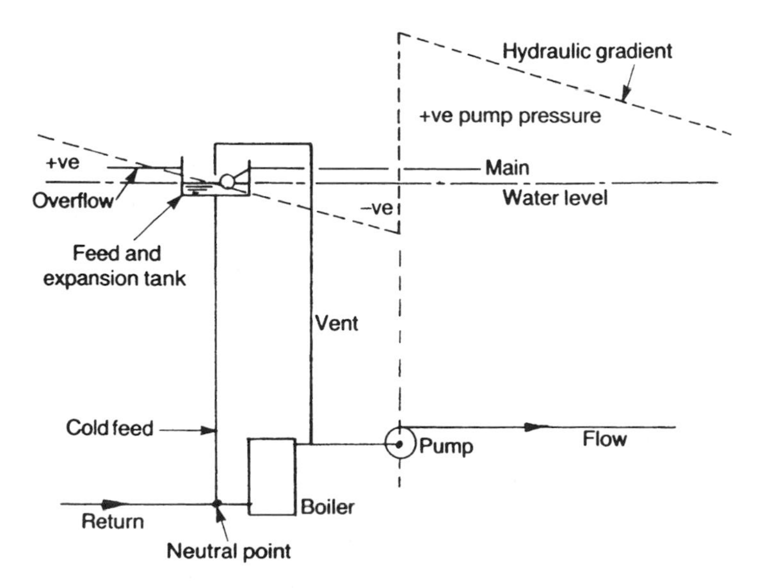

Figure 4.15 Neutral point in a heating system

the static head and the pump head, some of which generates suction pressure between the neutral point and the pump inlet.

The design of a pumped heating system is approached in the following way.

1. Calculate room heat losses.
2. Decide radiator and convector positions and then their sizes from manufacturers' literature.
3. Calculate the water flow rate for each heat emitter.
4. Design the pipework layout.
5. Mark the water flow rates on the pipework drawing and add them up all the way back to the boiler from the furthest heater, marking the drawing with each value.
6. Choose pipe sizes from a chart or data, using an estimate of pressure loss rate and maximum allowable water velocity.
7. Calculate the pump head and water flow rate; compare with the pump manufacturer's stated performance curves and choose a suitable pump.

The temperature of water flowing through a heat emitter or along a pipe will drop from its flow temperature t_f °C to its return temperature t_r °C and lose heat Q W. If there is one heater in a room, Q will be equal to the room steady-state heat loss. Heat losses from distribution pipework may initially be assumed to be 10% of the room heat loss, and the radiator water flow rate should be increased by this amount.

The specific heat capacity (SHC) of water is 4.19 kJ/kg K. The heat balance equation is

$$\text{heat lost by water} = \text{radiator heat output } Q + \text{pipe heat loss}$$

or

$$\text{water flow rate } q \times \text{SHC} \times \text{temperature drop} = Q + 10\% \times Q$$

$$\text{water flow rate } q = \frac{1.1\ Q}{\text{SHC}(t_f - t_r)} \text{ kg/s}$$

The radiator manufacturer's heat output data will be for a fixed temperature difference between the mean water temperature and the room air temperature. Usually a figure of 55 K is used. Comparison of design conditions and literature data can be made using the following equations:

$$\text{radiator mean water temperature MWT} = \frac{t_f + t_r}{2}$$

$$\text{radiator heat output at 55 K difference} = \frac{\text{room heat loss}}{\text{temperature correction factor}}$$

where

$$\text{temperature correction factor} = \left[\frac{(t_f + t_r)/2 - t_a}{55}\right]^{1.30}$$

and t_a is the room air temperature (°C).

EXAMPLE 4.1

A double-panel radiator is to be installed in a room where the air temperature is 22 °C and the heat loss is 3400 W. Water flow and return temperatures are to be 90 °C and 80 °C respectively. An extract from a radiator manufacturer's catalogue for a temperature difference of 55 K is given in Table 4.2. Select a suitable radiator and find the water flow rate through it.

Table 4.2 Heat output from steel double-panel radiators of 500 mm and 700 mm height

	Heat output (kW) for 55 K difference	
Radiator length (m)	500 mm	700 mm
1.720	2.00	2.60
1.920	2.30	2.90
2.200	2.60	3.25
2.400	2.85	3.60
2.600	3.10	3.90

$$\text{temperature correction factor} = \left[\frac{(90+80)/2-22}{55}\right]^{1.30}$$

$$= 1.193$$

$$\text{radiator heat output at 55 K difference} = \frac{3400 \text{ W}}{1.193} = 2850 \text{ W}$$

Either the 2.4 m × 500 mm radiator or the 1.92 m × 700 mm radiator would be satisfactory, depending upon the wall space available. The water flow rate is

$$q = \frac{1.1 \times 3.4}{4.19 \times (90-80)}$$

$$= 0.0893 \text{ kg/s}$$

A first estimate of pipe sizes can be made using an average pressure loss rate $\Delta p/EL$ for the whole system (EL is the equivalent length). Either an arbitrary figure of say 300 N/m^3 is chosen, or a figure is evaluated from the expected pump head. The index circuit length is found from the total flow and return pipe length from the boiler to the furthest radiator. Pipe fittings increase frictional resistance and an initial 25% increase in measured pipe length is made in order to find the equivalent length of the system:

$$\text{pump head } \Delta p \frac{\text{N}}{\text{m}^2} = EL \text{ m} \times \frac{\Delta p}{EL} \frac{\text{N}}{\text{m}^3}$$

This is often converted into head H m water gauge:

$$\text{pump head } H = \Delta p \frac{\text{N}}{\text{m}^2} \times \frac{\text{m}^3}{\rho \text{ kg}} \times \frac{\text{kg}}{g \text{ N}}$$

$$= \frac{\Delta p}{9.807 \times 1000} \text{ m}$$

$$= \frac{\Delta p}{9807} \text{ m}$$

Water flow rate through the pump is the sum of all the radiator water flow rates. Figure 4.16 shows typical pump performance curves. Figure 4.17 shows an arrangement of two centrifugal pumps that could be used in this type of application. The rotary valve at each pipe junction is used to isolate one or both pumps. The delivery pipe manifold has a non-return valve on each pump connection to stop back circulation through the stationary pump. The allowed pressure loss rate can be assessed from the pump characteristic curves. For example, if pump B is to be used at a flow rate of 1 kg/s, the corresponding head developed is 3 m. This is equal to a pressure

$$\Delta p = 9807\ H$$

$$= 9807 \times 3 \text{ m}$$

$$= 29\ 421 \text{ N/m}^2$$

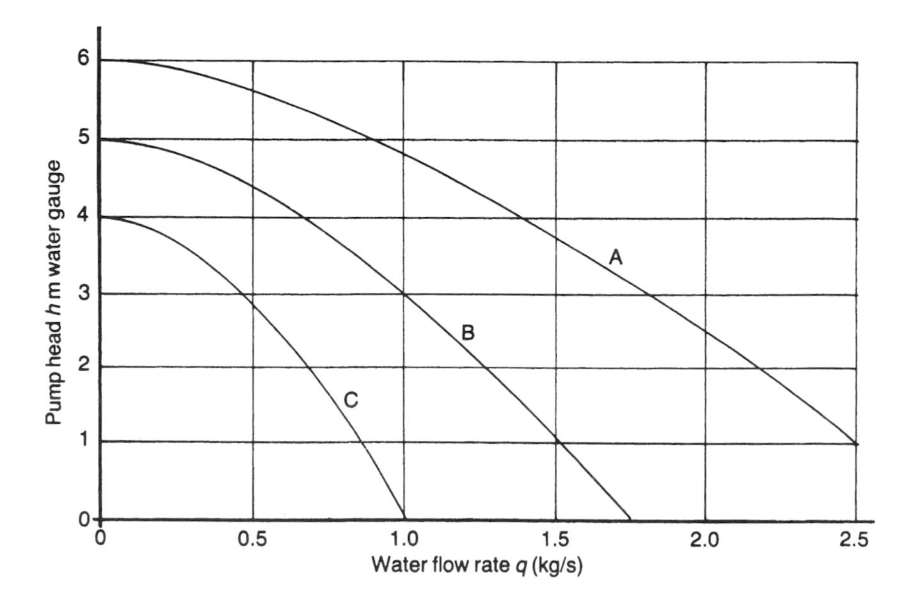

Figure 4.16 Pump performance curves

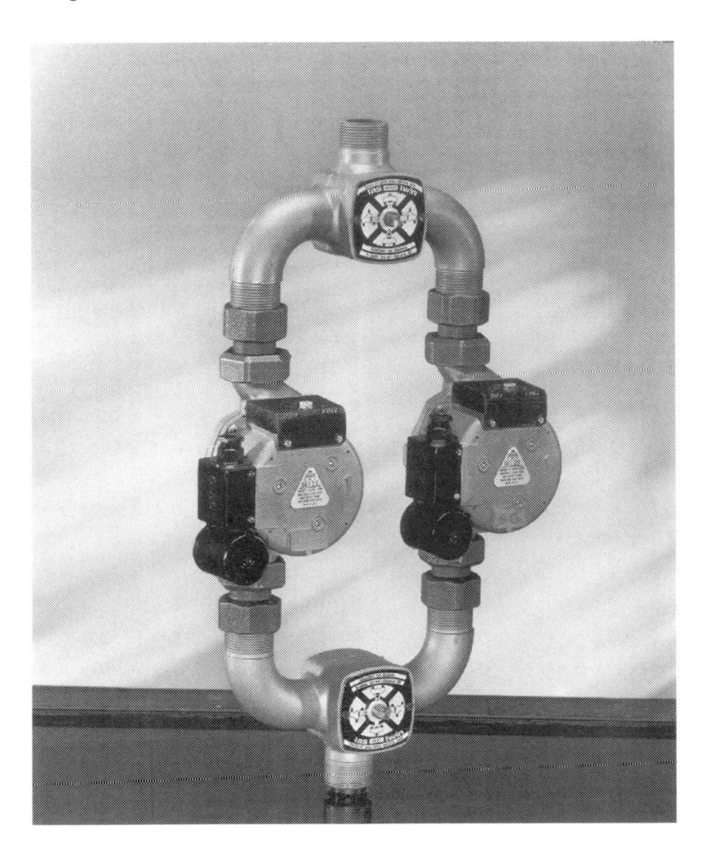

Figure 4.17 Twin centrifugal pump set (reproduced by courtesy of Potterton Myson Ltd)

As

$$29\,421\ \frac{\mathrm{N}}{\mathrm{m}^2} = EL\ \mathrm{m} \times \frac{\Delta p}{EL}\ \frac{\mathrm{N}}{\mathrm{m}^3}$$

$$\text{allowed}\ \frac{\Delta p}{EL} = \frac{29\,421}{EL}\ \mathrm{N/m^3}$$

If the measured index circuit is 50 m of pipework

$$EL = 50\ \mathrm{m} \times 1.25$$

$$= 62.5\ \mathrm{m}$$

and

$$\frac{\Delta p}{EL} = \frac{29\,421}{62.5} = 471\ \mathrm{N/m^3}$$

This would be the maximum pressure loss rate, averaged over all the pipework, and pipe sizes would be read from the 460 $\mathrm{N/m^3}$ line in Table 4.3 to ensure that the available pump head was not exceeded (Moss, 1996).

Table 4.3 Flow of water in copper pipes of various diameters

$\Delta p/EL$ (N/m^3)	Water flow rate q (kg/s) for diameters of					
	6 mm	15 mm	22 mm	28 mm	35 mm	42 mm
200	0.013	0.065	0.174	0.381	0.656	1.060 $v = 1.0$
260	0.015	0.075	0.202	0.441	0.760	1.230
300	0.016	0.081	0.219	0.478	0.823	1.330
360	0.018	0.090	0.242	0.529	0.910	1.470
400	0.019	0.096	0.257	0.561	0.965	1.560
460	0.020	0.104	0.278	0.607	1.040	1.680 $v = 1.50$
500 $v = 0.50$	0.021	0.109	0.291	0.635	1.090	1.760
560	0.023	0.116	0.310	0.677	1.160	1.880
600	0.024	0.120	0.323	0.703	1.210	1.950
660	0.025	0.127	0.340	0.741	1.270	2.050
700	0.026	0.131	0.352	0.766	1.320	2.120
760	0.027	0.138	0.368	0.802	1.380	2.220
800	0.028	0.142	0.379	0.825	1.420	2.280

v, water velocity (m/s); $\Delta p/EL$, rate of pressure loss due to friction.
Source: Reproduced from *CIBSE Guide* (CIBSE, 1986) by permission of the Chartered Institution of Building Services Engineers.

EXAMPLE 4.2

Figure 4.18 shows the arrangement of a two-pipe low-temperature hot-water heating system serving three radiators. The pipe dimensions indicated apply to both flow and return pipes. The frictional resistance of the pipe fittings amounts to 25% of the measured pipe length. Flow and return water temperatures at the boiler are to be 85 °C and 72 °C respectively. Each radiator is situated in a room air temperature of 20 °C. The heat outputs of radiators 1, 2 and 3 are 1 kW, 2 kW and 3 kW respectively. Pump C of Fig. 4.16 is to be used. Find the pipe sizes for the system.

For radiator 1

$$q_1 = \frac{1.1 \times 1}{4.19 \times (85 - 72)} = 0.02 \text{ kg/s}$$

Similarly, for radiators 2 and 3

$$q_2 = 0.04 \text{ kg/s}$$

$$q_3 = 0.06 \text{ kg/s}$$

Water flow in the distribution pipework will be

$$q_X = 0.12 \text{ kg/s}$$

$$q_Y = 0.10 \text{ kg/s}$$

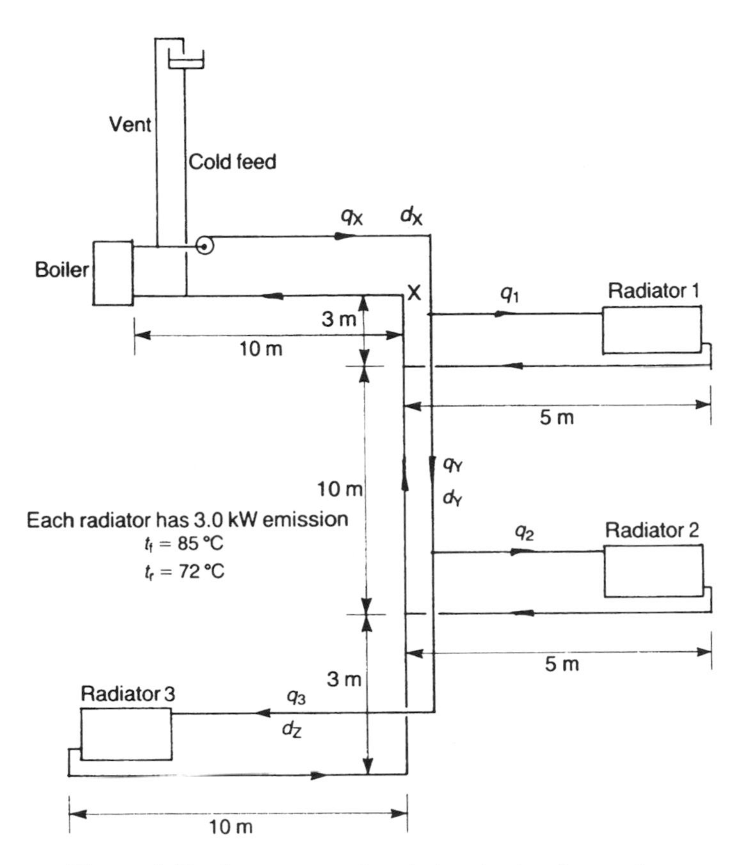

Figure 4.18 Low-temperature hot-water heating system

Water flow rate through the pump is

$$q_X = 0.12 \text{ kg/s}$$

Available pump head, from Fig. 4.16, is

$$H = 6 \text{ m}$$

Pump pressure rise is

$$\Delta p = 9807 \times 6 \text{ N/m}^2$$

$$= 58\ 842 \text{ N/m}^2$$

The index circuit is from the boiler to radiator 3; thus the measured pipe length is

$$2(10 + 3 + 10 + 3 + 10) \text{ m} = 72 \text{ m}$$

Therefore

$$EL = 1.25 \times 72 \text{ m}$$

$$= 90 \text{ m}$$

and

$$58\ 842\ \frac{N}{m^2} = 90\ m \times \frac{\Delta p}{EL}\frac{N}{m^3}$$

$$\text{maximum available}\ \frac{\Delta p}{EL} = \frac{58\ 842}{90}\ N/m^3$$

$$= 653.8\ N/m^3$$

Thus, from Table 4.2, using $\Delta p/EL = 600\ N/m^3$, the pipe sizes are

$$d_X = 15\ mm$$

$$d_Y = 15\ mm$$

$$d_Z = 15\ mm$$

Note that the pressure loss rates in pipes Y and Z are 460 N/m^3 and less than 200 N/m^3 respectively. Not all the allowable pump head of 6 m will be absorbed in pipe friction losses. A gate valve beside the pump will be partially closed to increase the resistance of the system. Each radiator has two handwheel valves, one for temporary adjustments by the occupant and the other for flow regulation by the commissioning engineer.

Oil-firing equipment

Fuel oil is graded in the Redwood no. 1 viscosity test according to its time of flow through a calibrated orifice at 38 °C. Vaporizing and wall-flame burners in boilers of up to 35 kW heat output use 28 s oil, pressure jet burners use gas oil class D (34 s), and industrial boiler plant uses grade E (250 s), grade F (1000 s) and grade G (3500 s). Power stations may use 6000 s residual oil, heated to make it flow. This is the tar residue from crude oil distillation and can only be burnt economically on such a large scale.

Figure 4.19 shows a typical domestic oil storage and pipeline installation. In the UK domestic oils can be stored in outdoor tanks. Grades E, F and G require immersion heaters in the tank and pipeline heating to ensure flow.

Wall-flame burners have a rotating nozzle, which sprays oil onto peripheral plates around the inside of a water-cooled vertical cylindrical combustion chamber. An electric spark ignites the oil impinging on the plates, establishing a ring of flame around the walls of the boiler. Correct oil flow rate from the reservoir is controlled by a ball valve.

Vaporizing burners consist of a vertical cylinder that is heated by the flame and evaporates further oil fed into its base from the reservoir flow control.

Pressure jet burners are usually confined to boilers in plant rooms as they produce more noise. Oil is pumped at high pressure through a fine nozzle, forming a conical spray in the furnace. Combustion air is blown into this oil mist from a centrifugal fan. The turbulent interaction of oil and air causes further atomization of the oil droplets, and the mixture is ignited by an electric spark.

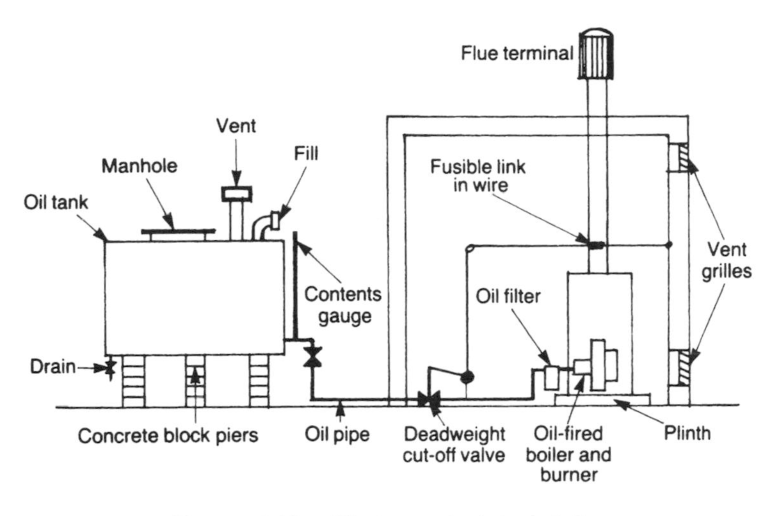

Figure 4.19 Oil storage tank installation

Combustion

Combustion is an exothermic chemical reaction that liberates heat. Fuel must be intimately mixed with sufficient oxygen and raised to a temperature high enough for combustion to be maintained. All the carbon and hydrogen in the fuel are burnt into gaseous products that can be safely vented into the atmosphere. Hydrocarbon fuels are highly energy-intensive. They require little storage volume and their combustion is controllable.

The constituents of dry air are 21% oxygen, 79% nitrogen and less than 1% other chemicals such as carbon dioxide, carbon monoxide, nitrous oxides and rare gases, measured by volume. Nitrogen is inert and takes no part in the chemistry of combustion, but it is heated in its passage through the furnace. Typical chemical compositions of fuels are given in Table 4.4.

Table 4.4 Fuel data

	Anthracite	Gas oil	Natural gas
Moisture (%)	8.0	0.05	
Ash (%)	8.0	0.02	
Carbon (%)	78.0	86.0	
Hydrogen (%)	2.4	13.3	
Nitrogen (%)	0.9		2.7
Sulphur (%)	1.0	0.75	
Oxygen (%)	1.5		
Methane (%)			90.0
Hydrocarbons (%)			6.7
Carbon dioxide (%)			0.6

Source: Reproduced from *CIBSE Guide* (CIBSE, 1986) by permission of the Chartered Institution of Building Services Engineers.

The quantity of air required for complete combustion and the composition of the products can be evaluated from the fuel chemistry. For methane (CH_4) the complete volumetric analysis would be

$$\text{methane} + \text{air} \rightarrow \text{carbon dioxide} + \text{water vapour} + \text{nitrogen}$$

The chemical symbols for these are as follows: oxygen, O_2; nitrogen, N_2; carbon dioxide, CO_2; water vapour, H_2O. Therefore (after complete combustion) we have

$$CH_4 + 2O_2 + N_2 \rightarrow CO_2 + 2H_2O + N_2$$

This means that one volume of CH_4, when reacted with two volumes of O_2 during complete combustion, will produce one volume of CO_2 and two volumes of water vapour. All measurements are at the same temperature and pressure. It is assumed that the water vapour is not condensed.

Some condensation is inevitable, however, and when sulphur (S) is present in the fuel, it combines with some of the O_2 to form sulphur dioxide (SO_2). If the gaseous SO_2 comes into contact with condensing water vapour and further O_2, weak sulphuric acid (H_2SO_4) may be formed in the flue. Coagulation of liquid H_2SO_4 and carbon particles from chimney surfaces leads to the discharge of acid smuts into the atmosphere, causing local damage to washing, cars and stonework. Acidic corrosion of the boiler and chimney greatly reduce their service period. The flue gas temperature is kept above the acid dew-point of about 50 °C to avoid such problems.

It can be seen from the methane combustion equation that 2 m^3 of O_2 are required to burn 1 m^3 of CH_3 completely. This O_2 is contained in $2/0.21 = 9.52$ m^3 of air, and this air contains $9.52 - 2 = 7.52$ m^3 N_2.

In order to ensure complete combustion under all operating conditions and to allow for deterioration of boiler efficiency between servicing, excess air is admitted. This ranges from 30% for a domestic pressure jet oil burner down to a few per cent in power station boilers where continuous monitoring and close control are essential. Excess O_2 from the excess air appears in the flue gas analyses. Measurement of O_2 and CO_2 levels reveals the quantity of excess air.

The presence of carbon monoxide (CO) in the flue gas indicates that some of the carbon in the fuel has not been completely burnt into CO_2 and that more combustion air is needed. The theoretically correct air-to-fuel ratio is the stoichiometric ratio. Figure 4.20 shows the variation of flue gas constituents with the air-to-fuel ratio.

The CO_2 content of oil-fired boiler plant flues will be about 12% at 30% excess air, the combustion air volume required per kilogram of fuel burnt will be about 14.6 m^3 and the flue gas temperature leaving the boiler will be about 200 °C. For domestic natural draught gas-fired boilers, excess air may be 60%, the flue gas temperature will be 165 °C and the CO_2 content will be around 7.5%.

Samples of flue gas taken during commissioning and routine servicing are tested for CO_2 and O_2 content by absorption into chemical solutions. The Orsat apparatus is typical. The smoke content is measured by drawing a sample through filter paper and comparing the discoloration with known values.

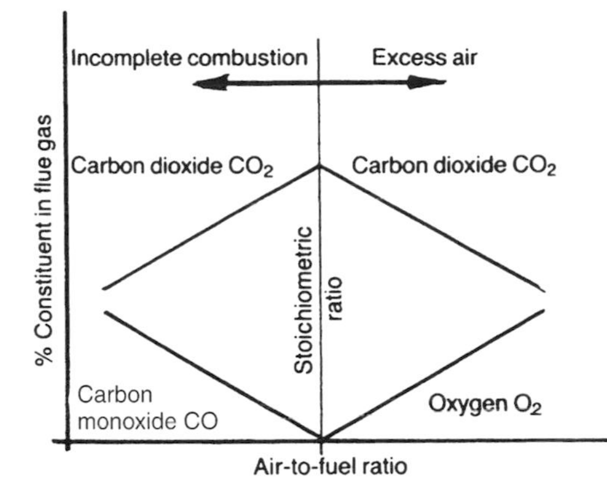

Figure 4.20 Variation of flue gas constituents with air-to-fuel ratio

Flues

Flue systems for oil-fired boilers are of either the conventional brick chimney or free-standing pipe designs so as to discharge combustion products into the atmosphere and allow sufficient dilution so that fumes reaching ground level will not be noticeable. Chimneys from large boiler plant may be subject to minimum height specification by the local authority. Efflux velocity from the chimney can be increased by utilizing a venturi shape when fan assistance is used, effectively raising the chimney height.

The flue pipe diameter will be equal to the boiler outlet connection. Each boiler in a multiple installation should have its own flue. Flues must be kept warm to prevent acidic condensation, and are therefore constructed within the building. Some useful heat is reclaimed in this way. Figure 4.21 shows the necessary separation of flue pipe from combustible materials at an intersection with a floor. External free-standing pipes are constructed of double-walled asbestos or stainless steel to reduce heat loss. Figure 4.22 shows two suitable arrangements.

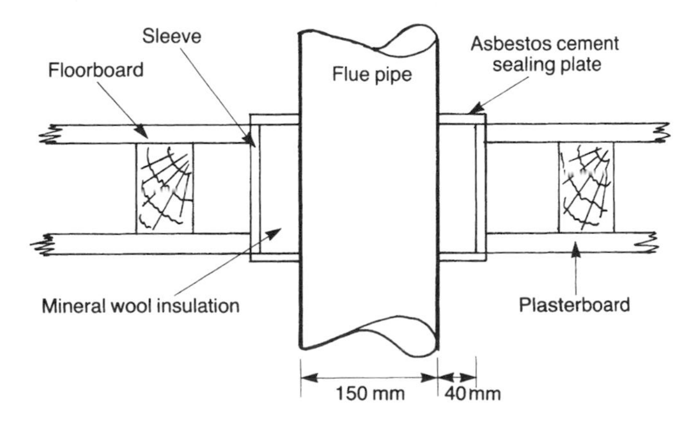

Figure 4.21 Separation of flue pipe from combustible materials in a floor

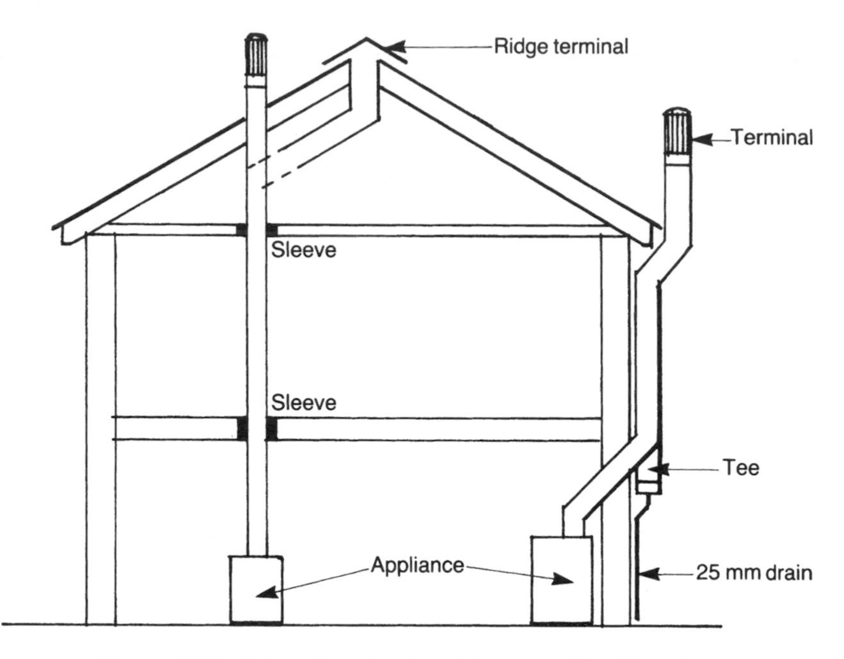

Figure 4.22 Internal and external free-standing flue pipes

Performance testing

A radiator heating system is subjected to a water pressure test at greater than its normal working pressure prior to thermal insulation. To check its thermal performance, an assessment is made of the expected room temperature for the external conditions prevailing during the test. Careful measurements of the external air temperatures at regular intervals, the amount of solar radiation, the wind speed, the occupancy, the use of internal lighting and the sources of heat gains are needed.

Under steady-state conditions

heat loss from room = heat output from radiator

$$UA\ (t_{ei} - t_{eo}) = \text{constant} \times \left[\frac{(t_f + t_r)/2 - t_{ai}}{55}\right]^{1.3}$$

The radiator constant is its heat emission for a 55 K temperature difference. The room heat loss is a series of constants multiplied by $(t_{ei} - t_{eo})$ K, including ventilation heat loss, and can be characterized by the air temperature difference $(t_{ai} - t_{ao})$ K. Thus $(t_f + t_r)/2$ can be replaced by the mean water temperature t_m °C. Therefore the heat balance equation simplifies to

$$(t_{ai} - t_{ao})_1 = C(t_m - t_{ai})_1^{1.3}$$

and

$$(t_{ai} - t_{ao})_2 = C(t_m - t_{ai})_2^{1.3}$$

where subscripts 1 and 2 refer to the test day and design conditions respectively. These equations can be divided:

$$\frac{(t_{ai} - t_{ao})_1}{(t_{ai} - t_{ao})_2} = \left[\frac{(t_m - t_{ai})_1}{(t_m - t_{ai})_2}\right]^{1.3}$$

The left-hand side of the equation represents the variation of building heat loss with temperature, and the right-hand side represents the heating system performance at various room temperatures. A graph of each side against a base of room air temperature can be drawn to find the room air temperature that satisfies both sides of the equation.

EXAMPLE 4.3

A radiator heating system is designed to produce an internal air temperature of 23 °C at an outside air temperature of −2 °C with a mean water temperature of 85 °C. It was tested on a calm cloudy day, before occupation, when the external air temperature remained stable at 5 °C for 2 days. Water flow and return temperatures at the boiler were found to be 88 °C and 76 °C respectively during the test. Room internal air temperature remained stable at 26 °C. State whether the heating system performance met its design conditions.

On the test day

$$t_m = \frac{88 + 76}{2}\ ^\circ\text{C} = 82\ ^\circ\text{C}$$

and

$$\frac{t_{ai} - 5}{23 - (-2)} = \left(\frac{82 - t_{ai}}{85 - 23}\right)^{1.3}$$

where t_{ai} is the expected internal air temperature on the test day. Then

$$\frac{t_{ai} - 5}{25} = \left(\frac{82 - t_{ai}}{62}\right)^{1.3}$$

Table 4.5 Radiator heat output test in Example 4.3

	Heat loss	Radiator heat output	
t_{ai}(°C)	$(t_{ai} - 5)/25$	$(82 - t_{ai})/62$	$[(82 - t_{ai})/62]^{1.3}$
30.0	1.0	0.839	0.80
25.0	0.8	0.919	0.90
20.0	0.6	1.000	1.000

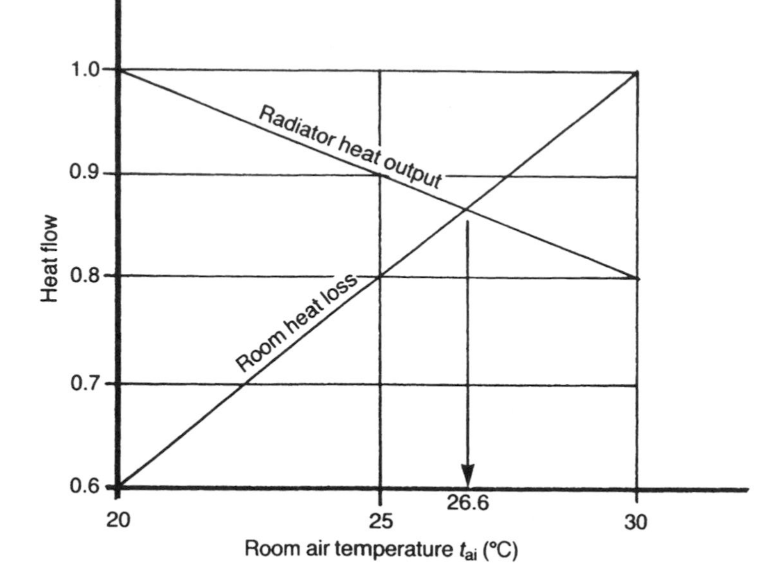

Figure 4.23 Variation of room heat loss and radiator heat output with room air temperature in Example 4.3

Assume a range of values for the internal air temperature of say 30, 25 and 20 °C. Evaluate each side of the equation (Table 4.5) and plot against t_{ai} (Fig. 4.23). The graphical solution reveals that an internal air temperature of 26.6 °C satisfies both equations. This can be verified by substituting 26.6 °C for t_{ai} in the heat balance equation. Thus the measured temperature on the test day is sufficiently close to the theoretically expected figure to say that the heating system meets its design specification.

Electrical power generation

Electricity is generated by alternators driven by steam turbines in power stations. The largest alternators produce 500 MW of electrical power at 33 kV. The steam is produced in a boiler heated by the combustion of coal or residual fuel oil, which could otherwise only be used for making tar. The oil is heated to make it flow through distribution pipework.

Nuclear power stations produce heat by a fission reaction and the active core is cooled by pressurized water (pressurized water reactor (PWR)), carbon dioxide gas (high temperature gas-cooled reactor (HTGR)), liquid sodium (fast breeder reactor) or heavy water (Canadian deuterium (CANDU) system). This fluid then transfers its heat to water, boiling it into steam to drive conventional turbines.

Smaller alternators are driven by methane combustion in gas turbine engines or by diesel engines. A large modern power station has four separate boiler–turbine–alternator sets, producing a total of 2000 MW at a maximum of 38% overall efficiency. Figure 4.24 shows the energy flows in a conventional power station.

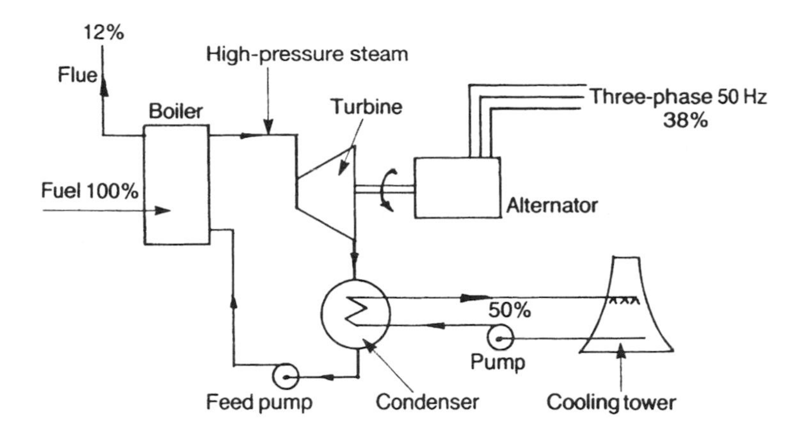

Figure 4.24 Conventional 2000 MW power station

Approximately half the input fuel's energy is dissipated in natural-draught cooling towers or sea water, depending on the plant location. Steam leaves the turbine at the lowest attainable subatmospheric pressure so that as much power as possible is extracted from it as it passes through the turbines. The temperature of the cooling water may be as low as 35 °C, which is of little practical use unless a mechanical heat pump is employed to generate a fluid at 60–90 °C. The heat could then be pumped to dwellings. Power stations are normally sited away from centres of population and heat transport costs are high.

During the next 25–100 years, the UK is going to have to make more efficient use of its indigenous hydrocarbon reserves, extend nuclear power generation capacity and develop alternative production methods such as tidal, wave, solar, wind, geothermal and hydroelectric plants.

Combined heat and power

Existing power stations generate electricity only, at as high an efficiency as possible. Combined heat and power (CHP) stations produce less electricity and more heat but improve overall fuel efficiency to about 50%, as indicated in Fig. 4.25.

Future CHP plants will be smaller than the present electricity-only plants and will be sited close to centres of industry and population. Coal-fired boilers will be used where practical. Fuel and ash will be mechanically handled and flue gases filtered to remove dust and impurities (Horlock, 1987).

District heating

District medium- or high-pressure hot-water heating, employing two-, three- or four-pipe underground distribution systems, will provide heat primarily to the largest and most consistent users, such as hospitals, factory estates and city centres. Further custom will be won from existing buildings by straight price

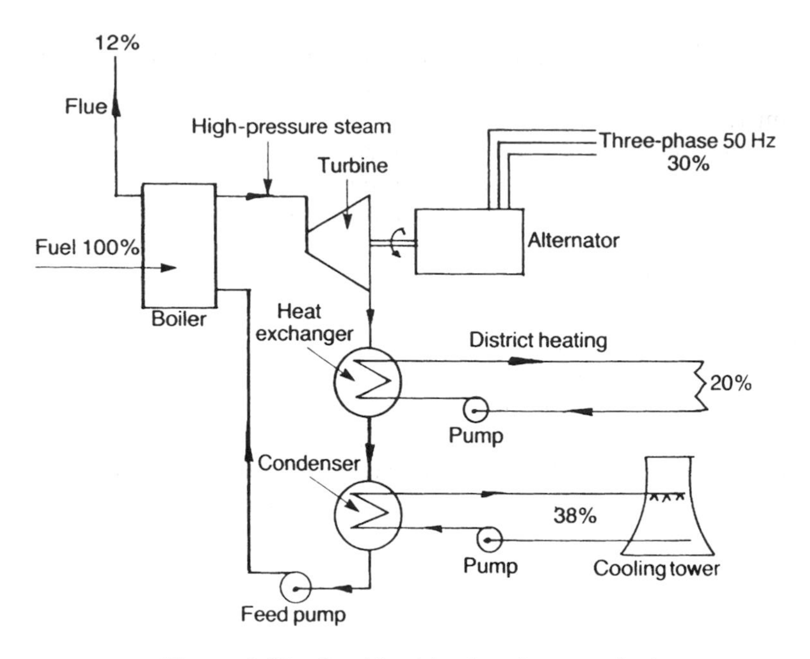

Figure 4.25 Combined heat and power plant

competition. The street distribution layout is indicated in Fig. 4.26. Flow and return pipes will be well insulated and may be installed inside one large-diameter pipe, which will form the structural duct and moisture barrier.

The CHP plant generates electricity for the locality and is connected into the national grid. It should also incinerate local refuse, utilizing the heat produced, and recycle materials such as metals and glass. It will provide hot water for sanitary appliances and air conditioning and, as these will be summer as well as winter heat loads, a method of separating them from the heating system will be

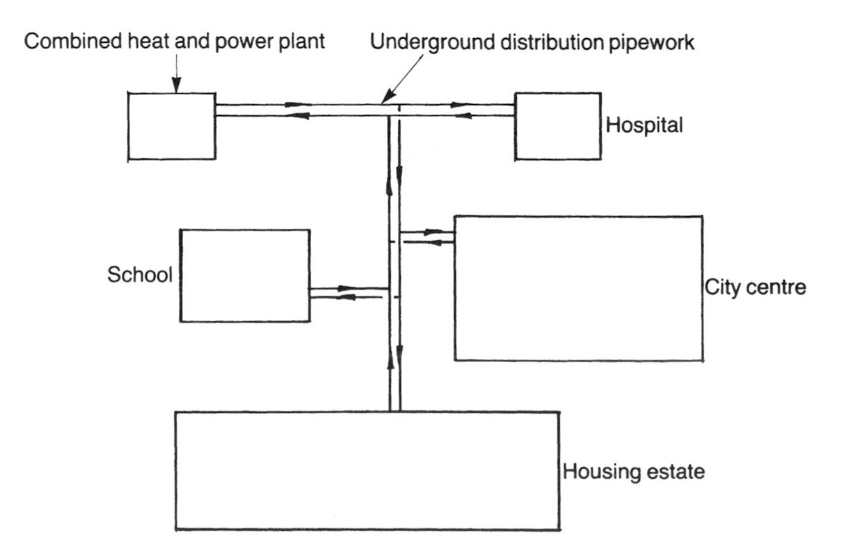

Figure 4.26 Medium- or high-pressure hot-water district heating system

used. This can be done with the three- and four-pipe arrangements shown in Figs 4.27 and 4.28 to economize on pump running costs and pipe heat losses during the summer.

The supply of heat to each dwelling will be controlled by an electric motorized valve, actuated by a temperature sensor in the heat exchanger, which will enable existing low-pressure hot-water systems to be connected. A heat meter, consisting of a water flow meter and flow and return temperature recorders, will continuously integrate the energy used, and quarterly bills could be issued through a directly linked computer.

Medium- and high-temperature hot-water heating systems are sealed from the atmosphere. Pressurization methods involve restraining thermal expansion, charging with air or nitrogen, or making use of the static head of tall buildings. As the boiling point of water increases with increasing pressure, high flow temperatures can be used. This permits a large drop in temperature from flow to return (50 °C or more), and water flow rates can be reduced compared with low-pressure hot-water open systems. Pipe sizes are smaller and the system is more economical to install when used on a large scale.

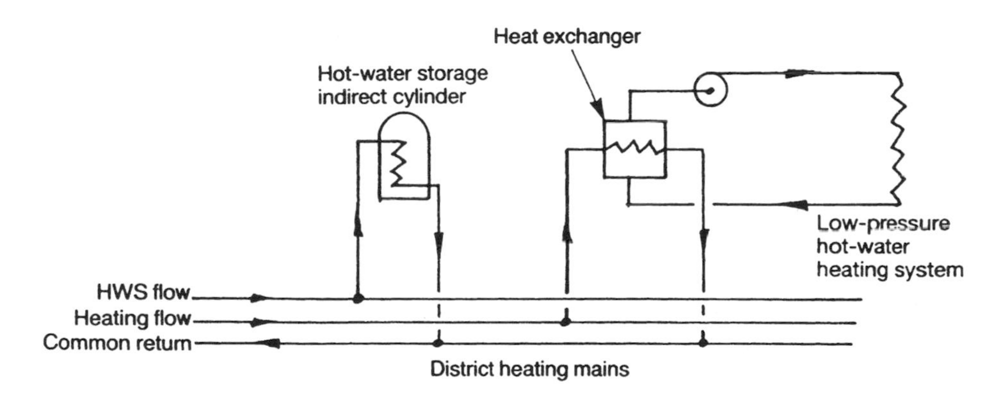

Figure 4.27 Three-pipe district heating system

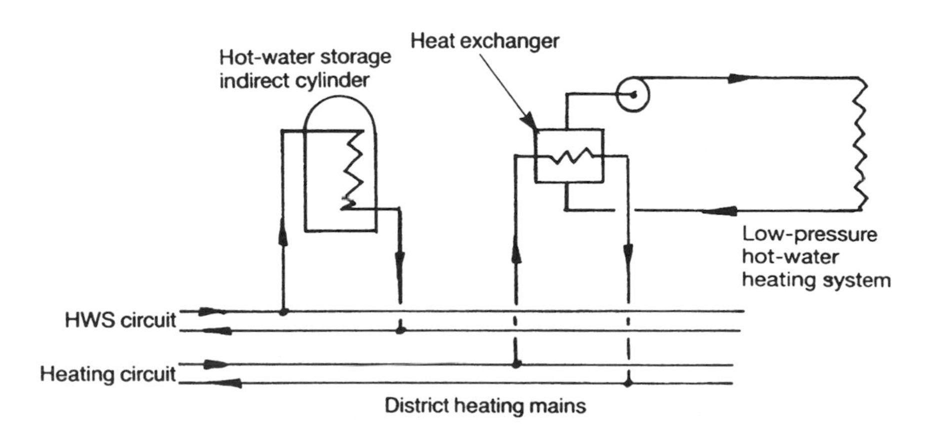

Figure 4.28 Four-pipe district heating system

District cooling from a central refrigeration plant serving air-conditioning units in commercial buildings can be developed alongside a CHP scheme. Underground chilled water pipework will be separate from the heat network, and space, cost and acoustic advantages could be gained in comparison with individual systems. A higher standard of service should be available from centralized services, with fewer breakdowns and closer control of pollution.

Building energy management systems

Computer-based remote control and continuous monitoring of energy-using systems, such as heating, air conditioning, electrical power, lighting and transportation, provide a higher standard of service than can be achieved manually (Fig. 4.29). The following types of control can be used.

1. The programmable logic controller (PLC) is a dedicated microprocessor programmed to operate a particular plant item such as a boiler, refrigeration compressor or passenger lift.
2. The energy management system (EMS) is a dedicated microprocessor that is linked to all the energy- and power-using systems such as heating, air conditioning, electrical power, lighting, lifts, diesel generators and air compressors. It may appear as a metal box on a wall of the plant room, having numbered buttons and a single line of screen display for maintenance staff to use for carrying out a limited range of routine changes. Such a unit may serve as an outstation that is either intelligent, having its own microprocessor, or dumb, merely passing data elsewhere.
3. A building energy management system (BEMS) is a supervisory computer that is networked to microprocessor outstations, which control particular plant such as heating and refrigeration equipment. All the energy-using systems within one building are accessed from the supervisor computer, which has hard disk data storage, a display unit, a keyboard, a printer and mimic diagrams of all the services.

   Additional buildings on the same site can be wired into the same BEMS by means of a low-cost cable. Remote buildings or sites are linked to the supervisor through a telephone modem. A modem is a **mod**ulator-**dem**odulator box, which converts the digital signals used by the computer into telecommunication signals suitable for transmission by the telephone network to anywhere in the world.
4. The plant management system (PMS) is used to control a large plant room such as an electrical power generator or district heating station. A PMS can be anything from a small dedicated PLC to an extensive supervisory computer system.
5. The building management system (BMS) is used for all the functions carried out by the building including the energy services, security monitoring, fire and smoke detection, alarms, maintenance scheduling, status reporting and communications. Types of BMS range from systems serving one small office, shop or factory to systems serving government departments and international shopping chains, which carry out financial audits, stocktaking and ordering of supplies each night utilizing telecommunications. Suppliers of, say, refrigera-

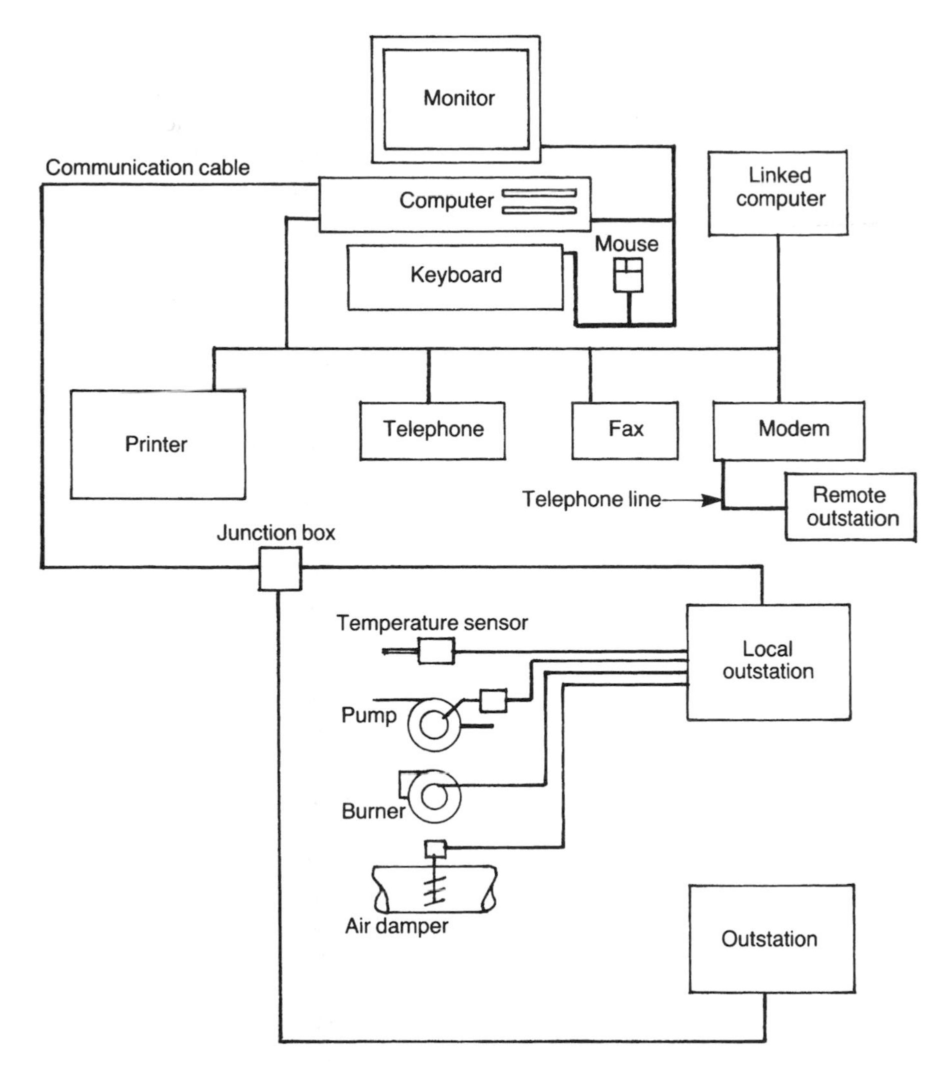

Figure 4.29 Building energy management system

tion equipment maintain links with all their installations in clients' buildings and are informed of faults as they occur and often prior to clients being aware of the problem.

An outstation is a microprocessor located close to the plant that it is controlling and is a channel of communication with the supervisor. An intelligent outstation has a memory and processing capability that enables it to make decisions on control and to store status information. A dumb outstation is a convenient point for collecting local data such as the room air temperature and whether a boiler is running or not, which is then packaged into signals for transmission to the supervisor in digital code. Each outstation has its own numbered address so that the supervisor can read the data from that source only at a discrete time.

The supervisor is the main computer, which oversees all the outstations, PLCs and modems, contacting them through a dedicated wiring system using up to 10 V and handling only digital code. Such communication can be made every few seconds, and accessing the data can take seconds or minutes depending on the quantity of data and the complexity of the whole system, i.e. the number of measurements and control signals transmitted.

The engineer in charge of the BEMS receives displayed and printed reports from the supervisor and has mimic diagrams (Fig. 4.30) of the plant, which enable identification of each pump, fan, valve and sensor together with the set points of the controllers that should be maintained. Alarm or warning status is indicated by means of flashing symbols and buzzers, indicating that corrective action by the engineer is required. Plant status, such as the percentage opening of a control valve and whether a fan is running, is recorded, but only the engineer can ascertain whether such information is correct, as some other component may have been manually switched off in the plant room by maintenance staff, or may have failed through fan belt breakage. Therefore a telephone line between the supervising engineer and the plant room staff is desirable to aid quick checking of facts.

The energy management engineer has seven main functions, which are accomplished by physical work, calculation and word processing:

1. supervision of the plant;
2. choosing the energy supply tariff;
3. reading energy consumption meters and calculating consumption;
4. organizing maintenance work and keeping records;
5. liaising with all levels of staff and management;
6. producing energy management reports;
7. monitoring fire and security systems.

The BEMS can automate these functions as follows.

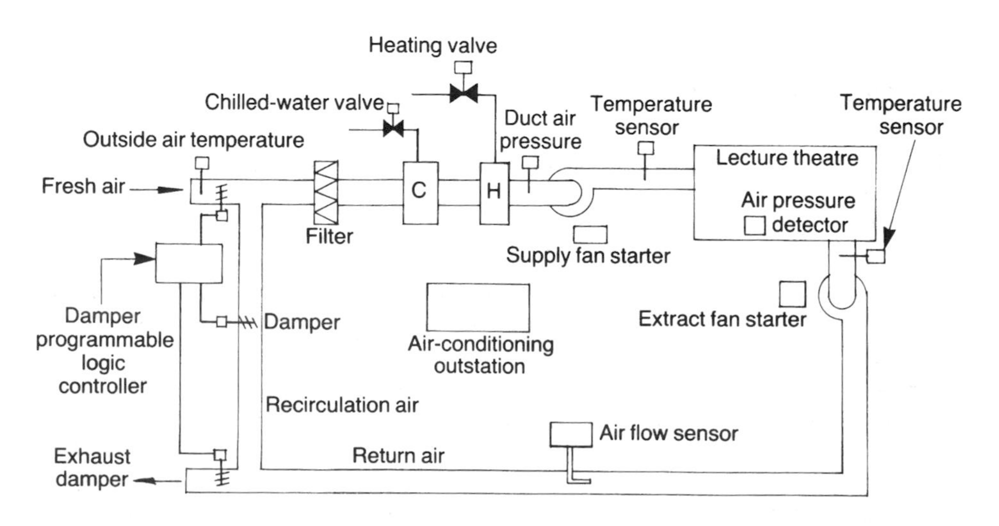

Figure 4.30 Mimic diagram on a BEMS screen

1. It can supervise the plant continually by means of analogue sensors, analogue-to-digital interfaces and communication cables.
2. It can perform remote reading of energy meters and of electricity, gas, oil and heat flow, integrate these with time to calculate total energy consumption, compare with desired values and control the plant to minimize consumption.
3. It can compare electricity and gas consumption rates with published tariffs and advise which would be most economical. It may be possible to switch off electrical loads automatically to keep to the lower-cost tariff.
4. It can continuously monitor fire detectors and security systems, such as door entry control and video camera operation, detect faults, inform about status and start alarm signals.
5. It can access BEMS control and monitor information at outstations by means of passwords, which disseminate information to, and allow restricted use by, different types of employee.
6. It can detect faults and alarm conditions as soon as they occur and display the warnings at the supervisory computer where they will be noticed without delay.
7. It can carry out normal control functions for energy-using systems and supervise what is going on.

Data which are sent to an outstation include the following:

1. measurement sensor data such as temperature, humidity, pressure, flow rate and boiler flue gas oxygen content;
2. control signals to or from valve or damper actuators;
3. plant operating status, which can be determined from the position of electrical switches which are open or closed, for example to check whether a pump is operational.

Connections are made to an outstation by means of a pair of low-voltage cables, multicore cable or fibre-optic cable (Levermore, 1992; Haines and Hittle, 1983; Coffin, 1992).

Geothermal heating

The Southampton geothermal heating system removes heat from salt water (brine), which is pumped up from 1.7 km depth by a unique down-hole pump. The well has been operational since 1981 and has had pumped circulation since 1988 (Southampton Geothermal Heating Company). The well head water temperature is 75 °C, which is sufficient to provide up to 2 MW of heating and hot-water services within the city centre. This renewable energy resource is the core of the district heating system.

The source of the warm water is a 16 m thick layer of porous Triassic Sherwood sandstone, which is maintained at 76 °C by the natural circulation of groundwater and heat flow from greater depths. The water has an acidic pH of 6.0 and high suspended solid, chloride salt and ammonia concentrations. The well is lined with a 245 mm diameter steel pipe. Water is extracted at the rate of 12 l/s by a down-hole pump that is driven from a down-hole turbine and is at a depth of 650 m. The turbine is rotated by water from a ground-level pump, which passes 31.6 l/s and consumes 192 kW of electrical power. This pump has a 250 kW three-phase electric motor, which is operated at a variable speed through a variable-frequency inverter. Figure 4.31 shows the well-head and the 250 kW pump.

The well water is passed through a titanium flat-plate heat exchanger, shown in Fig. 4.32, before being run into the storm drain at a temperature down to 30 °C. This discharge water temperature fluctuates owing to the variations in the district heating return water temperature. The use of diverting three-port flow control valves on the heating systems within the buildings served is discouraged, as these would increase the water temperature that is returned to the heat station. The use of the absorption heat pump (Chapter 5, p. 146) also influences the discharge temperature to the storm drain that directs the groundwater into the River Test estuary.

The heat station contains a 2 MW flat-plate heat exchanger, a diesel-powered engine which drives a 450 kWe (kWe = kW electrical power) electrical generator, a peak-load 2 MW oil-fired boiler, district water treatment and pressurization to 4 bar, district heating pumps, generator control and switchgear panels, a computerized heat monitoring and charging system and an absorption heat pump. The boiler also supplies high-temperature hot water for the absorption heat pump. The heat pump lowers the return water temperature from the district circuit so that more heat can be extracted from the flat-plate heat exchanger and the brine. The heat that is removed from the district circuit is put back into it after the water has passed through the flat-plate heat exchanger. This takes place at the higher-temperature part of the heat pump, in the condenser. The water jacket and the flue heat exchanger of the diesel engine produce 600 kW of heat, which is put into the district water circuit.

A two-pipe recirculatory heating pipe system distributes heat to the Civic Centre, Southampton Institute, BBC studios, shops, offices and hotels. Consumers pay charges from heat meters in their building. Each heat meter integrates the water flow rate passing through that consumer's building, the water specific heat capacity, along with flow and return temperatures, to evaluate remotely the MWh consumed. The consumer pays around £20/MWh (2 p/kWh) for the heat energy units and this includes a contribution towards the overhead costs and capital repayment to the supplier. The consumer does not pay maintenance costs for the heat service up to the heat meter, as with electric and gas services. Connection into the geothermal circuit may be free of capital charge to the consumer. It is planned to enlarge the heat station so that it will ultimately have a net output of 6.2 MWe and 11.2 MW of heat supply. When the existing gas- and oil-fired boilers in the Civic Centre and Institute are used, the heat supply system can have a capacity of 20.5 MW. This will be sufficient for a large part of the city centre within a 2 km radius.

Figure 4.33 shows a simplification of the general arrangement of the heating system. Valves V_1 and V_2 control the flow of hot water to the heat pump and the district system in accordance with the demand for heat. Valves V_3 and V_4 are manually operated when the heat pump is brought into use. The boiler and diesel engine both have an exhaust gas heat exchanger that heats the water that is pumped into the district circulation. Figure 4.34 shows the construction of the underground cased pre-insulated pipe.

Figure 4.31 Southampton geothermal well head, brine pump and filter (reproduced by courtesy of Southampton Geothermal Heating Company)

Figure 4.32 The titanium flat-plate heat exchanger transfers heat from the brine to the district water circulating through the Southampton geothermal heating system (reproduced by courtesy of Southampton Geothermal Heating Company)

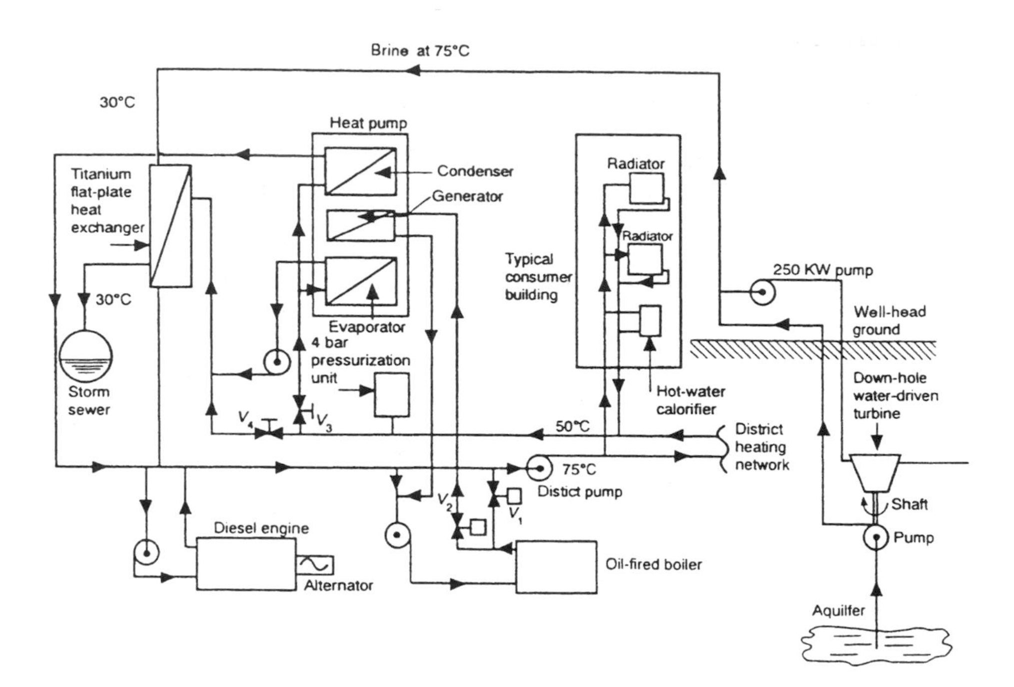

Figure 4.33 General arrangement of the Southampton geothermal heating system

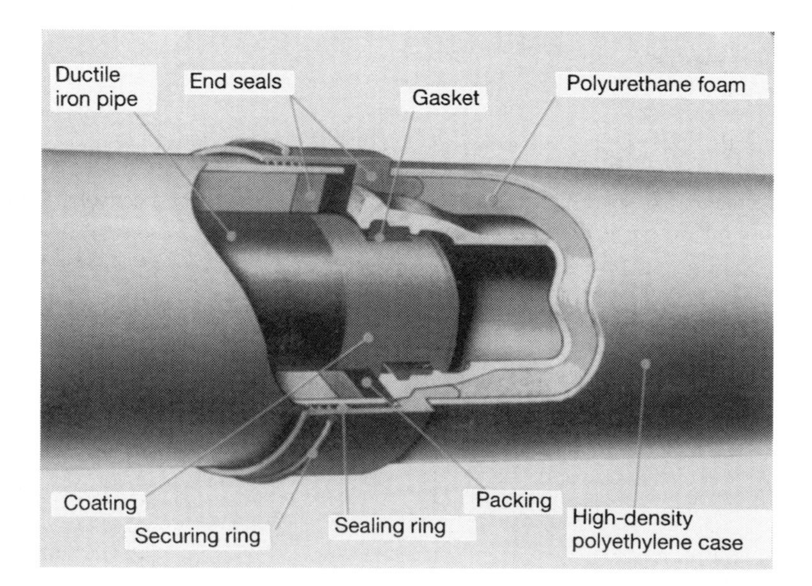

Figure 4.34 Underground district heating pipe (reproduced by courtesy of Southampton Geothermal Heating Company)

Questions

1. Sketch and describe two different types of heating system for each of the following applications: house, office, commercial garage, shop, warehouse and heavy engineering factory.

2. Why may the water in large heating systems be pressurized? Explain how pressurization systems work.

3. How do heating systems alter the mean radiant temperature of a room? Give examples.

4. What factors are included in the decision on the siting of a heat emitter? Give examples and illustrate your answer. What safety precautions are taken in buildings occupied by very young, elderly, infirm or disabled people?

5. How can radiant heating minimize fuel costs while providing comfortable conditions? Give examples.

6. Sketch the installation of a ducted warm-air heating system in a house and describe its operation.

7. List the characteristics of electrical heating systems and compare them with other fuel-based systems.

8. Outline the parameters considered when deciding whether to use a one- or two-pipe distribution arrangement for a radiator and convector-low pressure hot-water heating system.

9. Three rooms have heat losses of 2 kW, 4 kW and 5 kW respectively. Double-panel steel radiators are to be used on a two-pipe low-pressure hot-water system having flow and return temperatures of 85 °C and 72 °C respectively. Room air temperatures are to be 20 °C. Choose suitable radiators from Table 4.2 and calculate the water flow rate for each.

10. Sketch and describe a microbore heating installation serving hot-water radiators. State its advantages over alternative pipework systems.

11. A medium-pressure hot-water heating system is designed to provide a heat output of 100 kW with flow and return temperatures of 110 °C and 85 °C respectively. Calculate the pump water flow rate required in litres per second.

12. Find the dimensions of a double-panel steel radiator suitable for a room having an air temperature of 15 °C when the water flow and return temperatures are 86 °C and 72 °C respectively and the room heat loss is 4.25 kW.

13. The two-pipe heating system shown in Fig. 4.18 is to be installed in an office block where radiators 1, 2 and 3 represent areas with heat losses of 12 kW, 20 kW and 24 kW respectively. Water flow and return temperatures are to be 90 °C and 75 °C respectively. The pipe lengths shown are to be multiplied by 1.5. Pump A (Fig. 4.16) is to be used. Pipe heat losses amount to 10% of room heat losses. The friction loss in the pipes is equivalent to 25% of the measured length. Find the pipe sizes.

14. A hot-water radiator central heating system is commissioned and tested while the average outdoor air is 3 °C and there is intermittent sunshine and a moderate wind. The building is sparsely occupied. Water flow and return temperatures at the boiler are 90 °C and 80 °C respectively and the room average temperature is 27 °C. The heating system was designed to maintain the internal air at 22 °C at an external air temperature of −1 °C with flow and return temperatures of 85 °C and 73 °C respectively. State whether the heating system met its design specification and what factors influenced the test results.

15. List and discuss the merits of the methods used to generate electrical power.

16. Discuss the application of CHP systems in relation to density of heat usage, local and national government policy, possible plant sites, complexity of existing underground services, ground conditions, costs of competing fuels, type and age of buildings, traffic disruption during installation and better control of pollution. (The term 'density of heat usage' refers to the actual use of heat in megajoules per unit ground plan area, including all floors of buildings and appropriate industrial processes requiring the sort of heat to be sold.)

5 Ventilation and air conditioning

Key terms and concepts

Learning objectives

Study of this chapter will enable the reader to

1. recognize the physiological reasons for fresh air ventilation of buildings;
2. calculate fresh air requirements;
3. understand the basic design criteria for air movement control;
4. describe the four combinations of natural and mechanical ventilation;
5. describe the working principles of air-conditioning systems;
6. calculate ventilation air quantities;
7. understand psychrometric cycles for humid air;
8. calculate air-conditioning heating and cooling plant loads;
9. describe the various forms of air-conditioning system;
10. state where reciprocating piston, screw and centrifugal compressors are suitable;
11. understand the coefficient of performance;
12. explain the states of refrigerant occurring within a vapour compression refrigeration cycle;
13. explain the operation of refrigeration equipment serving an air-conditioning system;
14. comprehend the absorption refrigeration cycle;
15. explain how ventilation rates are measured;
16. choose suitable materials for air-conditioning ductwork;
17. understand the relationship between CFCs and the environment;
18. know the uses of CFCs, and good practice and handling procedures;
19. be able to discuss the problem of SBS;
20. know the symptoms, causes and possible cures for SBS;
21. relate the daily cyclic variation of air temperatures to the need for air conditioning.

Introduction

The reasons for ventilation lead into an understanding of the necessary combinations of natural and mechanical systems and air conditioning, which means full mechanical control of air movement through the building (Kut, 1993). The calculation of air changes and air flow rates from basic human requirements, and then for removal of heat gains, is fundamental. The sizing of air ducts and the calculation of heater and cooler loads utilizing the psychrometric chart of humid air properties is shown.

Systems of air conditioning ranging from small self-contained units to large commercial applications are described. Appropriate refrigeration elements are developed, from thermodynamic principles, into complete installations in a form that is easily understood. Both vapour compression and vapour absorption cycles are explained.

Sick building syndrome (SBS) has been attributed to air conditioning, but has been found to be due to an amalgam of possible causes, none of which is singly identifiable as the culprit. The possible causes and solutions relating to user complaints are discussed.

Chlorofluorocarbons (CFCs) were widely used in thermal insulation and refrigeration until their potential for environmental damage resulted in the Montréal Protocol agreement. The effects of this agreement on the building services industry are discussed.

Ventilation requirements

An attempt to calculate the quantity of ventilation air needed for habitation can be made by considering the CO_2 production P m^3/s of a sedentary adult, the concentration C_s% of CO_2% in outdoor air and the maximum threshold CO_2 concentration C_r% for the occupied space. Normally, $P = 4.7 \times 10^{-6}$ m^3/s per person, $C_s = 0.03\%$ and $C_r = 0.5\%$. If Q m^3/s is the rate of fresh air flowing through a room to provide acceptable conditions, a CO_2 balance can be made:

$$CO_2 \text{ increase in ventilation air} = CO_2 \text{ produced by occupant}$$

Now

$$\text{flow rate of } CO_2 \text{ entering room} = QC_s \text{ m}^3/\text{s}$$

$$\text{flow rate of } CO_2 \text{ leaving room} = QC_r \text{ m}^3/\text{s}$$

Therefore

$$QC_r - QC_s = P \text{ m}^3/\text{s}$$

Hence

$$Q(C_r - C_s) = P \text{ m}^3/\text{s}$$

and

$$Q = \frac{P}{C_r - C_s} \, \mathrm{m^3/s}$$

$$= \frac{4.7 \times 10^{-6}}{(0.5 - 0.03) \times 10^{-2}} \, \mathrm{m^3/s}$$

$$= 0.001 \, \frac{\mathrm{m^3}}{\mathrm{s}} \times \frac{10^3 \, \mathrm{l}}{1 \, \mathrm{m^3}}$$

$$= 1 \, \mathrm{l/s}$$

Other factors have a stronger influence on ventilation requirement: bodily heat production (about 100 W per person during sedentary occupation); moisture exhaled and evaporated from the skin (about 40 W per person for sedentary occupation); body odour; and fumes from smoking. These factors greatly outweigh the CO_2 requirement. The recommended fresh air supply per person is 10.4 l/s, which is increased by up to 50% in the event of expected heavy tobacco smoke. Building Research Establishment Digest 206: 1977 gives design curves for open plan and small offices of room height 2.7 m and floor space per person 4.5 m^2. A small office requires 2.25 air changes per hour of outdoor air, and this will normally be provided by natural ventilation.

The ventilation system should not produce monotonous draughts but preferably variable air speed and direction. Facilities for manual control of ventilation terminals allow sedentary workers some freedom of choice over their environment. Careful location of ventilation grilles and control of both the temperature and velocity of moving air in mechanical systems can ensure that neither cool nor hot draughts are caused. The maximum air velocity that can be perceived at neck level is related to its temperature. If the values given in Table 5.1 are not exceeded, then annoyance should be avoided.

Grille manufacturers' data reveal the length of the jet of air entering the room for particular air flow rates. Additionally, the air jet may rise or fall depending on whether it is warmer or cooler than the room air. Moving air currents tend to attach themselves to a stationary surface and follow its contours (Coanda effect). When this boundary layer flow either comes to the end of, say, a wall or hits a bluff body, such as a beam or luminaire, it may be suddenly detached and cause turbulent flow; this appears as a draught, which is an uncomfortable air movement.

Table 5.1 DIN criteria for air movement at the neck

Moving-air temperature (°C)	20	22	24	26
Maximum air speed (m/s)	0.10	0.20	0.32	0.48

Reproduced from the *IHVE Guide* (CIBSE, 1986 [IHVE, 1970]) by permission of the Chartered Institution of Building Services Engineers.

An air jet entering a room should be allowed to mix with the room air before entering the occupied space. This can be done either above head height with cooling systems or in circulation spaces at low level with heating systems. The design criteria for ventilation systems are as follows:

1. correct fresh air quantity (minimum normally 10.4 l/s per person);
2. avoidance of hot or cold draughts by design of the air inlet system;
3. some manual control over air movement;
4. mechanical ventilation to provide a minimum of four air changes per hour to ensure adequate flushing of all parts of rooms;
5. air change rates that can be increased to remove solar and other heat gains;
6. air cleanliness achieved by filtration of fresh air intake and recirculated room air.

Natural and mechanical systems

The four possible combinations of natural and mechanical ventilation are as follows.

1. Natural inlet and outlet: utilizing openable windows, air bricks, louvres, doorways and chimneys. Up to about three air changes per hour may be provided, but these depend upon prevailing wind direction and strength, the stack effect of rising warm air currents, and adventitious openings around doors and windows.
2. Natural inlet, mechanical outlet: mechanical extract fans in windows or roofs and ducted systems where the air is to be discharged away from the occupied space owing to its contamination with heat, fumes, smoke, water vapour or odour. This system can be used in dwellings, offices, factories or public buildings. A slight reduction in air static pressure is caused within the building, and external air flows inwards. This inflow is facilitated by air inlet grilles, sometimes situated behind radiators or convector heaters. There is no filtration of incoming impurities. This system is used particularly for toilet or kitchen extraction, smoke removal from public rooms and heat or fume removal from industrial premises. Heating of the incoming air is essential for winter use.
3. Mechanical inlet, natural outlet: air is blown into the building through a fan convector or ducted system to pressurize the internal atmosphere slightly with a heated air supply. The air leaks out of the building through adventitious openings and permanent air bricks or louvres. This system can be used for offices, factories, large public halls or underground boiler plant rooms.
4. Mechanical inlet and outlet: where natural ventilation openings would become unable to cope with large air flow rates without disturbing the architecture or causing uncontrollable draughts, full mechanical control of air movement is assumed. This may augment natural ventilation at times of peak occupancy or solar heat gain. When a building is to be sealed from the external environment, then a full air-conditioning system is used. Figure 5.1 shows the basic arrangement of a single-duct system.

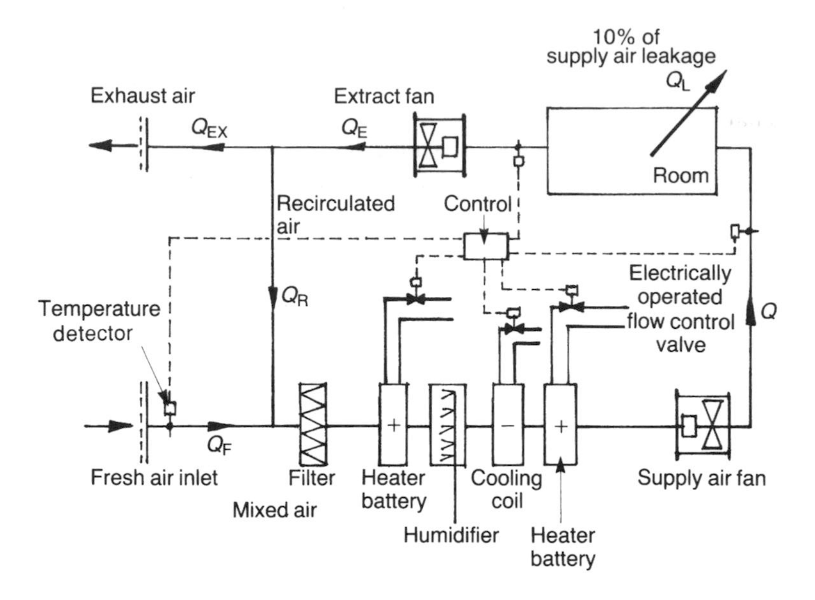

Figure 5.1 Single-duct air conditioning system

The designers of new buildings bear a heavy responsibility for the future environmental condition of the planet. A building that is a large user of primary energy, i.e. fossil fuel, to power its mechanical systems, usually air conditioning, is a charge on society for 50 or more years. Such buildings are unlikely to have their mechanical heating, cooling, lifts, lighting and computer power supply networks removed for reasons of economy by future tenants. Once the pattern of energy use is set for a building by the architect and original client, it remains that way until it is demolished.

The increasing need for comfortably habitable buildings in the temperate climate of the British Isles since the 1960s, led to demands for air conditioning. Warm weather data for the UK shows that an external air temperature of 25 °C d.b. is normally only exceeded during 1% of the summer period June to September (CIBSE, 1986, Figure A2.11). (Compare this to those areas of the world within 45 ° of latitude from the equator, where the outside air temperature exceeds 30 °C d.b.; people living and working in such areas would probably consider the use of air conditioning in the UK a waste of energy resources.) These 4 months total 120 days and include weekends, a public holiday and many people's annual holidays from their places of work. Therefore, only one day per summer is likely to have an outdoor air temperature that exceeds 25 °C d.b. When a long, hot summer is experienced in southern England, there can be several days which exceed 25 °C d.b., but this may not be repeated for a few years. In the rest of the UK, lower outdoor air temperatures are experienced most of the time. The UK has a basically maritime climate, with the Atlantic Ocean, English Channel and North Sea providing humid air flows at air temperatures that are modified by the evaporative cooling effect of the seas. Long, hot summers are produced by strong winds from eastern Europe. These winds have travelled across the hot and dry land mass of northern Europe and then been reduced in temperature due to the evaporation of water from the North Sea. Such evaporation of sea water is aided by high wind speed. Sea water is evaporated into the wind and a change of phase occurs in the water. In order to evaporate water into moisture that is carried by the air stream, the water is boiled into steam, a change of phase. This phase change can only take place with the removal of sensible heat from the air and the water, as the latent heat of evaporation has to be provided, just as in a steam boiler. Removal of sensible heat from the air, lowers its dry-bulb temperature. This is the principle of evaporative cooling that is employed in cooling towers and in the evaporative cooling heat exchangers that are used for indoor comfort cooling in hot, dry climates. The occasional long, hot summers in the UK coincide with the increase of outside air wet-bulb temperature that is associated with the evaporative cooling effect from the sea. The result is warm and humid weather in south-east England. It is the combined effect of the higher dry-and wet-bulb outdoor air temperatures which produces the humid air that many people find uncomfortable. The building designer has to decide how to address periods of discomfort due to warm outdoor weather conditions and explain the likely outcome to the client.

The principal source of indoor thermal discomfort in the UK is the manner in which modern buildings are designed and then filled with people and heat-producing equipment. Buildings that were constructed of masonry and brick, with small, openable windows, overhanging eaves from pitched tiled roofing, hinged wooden external or internal slatted shutters, permanent ventilation openings from air bricks, chimneys and roof vents, and north light windows for industrial use, have a large thermal storage mass to smooth out the fluctuations of the outdoor air temperature. Such buildings from the designs of the late nineteenth and early twentieth centuries, are better able to provide shaded and cool habitable zones than the exposed glazing high-rise buildings that have been popular since the 1960s. Deep core floor plan buildings that are unreachable by outdoor air to provide the required ventilation and exterior façades which are deliberately exposed to solar heat gain with no provision for external shading screens, trees or shadows cast from surrounding buildings, are designed to create interior discomfort. Current buildings are also packed to capacity with productive workers who each need a personal computer and permanent artificial illumination in order to carry out their tasks. The other mistake is to employ people to work in these glass boxes only during daylight hours that coincide with the peak cooling load times. The Mediterranean countries close many of their businesses during the afternoons, close the wooden window shutters and sleep during the hottest part of the day, coming out in the late afternoon to start work again.

Energy conservation engineers can assist the building maintenance manager to make the best use of what has been built. There will be many technical methods that are available to reduce the use of energy by the mechanical and electrical systems within the building, but at best, these are likely to reduce energy use by a maximum of 25%, often much less can be achieved. Retrofitting a building with energy-saving systems may accompany a change of use or an overall refurbishment. There has been a strong move in the direction of low-energy-use buildings within the UK during recent years. Greenfield sites in the mild climate of the British Isles have provided building designers with more options than are available in more extreme climates. The outdoor air temperatures that are used for the design of the heating and cooling systems in the UK are within the range of −3 °C d.b to 30 °C d.b. The minimum overnight outdoor air temperature may drop to −17 °C d.b. once per year during a severe winter. The extreme low and high outdoor air temperatures that can be experienced are of little importance to the designers of commercial and most industrial buildings, unless indoor air temperatures are of critical importance for the condition of products, human or animal safety, or industrial processes. The users of most buildings are expected to withstand colder or warmer indoor temperatures for a few days per year and not complain too much. In extreme cases, workplace agreements are made to allow work to be stopped for short periods so that the occupants of the building can move away from the workstation for recuperation. An example is a production factory where 10 min breaks in an 8 h shift are agreed for production-line workers to go outside the building when the indoor air temperature exceeds 30 °C d.b., which it does frequently during long, hot summers, due to the minimalist thermal insulation value of the building's walls and roofing.

The design team of a new building need to develop a method of solving the complex issues surrounding the task of providing a building which satisfies the:

basic needs of the owner;
architectural and local planning design philosophy;
requirements of legislation;
access, spatial, visual, aural and thermal comfort needs of the occupants;
use of energy during the 50-year service period of the building;
sources of energy that are available for the building and the maintainability of the whole complex.

Those matters which relate to the need, or otherwise, for air conditioning, can be summarized as follows:

1. the local design weather conditions;
2. the indoor design set points for zone air temperature and relative humidity;
3. the allowable variation in the indoor design air conditions;
4. the number of occasions during each year that divergence from the specified indoor design conditions are allowable;
5. the time periods when the building will be occupied by the main users and the service personnel;
6. the sources of primary energy available for the building, their long-term reliability, storage requirements and safety considerations;
7. whether renewable energy can be used on the site;
8. the building usage;
9. the means by which the building can be heated;
10. the outdoor air ventilation quantity requirements;
11. the peak energy requirements for heating and cooling;
12. the location and sizes of plant and service shaft spaces available;
13. whether natural ventilation or assisted natural ventilation can provide the required air flow through the building;
14. whether mechanical cooling systems are needed to maintain the specified peak design conditions;
15. the need to provide accurate indoor environmental conditions for equipment, material storage and handling, or an industrial process;
16. whether low-cost cooling systems can be used;

17. whether there is a real need to provide a mechanical means of air conditioning;
18. if there is a process requirement for closely conditioned air within the building;
19. energy recovery strategies available;
20. the maintainability of the mechanical services and how replacement plant can be provided without major structural works becoming necessary.

These considerations all have an impact on the building design team's decision-making process. The local weather conditions that create the maximum heating and cooling loads during the occupied part of the day, or night, will determine the size of the heating and cooling plant that are required. Occupancy times can sometimes be varied in order to minimize the plant capacity that is needed, for example by using the Mediterranean region off-peak working principle, however unpalatable this may seem to northern European practices. The architectural and engineering designers have the option to experiment with the thermal insulation value of the exterior envelope of the building in varying the area of glazing, walling and roofing and the thermal transmittance of each component. Life-cycle costing of each alternative thermal design will reveal the total cost of constructing and using the building for 50 years, with reasonable assumptions on price changes each year and the cost of refitting the building every 20 years because of improvements in technology.

Selection of the indoor design air set points for temperature and relative humidity will be determined by legislative and comfort standards. Short-term variances within an allowable range can minimize the use of energy at peak times. It may be possible to avoid the installation of mechanical cooling systems where the users of the building agree to accept regular, short-term divergence from the standard design conditions. The provision of ceiling-mounted or portable fans with comfort breaks and refrigerated drinks dispensers might allow a building to avoid the installation of mechanical cooling, depending upon the number and frequency of divergencies from the normal standards. Allowing the indoor air temperature to rise to 25 °C d.b. in summer before switching on the air-conditioning chiller during the afternoon, and whether the building is occupied, the time is before 3.30 p.m. and the outdoor air temperature is above 25 °C d.b; all these conditions can be accessed through the computer-based building management system, if one is provided. Automatic control programming can be set to minimize the use of electrical energy by the mechanical cooling system as a deliberate strategy by the designers and operators of the building. Dynamic thermal analysis software is used to model buildings and their services systems to assess the indoor air conditions that are expected to occur.

The provision of outdoor air ventilation is a legislative requirement based on the activities of and the numbers of people within the enclosure. The minimum quantity of outdoor air must be maintained for each person to comply with the standards. Outdoor air ventilation does more than provide air for breathing, it also flushes the building with outdoor air to remove heat and to control the concentrations of odours and atmospheric pollutants that are produced within the building. Toilet exhaust air flows often partly match the inflow rate of outdoor air in commercial buildings. Any balance of flow between the exhaust quantity and inflow of outdoor air may be allowed to leak though doors, permanent ventilators and other openings in the structure, such as gaps around window frames. The inclusion of carbon dioxide sensors in the return or exhaust air ducts allows the building management system to minimize the opening of the outside air intake motorized dampers and consequently reduce the heating or cooling load of the air-conditioning plant and save energy. Carbon dioxide in air that is removed from the occupied space is a direct assessment of the number of occupants and their activity level, allowing energy use to match the instantaneous load on the building. Other means of assessing room occupancy are available from infrared or ultrasonic sensors. If the designers of the building are able to know the divergence in the patterns of occupancy of rooms or spaces, then decisions can sometimes be taken to diversify the provision of lighting, heating and cooling zones to minimize the use of mechanical and electrical plant and systems, with benefits in the initial plant capacity and in the use of energy in the long term.

Natural ventilation can be used in atria and industrial buildings where the stack effect of height within the building is used to create air movement. Low-level outside air intakes and roof-level air extractors or openable ventilation units can be mechanically controlled to match the heating and cooling load on the building to the flow of air through the spaces. The avoidance of draughts around sedentary occupants will always be the main challenge. Low-level radiant heating from warmed floors or overhead radiant panels can offset high rates of air movement. Anyone who has sat within cathedral ceiling spaces will recognize the potential discomfort problems in cold weather within intermittently used high thermal-mass buildings, for example stone churches and sports halls. The use of natural ventilation in the UK is accompanied by the problem of allowing the uncontrolled ingress of the moisture that is present in the outside air. Any surface within the building that is below the dew-point temperature of the space will accumulate damp and create long-term mould growth. This becomes a comfort, health and maintenance cost if damage to the building is to be avoided. In climates where the outdoor air temperature rises above 20 °C d.b. on most days of the year, i.e. those within 40° latitude, the use of outdoor air natural ventilation is often out of the question. This is due to the combination of intense solar radiation heat gains in these regions and the lack of a natural cooling function by the outdoor air. Within the tropics, latitude 27°, the constant high moisture content of the outdoor air makes it unsuitable for natural ventilation practice in commercial buildings. Buildings in the UK that have an internal atrium have restored a historical precedent in turning the building inside-out. The exterior surfaces need less glazing as the occupants' view is directed inwards towards a planted open space that has natural daylight. The atrium can be used to return conditioned air back to the air-handling plant room without the need for a return air fan, collect exhaust air and expel it through roof openable vents and facilitate the removal of smoke during an emergency. Heat produced from the occupants, fluorescent lighting, computers and electrical equipment assists the upward flow of air away from the occupied zones. Atria are used in all climates from sub-zero through 40 °C d.b. environments and fully provide indoor office, retail shopping malls, casino, hotel and entertainment spaces throughout the world.

The options that are available to the designers in the selection of the method of controlling the environmental conditions within the building include the following.

1. Natural ventilation – applicable when the external climate, the use and design of the building permit it. Mild climate localities usually close to the coast

where the sea is warm; internal air conditions are allowed to vary widely and are directly related to the external weather conditions; central heating system; cooling may be provided from packaged direct expansion refrigeration units within each zone; manually operated internal or external shading blinds; passive solar architecture that may include thermal storage walls; chilled-water beams or flat panels may be installed at high level within offices to provide limited cooling (the chilled water in the beams and panels is maintained at a temperature that is above the room air dew-point so that there is no condensation on the exposed surfaces or within the ceilings, avoiding dehumidification and control of the zone relative humidity); chilled beams and panels provide no ventilation air.

2. Assisted natural ventilation – a development of natural ventilation, as above, where applicable. Mechanically operated ventilation louvres and exhaust air fans improve the control of air flow through the occupied spaces; the incoming outside air may be cooled with a water spray evaporative cooler in hot, dry climates; evaporative cooling is a low-cost means of cooling the incoming outside air, which is exhausted by either natural openings of doors and windows, or by exhaust air fans in confined zones; the incoming outdoor air may be cooled through a specific temperature range, say 10 K, to provide limited cooling by means of direct expansion or chilled water refrigeration plant.
3. Mechanical ventilation which only passes outside air through the zone. This usually applies to moderate climates such as the UK where minimal cooling is required; in climates where the outside air temperature exceeds 30 °C d.b. the flow rate of outside air is likely to be insufficient to provide enough cooling for zone temperature control in an air-conditioning system; where it is possible to locate the exhaust air duct alongside the incoming outside air duct, within the ceiling space or a plant room, an air-to-air flat-plate heat exchanger is used to transfer heat between the incoming and outgoing air streams; the outgoing exhaust air is already at the correct zone temperature, of around 22 °C d.b. (it cannot be recycled as it is vitiated with carbon dioxide, odours and atmospheric pollutants); heat transfer works throughout the year (in winter, the incoming outdoor air is preheated by up to 10 K from the outgoing exhaust air; in summer, the incoming outdoor air is precooled by up to 10 K from the outgoing exhaust air); heat transfer efficiency is around 55%; a similar heat transfer can be obtained from a recuperative heat wheel that transfers heat from the outgoing exhaust air to the incoming outside air, generating up to a 10 K temperature change in the outside air stream; the heat transfer medium of the wheel may be strips of mylar film.
4. Mechanical ventilation with recirculated room air. The maximum quantity of conditioned room air is recirculated to save energy use at the heating and refrigeration plant; the outside air motorized dampers are modulated from closed to fully open to control the zone air temperature without the use of the mechanical refrigeration plant for as long a time as possible (this provides low-cost cooling to the building); when the refrigeration plant has to be used, the outside and exhaust air motorized dampers are moved to their minimum outside air positions, often around 10% open, allowing the maximum use of recirculated room air; a range of ducted air-conditioning systems are in use, including single duct, dual duct, induction units, fan coil units and variable air volume systems.

The single-duct system works in the following way. Some of the air extracted from the room is exhausted to the atmosphere and as much as possible is recirculated to reduce running costs of heating and cooling plants. Incoming fresh air is filtered and mixed with that recirculated; it is then heated by a low-, medium- or high-pressure hot-water or steam finned pipe heat exchanger *or* an electric resistance element. The heated air is supplied through ducts to the room. The hot-water flow rate is controlled by a duct-mounted temperature detector in the extract air, which samples room conditions. The electrical signal from the temperature detector is received by the automatic control box and corrective action is taken to increase or reduce water flow rate at the electrically driven motorized valve at the heater battery.

During summer operation, chilled water from the refrigeration plant is circulated through the cooling coil and room temperature is controlled similarly.

A temperature detector in the fresh air duct will vary the set value of the extract duct air temperature – higher in summer, lower in winter – to minimize energy costs. A low-limit temperature detector will override the other controls, if necessary, to avoid injection of cold air to the room.

The building is slightly pressurized by extracting only about 95% of the supply air volume, allowing some conditioned air to leak outwards or exfiltrate.

Energy savings are maximized by recirculating as much of the conditioned room air as possible. Room air recirculation with economy-cycle motorized dampers can, sometimes, be retrofitted to existing systems as an energy conservation measure. In mild climates, such as in the UK full outside air systems are also used. These have no recirculation air ducts; either a flat-plate heat exchanger or run-around pipe coils can be installed to preheat and precool the incoming outside air to save energy. Such heat exchangers are around 55% efficient, which is not as good as recirculation.

EXAMPLE 5.1

A room 12 m × 10 m × 3 m high is to have a ventilation rate of 6 air changes per hour. Air enters from a duct at a velocity of 10 m/s. Find the air volume flow rate to the room and the dimensions of the square duct.

The air flow rate is given by

$$Q = \frac{N \text{ air changes}}{\text{hour}} \times \frac{V \text{ m}^3}{\text{air change}} \times \frac{1 \text{ h}}{3600 \text{ s}}$$

where room volume V m^3 = 1 air change. Hence

$$Q = \frac{NV}{3600} \text{ m}^3/\text{s}$$

$$= \frac{6 \times 12 \times 10 \times 3}{3600} \text{ m}^3/\text{s}$$

$$0.6 \text{ m}^3/\text{s}$$

Also,

$$Q \text{ m}^3/\text{s} = \text{duct cross-sectional area } A \text{ m}^2 \times \text{air velocity } v \text{ m/s}$$

Therefore

$$A = \frac{Q}{v} = \frac{0.6}{10} \text{ m}^2 = 0.06 \text{ m}^2$$

If the duct side is l m, then $A = l^2$ m^2. Therefore

$$l = \sqrt{A} \text{ m}$$
$$= \sqrt{0.06} \text{ m}$$
$$= 0.245 \text{ m}$$

EXAMPLE 5.2

A lecture theatre has dimensions 30 m × 20 m × 4 m high and has 50 occupants; 12 l/s of fresh air and 50 l/s of recirculated air are supplied to the theatre for each person. A single-duct ventilation system is used. If 10% of the supply volume leaks out of the theatre, calculate the room air change rate and the air volume flow rate in each duct.

$$\text{supply air quantity per person} = (12 + 50)\,\frac{\text{l}}{\text{s}} \times \frac{1 \text{ m}^3}{10^3 \text{ l}}$$
$$= 0.062 \text{ m}^3/\text{s}$$

Hence

$$Q = 0.062 \times 50 \text{ m}^3/\text{s}$$
$$= 3.1 \text{ m}^3/\text{s}$$

Now

$$Q = \frac{NV}{3600}$$

and hence

$$N = \frac{3600\,Q}{V}$$
$$= \frac{3600 \times 3.1}{30 \times 20 \times 4} \text{ air changes per hour}$$
$$= 4.65 \text{ air changes per hour}$$

The leakage from the theatre is

$$Q_1 = 10\% \times Q$$
$$= 0.1 \times 3.1 \text{ m}^3/\text{s}$$
$$= 0.31 \text{ m}^3/\text{s}$$

The quantity of air extracted from the theatre is

$$Q_e = 3.1 - 0.31 \text{ m}^3/\text{s}$$
$$= 2.79 \text{ m}^3/\text{s}$$

The quantity of fresh air entering the ductwork is

$$Q_f = \frac{12 \times 50}{10^3} \text{ m}^3/\text{s}$$
$$= 0.6 \text{ m}^3/\text{s}$$

The quantity of recirculated air is

$$Q_r = Q - Q_r$$
$$= 3.1 - 0.6 \text{ m}^3/\text{s}$$
$$= 2.5 \text{ m}^3/\text{s}$$

The exhaust air quantity is

$$Q_{ex} = Q_e - Q_r$$
$$= 2.79 - 2.5 \text{ m}^3/\text{s}$$
$$= 0.29 \text{ m}^3/\text{s}$$

EXAMPLE 5.3

There are 20 people in a gymnasium, each producing CO_2 at a rate of 12×10^{-6} m^3/s. If the maximum CO_2 level is not to exceed 0.25%, find the air supply rate necessary. The outdoor air CO_2 concentration is 0.05%.

$$Q = \frac{P}{C_r - C_s} \text{ m}^3/\text{s}$$
$$= \frac{20 \times 12 \times 10^{-6}}{(0.25 - 0.05) \times 10^{-2}} \text{ m}^3/\text{s}$$
$$= 0.12 \text{ m}^3/\text{s}$$

Removal of heat gains

Ventilation air is used to remove excess heat gains from buildings. Two types of heat gain are involved: sensible and latent.

Sensible heat gains result from solar radiation, conduction from outside to inside during hot weather, warm ventilation air, lighting, electrical machinery and equipment, people and industrial processes. Such heat gains affect the temperature of the air and the building construction.

Latent heat gains result from exhaled and evaporated moisture from people, moisture given out from industrial processes and humidifiers. These heat gains do not directly affect the temperature of the surroundings but take the form of transfers of moisture. They can be measured in weight of water vapour transferred or its latent heat equivalent in watts.

The latent heat of evaporation of water into air at a temperature of 20 °C and a barometric pressure of 1013.25 mb is 2453.61 kJ/kg. Thus the latent heat (LH) required to evaporate 60 g of water in this air is

$$\mathrm{LH} = 60\ \mathrm{g} \times \frac{1\ \mathrm{kg}}{10^3\ \mathrm{g}} \times 2453.61\ \frac{\mathrm{kJ}}{\mathrm{kg}}$$

$$= 147.22\ \mathrm{kJ}$$

If this evaporation takes place over, say, 1 h, the rate of latent heat transfer will be

$$\mathrm{LH} = 147.22\ \frac{\mathrm{kJ}}{\mathrm{h}} \times \frac{1\ \mathrm{h}}{3600\ \mathrm{s}} \times \frac{10^3\ \mathrm{J}}{\mathrm{kJ}} \times \frac{\mathrm{W\ s}}{\mathrm{J}}$$

$$= 40.9\ \mathrm{W}$$

This is the moisture output from a sedentary adult.

Removal of sensible heat gains to control room air temperature is carried out by cooling the ventilation supply air and increasing the air change rate to perhaps 20 changes per hour. Figure 5.2 shows this scheme. The temperature and moisture content of the supply air increase as it absorbs the sensible and latent heat gains until it reaches the desired room condition. The net sensible heat flow will be into the room in summer and in the outward direction in winter.

Rooms that are isolated from exterior building surfaces have internal heat gains from people and electrical equipment, producing a net heat gain throughout the year. The heat balance is as follows:

net sensible heat flow into room = sensible heat absorbed by ventilation air

Therefore

SH = air mass flow rate × specific heat capacity × temperature rise

where SH is the sensible heat. The specific heat capacity of humid air is 1.012 kJ/kg K. The volume flow rate of air is normally used for duct design.

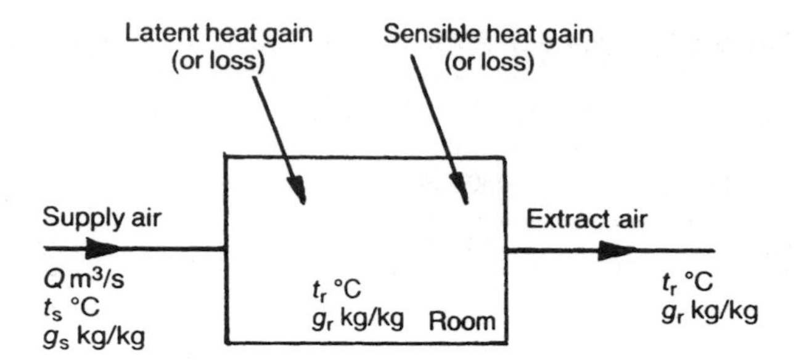

Figure 5.2 Schematic representation of heating, cooling and humidity control of an air-conditioned room

Therefore

$$Q\,\frac{\text{m}^3}{\text{s}} = \text{air mass flow rate}\,\frac{\text{kg}}{\text{s}} \times \frac{\text{m}^3}{\rho\ \text{kg}}$$

The density of air at 20 °C d.b. and 1013.25 mb is $\rho = 1.205\ \text{kg/m}^3$.

The supply air temperature t_s can have any value, and as density is inversely proportional to the absolute temperature, from the general gas laws,

$$\text{SH kW} = Q\,\frac{\text{m}^3}{\text{s}} \times 1.205\,\frac{\text{kg}}{\text{m}^3} \times \frac{(273+20)\ \text{K}}{(273+t_s)\ \text{K}} \times (t_r - t_s)\ \text{K} \times 1.012\,\frac{\text{kJ}}{\text{kg K}} \times \frac{\text{kWs}}{1\ \text{kJ}}$$

For summer cooling, t_r is greater than t_s during a net heat gain. For winter heating, t_r is less than t_s as there is a net heat loss, and so the temperature difference $t_s - t_r$ must be used in the equation. It is more convenient to rewrite the equation in the form

$$Q = \frac{\text{SH kW}}{t_r - t_s} \times \frac{273 + t_s}{357}\ \text{m}^3/\text{s}$$

EXAMPLE 5.4

An office has 20 occupants, 30 m² of windows, ten 65 W fluorescent tube light fittings, a photocopier with a power consumption of 1500 W and conduction heat gains during summer amounting to 1 kW. Solar heat gains are 450 W/m² of window area. The sensible heat output from each person is 110 W. The room air temperature is not to exceed 23 °C when the supply air is at 15 °C. Calculate the supply air flow rate required.

The sensible heat gains are summarized in Table 5.2. Then

$$Q = \frac{18.85}{23-15} \times \frac{273+15}{357}\ \text{m}^3/\text{s}$$

$$= 1.901\ \text{m}^3/\text{s}$$

Table 5.2 Summary of sensible heat gains in Example 5.4

Source	Quantity ($W/m^2 \times m^2$)	*SH*(W)
Windows	450 × 30	13 500
Occupants	110 × 20	2 200
Lights	65 × 10	650
Photocopier		1 500
Conduction		1 000
		SH gain = 18 850
		= 18.85 kW

Some engineers prefer to calculate the mass flow rate of air flow through the air-conditioned space. This is easily found by multiplying the volume flow rate by the density of air at that location, thus,

$$\text{supply air mass flow rate} = 1.901\ \frac{m^3}{s} \times 1.205\ \frac{kg}{m^3} = 2.29\ kg/s$$

Air mass flow rate does not change with temperature, volume flow does.

EXAMPLE 5.5

A room has a heat loss in winter of 15 kW and a supply air flow rate of 2 m^3/s. The room air temperature is to be maintained at 21 °C. Calculate the supply air temperature to be used.

For winter, the equation is

$$Q = \frac{SH}{t_s - t_r} \times \frac{273 + t_s}{357}$$

This can be rearranged to find the supply air temperature t_s required:

$$Q \times 357(t_s - t_r) = SH(273 + t_s)$$

$$Q \times 357t_s - Q \times 357t_r = 273SH + SH \times t_s$$

$$Q \times 357t_s - SH \times t_s = 273SH + Q \times 357t_r$$

Thus

$$t_s(Q \times 357 - SH) = 273SH + Q \times 357t_r$$

$$t_s = \frac{273SH + 357Qt_r}{357Q - SH}\ °C$$

In this example

$$t_s = \frac{273 \times 15 + 357 \times 2 \times 21}{357 \times 2 - 15}$$

$$= 27.3\ °C$$

The heated or cooled air supply may also have its humidity modified by an air-conditioning plant so that the percentage saturation of the room air is controlled. A water spray or steam injector can be used for humidification. The refrigeration cooling coil lowers the air temperature below its dew-point to condense the moisture out of the air.

Figure 5.3 shows how the properties of humid air are determined from a psychrometric chart. If g_r and g_s are the moisture contents of the room and the supply air respectively, then a heat balance can be written for the latent heat gain absorbed by the previously calculated supply air flow rate Q:

latent heat gain to room = latent heat equivalent of moisture removed by the conditioned air

$$\text{LH kW} = \text{moisture mass flow rate } \frac{\text{kg}}{\text{s}} \times \text{latent heat of evaporation } \frac{\text{kJ}}{\text{kg}}$$

$$= \text{air mass flow rate } \frac{\text{kg}}{\text{s}} \times \text{moisture absorbed } \frac{\text{kg H}_2\text{O}}{\text{kg dry air}} \times 2453.61\ \frac{\text{kJ}}{\text{kg H}_2\text{O}}$$

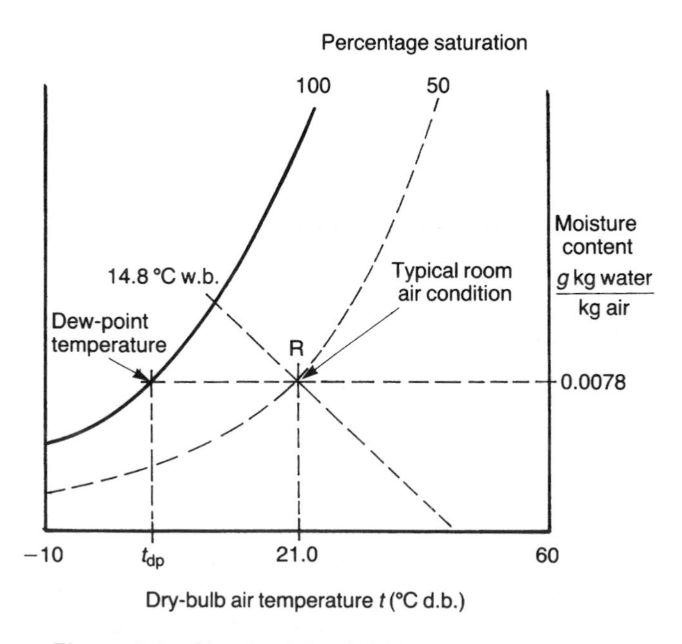

Figure 5.3 Sketch of the CIBSE psychrometric chart

$$\text{LH kW} = Q\,\frac{\text{m}^3}{\text{s}} \times 1.205\,\frac{\text{kg}}{\text{m}^3} \times \frac{273+20}{273+t_s} \times (g_r - g_s)\,\frac{\text{kg H}_2\text{O}}{\text{kg air}} \times 2453.61\,\frac{\text{kJ}}{\text{kg H}_2\text{O}}$$

$$Q = \frac{\text{LH}}{g_r - g_s} \times \frac{273+t_s}{866\,284}\ \text{m}^3/\text{s}$$

The denominator can be rounded to 860 000 with an error of less than 1.0%.

EXAMPLE 5.6

The people in the office in Example 5.4 each produce 40 W of latent heat. Find the supply air moisture content to be maintained, given that the room air is to be at 50% saturation and the corresponding moisture content g_r is 0.008 905 kg H_2O/kg air.

$$\text{LH} = 20\ \text{people} \times 40\,\frac{\text{W}}{\text{person}} \times \frac{1\ \text{kW}}{10^3\ \text{W}} = 0.8\ \text{kW}$$

$$t_s = 15\ °\text{C} \qquad Q = 1.9\ \text{m}^3/\text{s}$$

$$\text{air mass flow rate} = 1.9\ \text{m}^3/\text{s} \times 1.205\ \text{kg/m}^3 = 2.29\ \text{kg/s}$$

$$Q = \frac{\text{LH}}{g_r - g_s} \times \frac{273+t_s}{860\,000}\ \text{m}^3/\text{s}$$

we obtain

$$g_r - g_s = \frac{0.8 \times (273+15)}{1.901 \times 860\,000}\ \text{kg H}_2\text{O/kg air}$$

Hence

$$g_s = 0.008\,905 - 0.000\,141\ \text{kg H}_2\text{O/kg air}$$

$$= 0.008\,764\ \text{kg H}_2\text{O/kg air}$$

Psychrometric cycles

Heating and cooling processes in air-conditioning equipment can be represented on the psychrometric chart in the following manner.

Heating

Heating is performed by low- or high-pressure hot-water finned pipe heater battery, electric resistance heater or fuel-fired heat exchanger as shown in Fig. 5.4. SE is the specific enthalpy (the total heat content) of the air, as read from the chart.

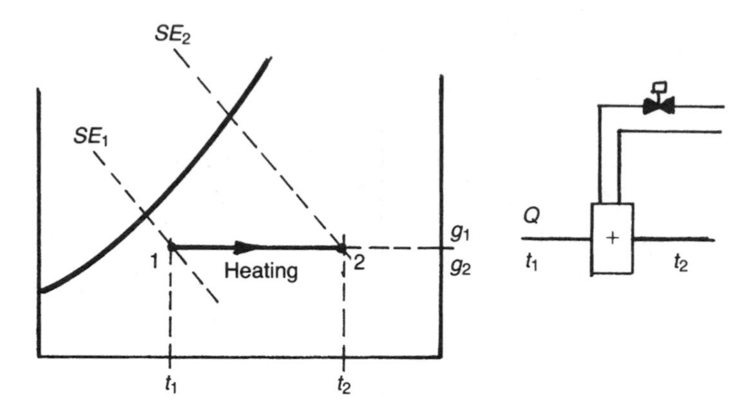

Figure 5.4 Air heating depicted on a psychrometric chart

Cooling

Cooling is performed by pumping chilled water, brine or refrigerant through a finned pipe battery. When the coolant temperature is below the air dew-point, condensation occurs and the air will be dehumidified. Figure 5.5 shows the two possible cycles.

Mixing

Mixing of two airstreams occurs when the fresh-air intake joins the recirculated room air. The quantity of each airstream is regulated by multileaf dampers operated by electric motors under the direction of an automatic control system. Varying the intake of fresh air between the minimum amount during peak summer and winter conditions and 100% when free atmospheric cooling can be achieved during mild weather and summer evenings can result in minimizing the energy costs of the heating and refrigeration plants. Figure 5.6 shows the operational process.

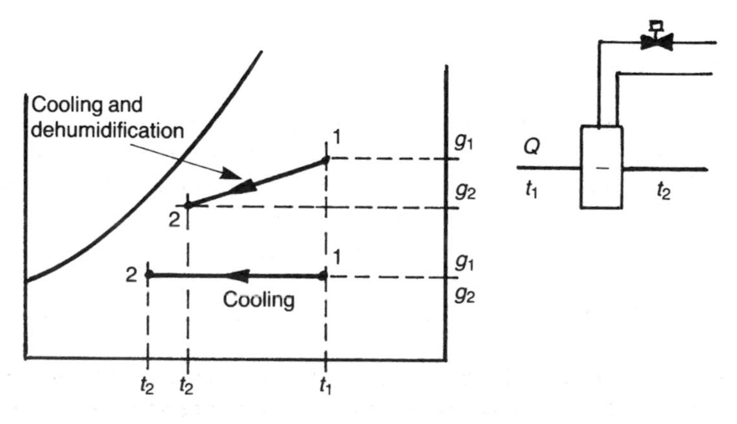

Figure 5.5 Cooling and dehumidification

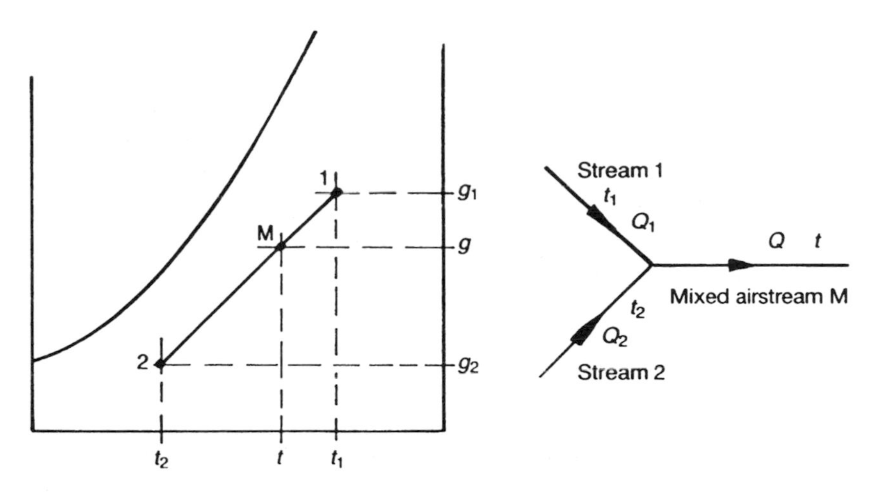

Figure 5.6 Psychrometric cycle for the mixing of two airstreams

The mass flow balance for the junction is

mass flow of stream 1 + mass flow of stream 2 = mixed mass flow

or

$$Q_1\rho_1 + Q_2\rho_2 = Q\rho$$

The enthalpy balance, taking the specific heat capacity as constant, is

$$Q_1t_1 + Q_2t_2 = Qt$$

Dividing through by Q gives

$$\frac{Q_1}{Q}t_1 + \frac{Q_2}{Q}t_2 = t$$

The mixed air temperature and moisture content lie on the straight line connecting the two entry conditions and can be found by the volume flow rate proportions as indicated by the equation.

Humidification

In winter, incoming fresh air with a low moisture content can be humidified by steam injection, banks of water sprays, evaporation from a heated water tank, a spinning disc atomizer or a soaked porous plastic sponge. A preheater low-pressure hot-water coil usually precedes the humidifier to increase the water-holding capacity of the air. This also offsets the reduction in temperature of the air owing to transference of some of its sensible heat into latent energy, which is needed for the evaporation process. Figure 5.7 shows such an arrangement.

A temperature sensor in the humidified air is used as a dew-point control by modulating the preheater power to produce air at a consistent moisture content throughout the winter. For comfort air conditioning, the room percentage

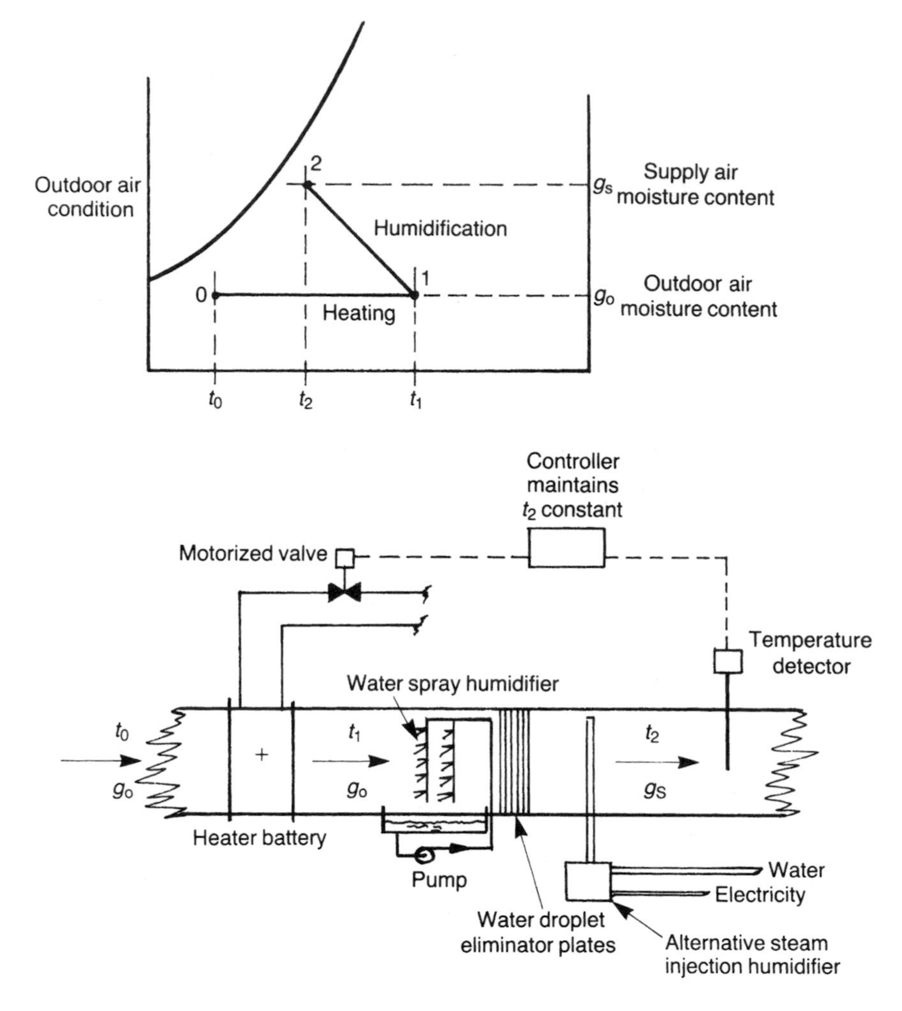

Figure 5.7 Preheating and humidification

saturation will be $50\% \pm 10\%$. This permits a wide range of humidifier performance characteristics.

The humidification process often follows a line of constant wet-bulb temperature. The water spray temperature is varied to alter the slope of the line on the psychrometric chart. A complete psychrometric cycle for a single-duct system during winter operation is shown in Fig. 5.8. The preheating and humidification stages have been omitted, as close humidity control is deemed not to be needed in this case. A typical summer cycle is shown in Fig. 5.9.

Some reheating of the cooled and dehumidified air will be necessary because of practical limitations of cooling coil design. Part of the boiler plant remains operational during the summer. Reheating can be avoided by using a cooling coil bypass which mixes air M and air C to produce the correct supply condition. Heating and refrigeration plant capacities are found from the enthalpy changes

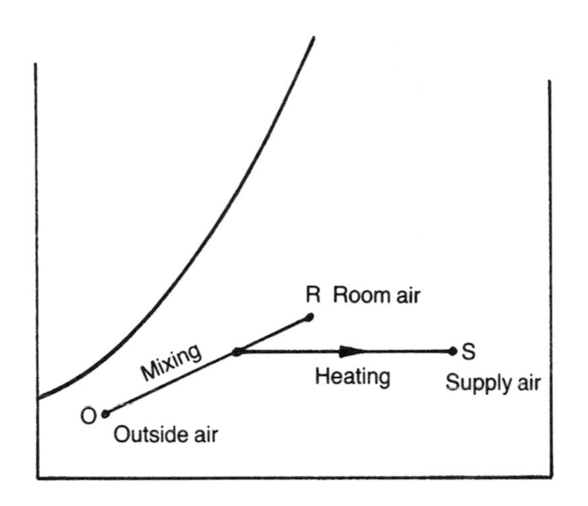

Figure 5.8 Winter psychrometric cycle for a single-duct system

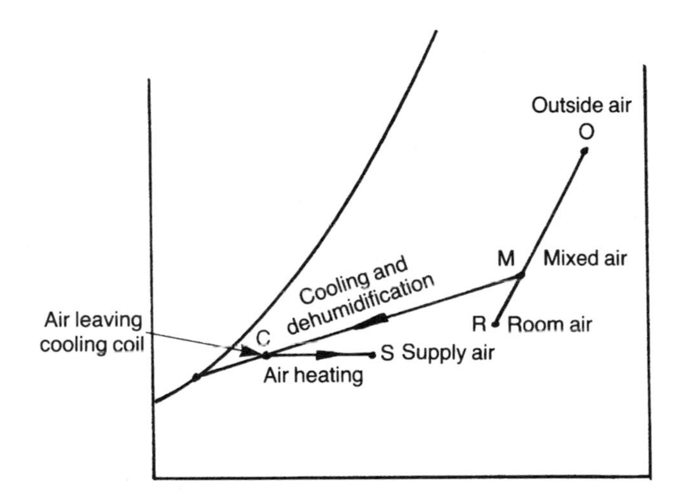

Figure 5.9 Summer psychrometric cycle for a single-duct system

and specific volume, read from the chart, and the air volume flow rate:

$$\text{heat transfer rate} = Q\,\frac{\text{m}^3}{\text{s}} \times \frac{\text{kg}}{v_s\,\text{m}^3} \times (\text{SE}_1 - \text{SE}_2)\,\frac{\text{kJ}}{\text{kg}} \times \frac{\text{kWs}}{\text{kJ}}$$

EXAMPLE 5.7

Outside air at −5 °C d.b., 80% saturation enters a preheater battery and leaves at 24 °C d.b. The air volume flow rate is 2 m^3/s. Find (a) the outdoor air wet-bulb temperature and specific volume, (b) the heated air moisture content and percentage saturation, and (c) the heater battery power.

From the CIBSE psychrometric chart:

(a) $-5.9\ ^\circ\text{C}$ and $0.7615\ \text{m}^3/\text{kg}$;

(b) 0.001 98 kg H_2O/kg air and 10%;

(c) $$\text{heater duty} = 2\ \frac{\text{m}^3}{\text{s}} \times \frac{\text{kg}}{0.7615\ \text{m}^3} \times [29 - (-0.1)]\ \frac{\text{kJ}}{\text{kg}} \times \frac{\text{kWs}}{\text{kJ}}$$

$$= 76.428\ \text{kW}$$

EXAMPLE 5.8

A cooling coil has water passing through it at a mean temperature of 10 °C; an air flow of 1.5 m^3/s air enters the coil at 28 °C d.b., 23 °C w.b. and leaves at 15 °C d.b. (a) Find the leaving air wet-bulb temperature and percentage saturation; (b) calculate the refrigeration capacity of the coil.

(a) 14.2 °C w.b. and 91% saturation;

(b) $$\text{refrigeration capacity} = 1.50\ \frac{\text{m}^3}{\text{s}} \times \frac{\text{kg}}{0.874\ \text{m}^3} \times (68 - 40)\ \frac{\text{kJ}}{\text{kg}} \times \frac{\text{kWs}}{\text{kJ}}$$

$$= 48.055\ \text{kW}$$

EXAMPLE 5.9

2 m^3/s of recirculated room air at 22 °C d.b., 50% saturation is mixed with 0.5 m^3/s of incoming fresh air at 10 °C d.b., 6 °C w.b. Calculate the mixed air dry-bulb temperature. Plot the process on a psychrometric chart and find the mixed air moisture content.

Using the equation

$$t = \frac{Q_1}{Q} t_1 + \frac{Q_2}{Q} t_2$$

$$t = \frac{2}{2.5} \times 22 + \frac{0.50}{2.5} \times 10$$

$$= 19.6\ ^\circ\text{C d.b.}$$

From the chart

$$g = 0.0076\ \text{kg H}_2\text{O/kg air}$$

Air-conditioning systems

Single duct

The single-duct system (Fig. 5.10) is used for a large room such as an atrium, a banking hall, a swimming pool, or a lecture, entertainment or operating theatre. It can be applied to groups of rooms with a similar demand for air conditioning, such as offices facing the same side of the building. A terminal heater battery under the control of a temperature sensor within the room can be employed to provide individual room conditions.

A variable air volume (VAV) system has either an air volume control damper or a centrifugal fan in the terminal unit to control the quantity of air flowing into the room in response to signals from a room air temperature sensor. Air is sent to the terminal units at a constant temperature by the single-duct central plant, according to external weather conditions. A reducing demand for heating or cooling detected by the room sensor causes the damper to throttle the air supply or the fan to reduce speed until either the room temperature stabilizes or the minimum air flow setting is reached.

Air flow from the diffuser is often blown across the ceiling to avoid directing jets at the occupants. As a result of the Coanda effect the air stream forms a boundary layer along the ceiling and entrains room air to produce thorough mixing and temperature stabilization before it reaches the occupied part of the room. When the VAV unit reduces air flow, there may be insufficient velocity to maintain the boundary layer, and in summer cool air can dump or drop from the ceiling onto the occupants, resulting in complaints of cool draughts.

Dual duct

In order to provide for wide-ranging demands for heating and cooling in multiroom buildings, the dual-duct system, as shown in Fig. 5.11, is used. Air flow in the two supply ducts may, of necessity, be at a high velocity (10–20 m/s) to fit into service ducts of limited size. Air turbulence and fan noise are prevented from entering the conditioned room by an acoustic silencer.

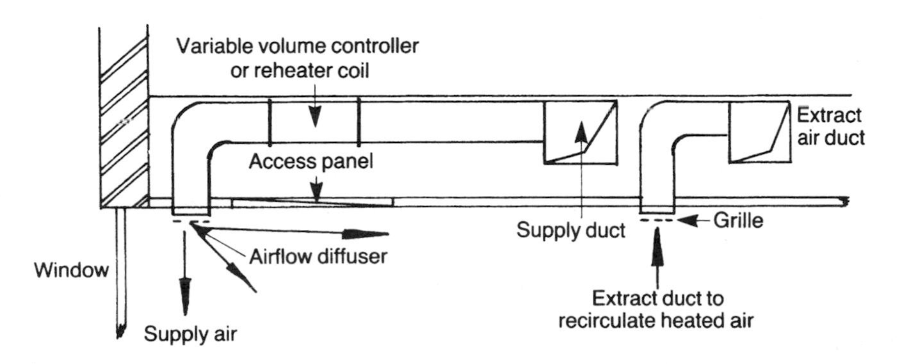

Figure 5.10 Single-duct all-air installation in a false ceiling

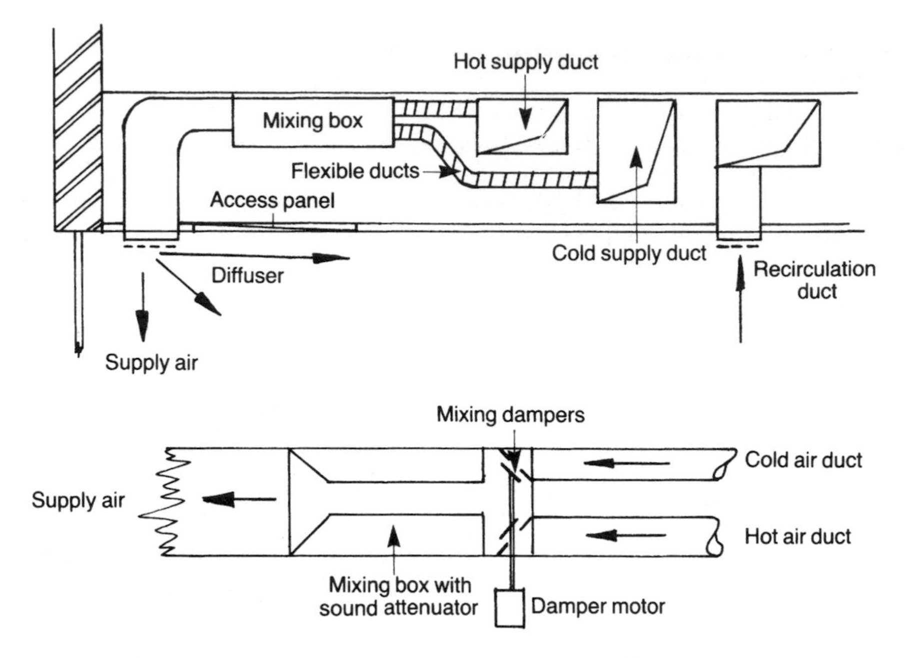

Figure 5.11 Dual-duct installation in a false ceiling and detail of the mixing box

In summer, the hot duct will be for mixed fresh and recirculated air, while the cold duct is for cooled and dehumidified air. The two streams are mixed in variable proportions by dampers controlled from a room air temperature detector. During winter, the cold duct will contain the untreated mixed air and the air in the hot duct will be raised in temperature in the plant room. The system is used for comfort air conditioning as it does not provide close humidity control. It reacts quickly to changes in demand for heating or cooling when, for example, there is a large influx of people or a rapid increase in solar gain.

Induction

Induction is a less costly alternative to the all-air single- and dual-duct systems for multiroom applications. The central air-conditioning plant handles only fresh air, perhaps only 25% of the supply air quantity for an equivalent single-duct system. All the humidity control, and also some of the heating and cooling for the building, is achieved by conditioning the fresh air intake in the plant room.

Primary fresh air is injected through nozzles into the induction unit in each room. These units may be in the floor, in the ceiling void or under the window-sill. Because of the high-velocity jets, the local atmospheric pressure within the unit is lowered and air is induced into it from the room. The induced air may enter at three or four times the volume flow rate of primary air, and it flows through a finned pipe bank and dust filter before mixing with primary air and being supplied to the room.

The secondary air flow rate can be manually adjusted using a damper. Either hot or chilled water is passed through the room coil depending upon demand. A two-, three- or four-pipe distribution system will be used. The two-pipe system requires a change-over date from heating to cooling plant operation, but a three-way valve can blend hot and chilled water from the three-pipe arrangement. The third alternative has separate hot- and chilled-water pipe coils and pipework.

The extract ductwork and fan removes 90% of the primary air supply and exhausts it to the atmosphere. All recirculation is kept within the room and this greatly reduces duct costs and service duct space requirements. Figure 5.12 shows a typical installation in an office.

Fan coil units

Heating and cooling loads that prove to be too great for induction units can be dealt with by separate fan and coil units fitted into the false ceiling of each room or building module. Better air filtration can be achieved than with the induction unit. A removable access hatch below the unit is required to facilitate motor and filter maintenance.

Care is taken to match the fan-generated noise to the required acoustic environment. As with the other systems, the extracted air can be taken through ventilated luminaires to remove the lighting heat output at source and avoid overheating the room. The supply and extract ducts only carry the fresh air. All recirculation is confined to the room. A typical layout is shown in Fig. 5.13.

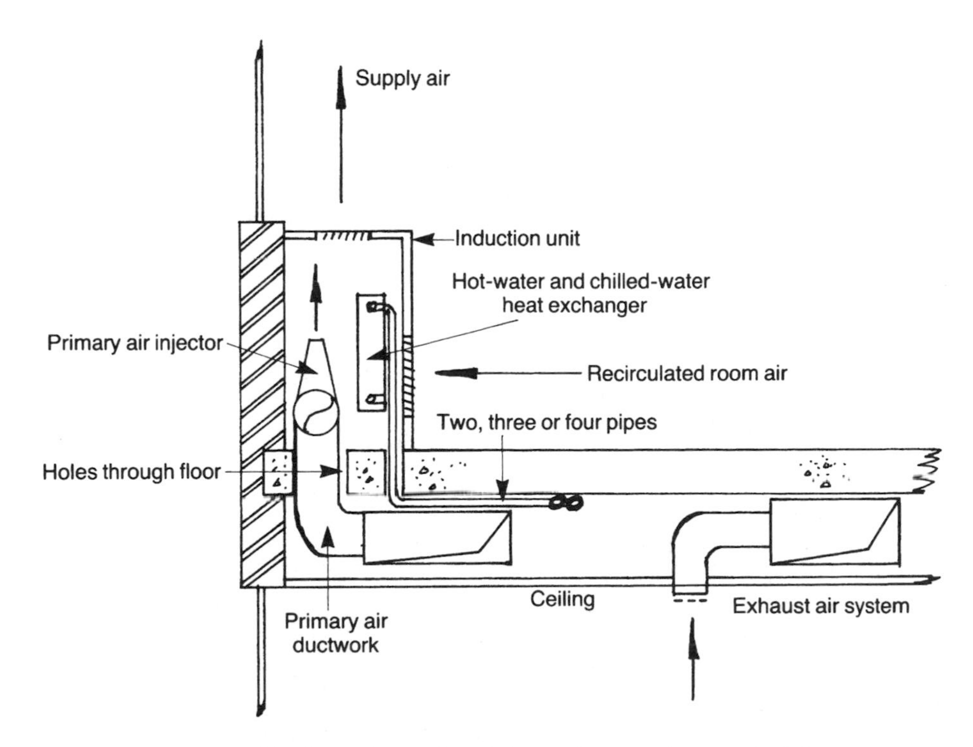

Figure 5.12 Induction unit installation in a multi-storey building

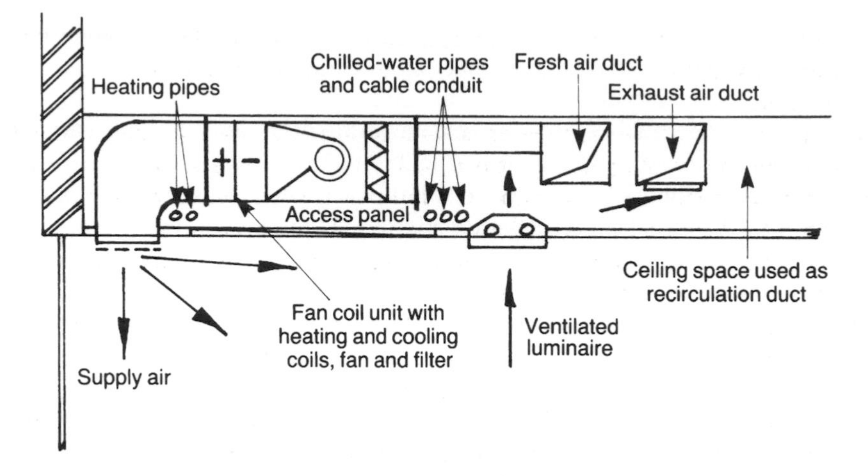

Figure 5.13 Fan coil unit installation in a false ceiling

Packaged unit

A packaged unit is a self-contained air-conditioning unit comprising a hermetically sealed refrigeration compressor, a refrigerant evaporator coil to cool room air, a hot-water or electric resistance heater battery, a filter, a water- or air-cooled refrigerant condenser and automatic controls. Packaged units can either be completely self-contained, needing only a supply of electricity, or piped to central heating and condenser cooling-water plant. Small units are fitted into

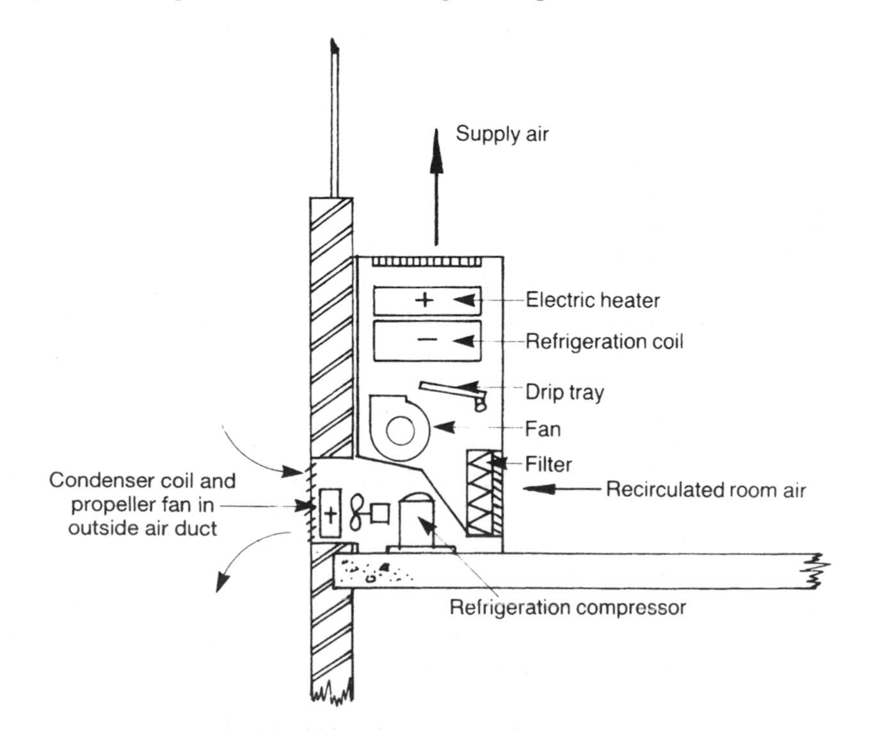

Figure 5.14 Packaged air-conditioning unit

an external wall and have a change-over valve to reverse the refrigerant flow direction. This enables the unit to cool the internal air in summer and the external air in winter.

Heat rejected from the condenser is used to heat the internal environment in winter. In this mode of operation it is called a heat pump. A separate ventilation system may be needed. Compressor and fan noise levels are compared with the acceptable background acoustic environment. Maintenance requirements are filter cleaning, bearing lubrication and replacement of the compressor when it becomes too noisy or breaks down.

Split system units have a separate condenser installed outside the building. Two refrigerant pipes of small diameter connect the internal and external equipment boxes. This allows greater flexibility in siting the noise-producing compressor. Ducted models provide conditioning and ventilation and are often sited on flat roofs. Figure 5.14 shows a typical through-the-wall installation.

Vapour compression refrigeration

The electrically driven vapour compression refrigeration system is the principal type used. Its rival, the absorption cycle, burns gas to produce cooling but has a coefficient of performance of around 1, whereas vapour compression has a coefficient of performance in the range 2–5, and so it is cheaper to operate. Compressor types are as follows.

1. Single- or multicylinder reciprocating piston compressor with spring-loaded valves: domestic refrigerators and small air conditioners have hermetically sealed motor–compressor units which are sealed for their service period, i.e. about 10 years.

   A condensing unit comprises a sealed compressor, a refrigerant condenser, a liquid receiver, pipework and controls. Refrigerant pipework is installed on site from this unit to a finned-pipe forced-draught air-cooling coil in the air-conditioning system.

   Large air-cooling plant comprises a multicylinder in-line or V-formation compressor, a shell and tube refrigerant to a water evaporator producing chilled water, and a shell and tube refrigerant to a water condenser where the refrigerant vapour is condensed into liquid and the heat given out is carried away by a water circuit to a cooling tower on the roof.
2. The centrifugal compressor is used in large chilled-water plants where the noise and vibration produced by the reciprocating type would be unacceptable. A centrifugal impeller of small diameter is driven through a step-up gearbox from a three-phase electric motor. The lack of vibration and compactness of the very high-speed compressor makes siting the plant easier.
3. The screw compressor has two meshed gears, which compress the refrigerant in the spaces between the helical screws. One gear is driven by an electric motor through a step-up gearbox. The compressor operates at high speed and has very low noise and vibration levels.

The operation of a vapour compression refrigeration plant is shown in Fig. 5.15.

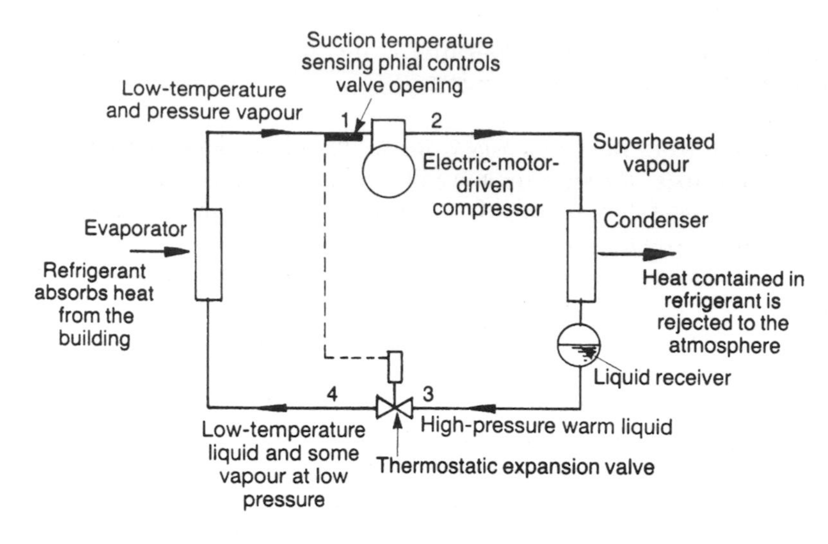

Figure 5.15 Vapour compression refrigeration system

Refrigerants commonly used are non-toxic fluorinated hydrocarbon fluids with high latent heat. Refrigeration plant with a capacity of up to about 175 kW uses refrigerant 12 (R12, which is CCl_2F_2), which boils at −29.8 °C in the atmosphere. In a typical system it will be evaporated at 5 °C under a pressure of 3.6 bar and condensed at 40 °C at 9.6 bar. Larger plant uses R22 ($CHClF_2$), which has a greater refrigerating effect per kilogram but is more expensive.

The coefficient of performance (COP) is an expression of cycle efficiency and is found from

$$\text{COP} = \frac{\text{heat absorbed by refrigerant in the evaporator W}}{\text{power consumption by the compressor W}}$$

The vapour compression cycle can be represented on a pressure–enthalpy diagram for the refrigerant as shown in Fig. 5.16. Referring to Figs 5.15 and 5.16, compression 1–2 raises the temperature of the refrigerant dry superheated

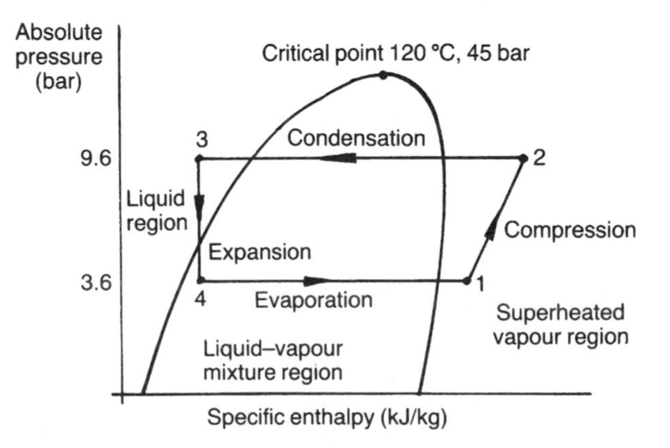

Figure 5.16 Pressure–enthalpy diagram for a refrigerant showing the vapour compression cycle

vapour from about 20 °C to 60 °C, where it can then be cooled and condensed at a sufficiently high temperature to reject the excess heat from the building to the hot external environment.

It condenses at 40 °C and collects in the liquid receiver. This warm high-pressure liquid passes through an uninsulated pipe so that it is subcooled to below its saturation temperature (about 20 °C) at the expansion valve located alongside the evaporator.

The pressure rise produced through the compressor is dissipated in friction through the fine orifice in the valve. Such a sudden pressure drop is almost an adiabatic thermodynamic process. This is represented by the vertical line 3–4 on the pressure–enthalpy diagram. Some heat loss from the valve body takes place, so that 3–4 will be slightly curved. Condition 4 is at the lower pressure of the evaporation process, where the refrigerant temperature has dropped to 5 °C. Some of the refrigerant liquid has flashed into vapour.

The liquid and flash vapour mixture flows through the evaporator, where it is completely boiled into vapour and is then given a small degree of superheat (path 4–1). It then enters the compressor as dry low-pressure superheated vapour at 20 °C. A suction temperature-sensing phial controls the refrigerant flow rate by means of a liquid-filled bellows on the thermostatic expansion valve. This matches refrigerant flow to the refrigerating effect required by the air-conditioning system and ensures that liquid droplets are not carried into the compressor, where they could cause damage.

Large plant has refrigerant pressure controllers, which reduce compressor performance by unloading some cylinders. Lubricating oil contaminates the refrigerant leaving the compressor. This is separated gravitationally and returned to the crankcase.

The pressure–enthalpy diagram depicts the reverse Carnot cycle. In the other direction, the cycle is for an internal combustion engine. The theoretical ideal coefficient of performance is given by

$$\text{ideal COP} = \frac{T_1}{T_2 - T_1}$$

where T_1 is the evaporation absolute temperature (K) and T_2 is the condensation absolute temperature (K). With the temperatures previously used,

$$\text{ideal COP} = \frac{273 + 5}{(273 + 40) - (273 + 5)}$$

$$= 7.94$$

Friction losses from fluid turbulence, heat transfers to the surroundings, and mechanical and electrical losses in the compressor all reduce this value to 2–4 in commercial equipment. The electricity consumption of fans and pumps adds to the running costs.

Figure 5.17 shows the installation of a chilled-water refrigeration plant serving one of the cooling coils in an air-conditioning system in a large building. Each air-handling system has filters, heaters, humidifiers and recirculation ducts appropriate to the design. There are several separate air-handling plants, each

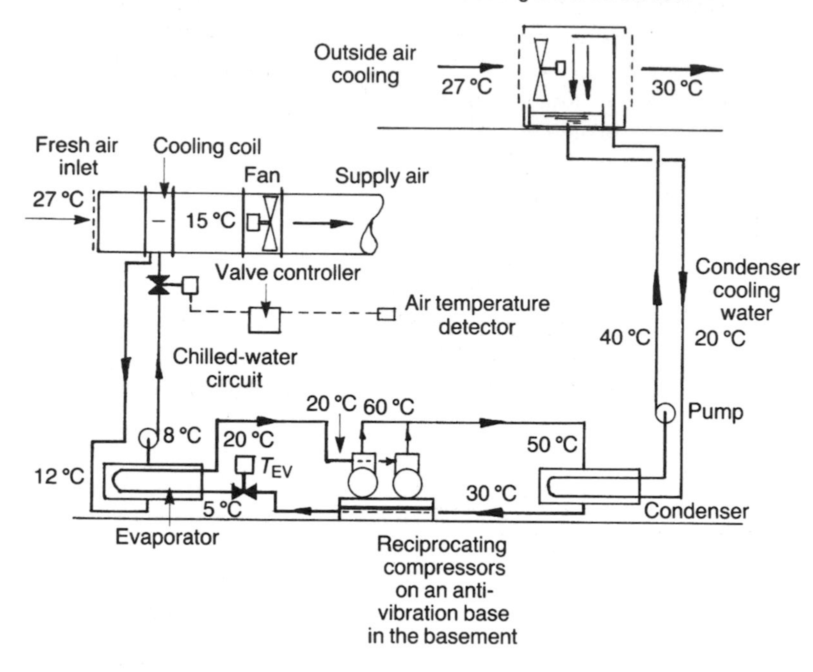

Figure 5.17 A refrigeration plant serving an air-conditioning system, showing typical fluid temperatures

serving its own zone. Zones are decided by the similarity of demand for conditioning. The south-facing orientation has a cyclic requirement for cooling that is distinct from that of the other sides of a building. Internal areas require cooling throughout the year. These differing needs are often met by having separate zones for each area.

Absorption refrigeration cycle

An example of a two-drum absorption refrigeration cycle is shown in Fig. 5.18. The input heat source may be gas, steam or hot water from a district heating scheme, or rejected heat from a gas turbine electricity-generating set.

The generator (1) contains a concentrated solution of lithium bromide salt in water. Pure water is boiled off this solution and condenses on the cooling-water pipes, which are connected to an external cooling tower. The generator drum pressure is subatmospheric at 0.07 bar, with boiling and condensation taking place at 38 °C. Water leaves the condenser (2), and then passes through an expansion valve (3), where its pressure and temperature are lowered to 0.01 bar and 7 °C. It then completely evaporates while being sprayed over water pipes in the evaporator. These pipes are the chilled-water circuit at 6–10 °C, which supplies the refrigeration for the air-conditioning cooling coils. Water vapour in the evaporator drum (4) is sucked into a weak lithium bromide solution by the salt's affinity for water. Latent heat given up by the water vapour as it condenses

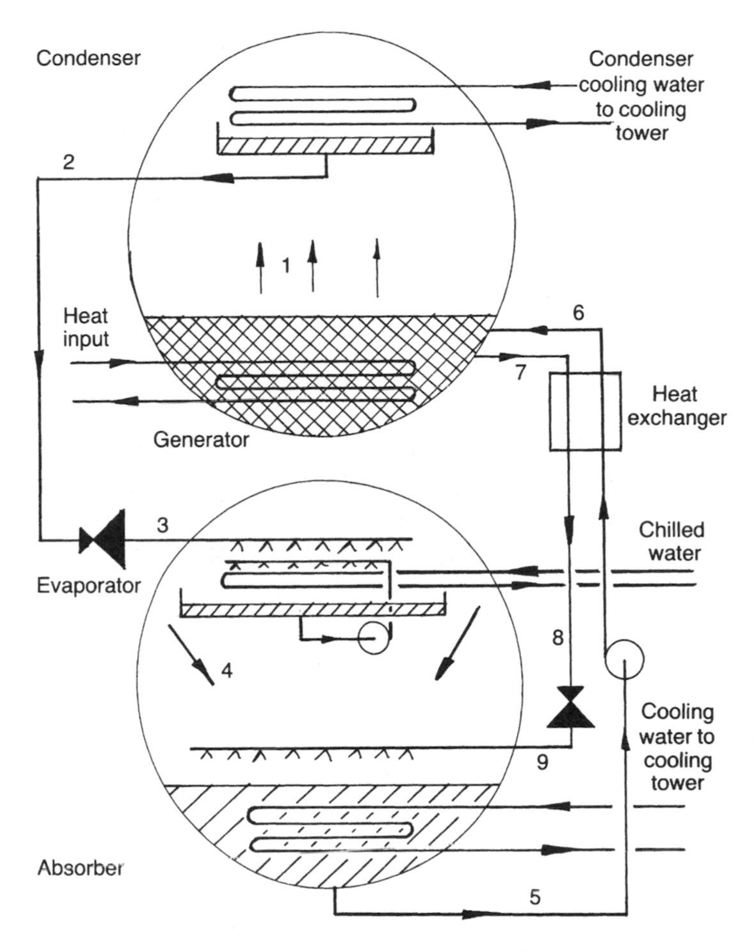

Figure 5.18 Two-drum absorption refrigeration cycle

into the solution is removed from the cooling tower by cooling-water pipes. The weak solution (5) is pumped back into the higher-pressure generator drum to complete the cycle. This pump is the only moving part of some systems. Concentrated salt solution (6) is passed down to the absorber via a heat exchanger and pressure-reducing valve to replace the salt removed. The production of chilled water is equivalent to about half the heat input to the generator.

The gas domestic refrigerator works on an absorption system using liquid evaporation to provide the cooling effect. Figure 5.19 shows the features of modern equipment. A solution of ammonia in water is heated in the boiler (A) by a small gas flame. This is the only energy input. Ammonia gas is driven off the solution and then condensed to a liquid in the air-cooled condenser (B) outside the refrigerator cabinet. The liquid ammonia then passes, with some hydrogen, into the evaporator (C) inside the refrigerator cabinet. This is the ice box. The ammonia completely evaporates while absorbing the heat from the cabinet. The two gases are then led into the absorber (D), where the ammonia is absorbed by a weak solution trickling down the absorber. The strong ammonia solution produced is then driven back into the boiler, while the hydrogen gas, which is not

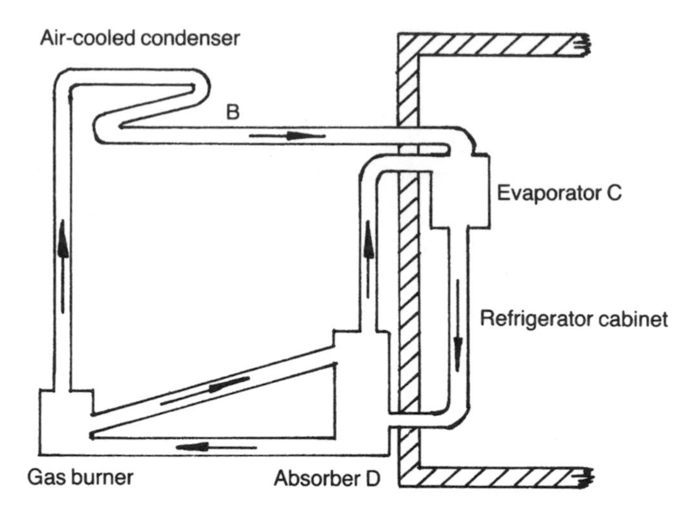

Figure 5.19 Gas-fired domestic absorption refrigeration system

absorbed, passes to the evaporator. The weak solution trickling down the absorber is provided from the boiler. Both types of absorption refrigerator provide cooling from a source of heat, they have few or no moving parts and they only require low-power pumps. This makes the cycle suitable for solar power input. The equipment is vibrationless and very quiet in operation.

Ventilation rate measurement

Measurement of room ventilation rate may be required for research into the energy consumption of heated buildings or to carry out commissioning tests on warm-air heating, ventilation or air-conditioning systems. Three basic methods are available.

Smoke

Provided that smoke detectors and alarms are deactivated and suitable warning is given, the room can be filled with smoke and the ventilation system switched on. The time to clear the smoke is used to calculate the air change rate and volume flow rate. Smoke candles or an oil-burning generator are used.

Anemometer

The air velocity through each ventilation grille and any obvious gaps around doors and windows is measured using a suitable anemometer: a rotating vane for large grilles, and a thermistor, mini-vane or pitot-static tube for small airways. The air flow rates into and out of the room are calculated from the airway areas and the average air velocities through them.

Tracer gas

A non-toxic tracer gas (nitrous oxide or helium) is released into the room and thoroughly mixed with portable fans to fill the complete volume. Samples of room air are taken at intervals and passed through an analyser, which measures the concentration of tracer gas. The room air change rate is calculated from two known concentrations and the time interval between them. This technique can be used for naturally ventilated buildings and produces accurate results.

The katharometer measures air electrical conductivity and gives an output of percentage concentration of tracer in air. An infrared analyser uses a source of infrared radiation and passes it down two tubes to receiving photocells. One tube contains a reference gas and the other the sample of room air. The different gases absorb different amounts of radiation, and the variation in the signals from the photocells is calibrated as the percentage of tracer gas in the air.

Consider the injection of tracer gas into a room such that its concentration is C_r%. A stirring fan is used. Ventilation of the room is at the rate of N air changes per hour. The tracer gas concentration falls to C_r% during time interval τ. The room air change rate can be found as follows:

$$N = \frac{1}{\tau}\ln\left(\frac{C_r}{C_\tau}\right)$$

so that

$$N\tau = \ln\left(\frac{C_r}{C_\tau}\right)$$

and hence

$$e_{N\tau} = C_r/C_\tau$$

Thus the concentration C_τ at time τ is given by

$$C_\tau = C_r \times e^{-N\tau}$$

EXAMPLE 5.10

Nitrous oxide tracer gas is admitted into a building and mixed with the internal air to achieve a 10% concentration. After 30 min, the tracer gas concentration has fallen to 5%. Calculate the air change rate per hour.

$$N\,\frac{\text{air changes}}{\text{h}} = \frac{1}{\tau \text{h}} \times \ln\left(\frac{C_r}{C_\tau}\right)$$

Now

$$\tau = 30\,\text{min} = 0.5\ \text{h}$$

$$C = 10\%$$

and

$$C_T = 5\%$$

Hence

$$N = \frac{1}{0.5} \ln\left(\frac{10}{5}\right)$$

$$= 2 \ln 2$$

$$= 1.386 \text{ air changes per hour}$$

Materials for ventilation ductwork

The materials used for ventilation ductwork are listed in Table 5.3. Thin-gauge galvanized mild steel sheet ducts are the most popular because of their low cost. Prefabricated ducts and fittings allow rapid site erection. Circular, rectangular, flat or spirally wound circular ducts are generally used. Joints are pop-riveted and sealed with waterproof adhesive tape, hard-setting butyl bandage or heat-shrunk plastic sleeves. Large ducts have bolted angle-iron flanges, which also act as support brackets. Stiffening steel strips or tented sheets are used to reduce the

Table 5.3 Ventilation ductwork materials

Material	Application	Jointing technique
Galvanized mild steel sheet	All ductwork	Riveted slip joints, machine-formed snap-lock, flanged, butyl cement bandage, heat-shrunk sleeve
UPVC and polypropylene	Prefabricated systems for housing and toilet extract	Flanged, socket and spigot
Resin-bonded glass fibre	Low-velocity and domestic warm air heating	Butt, sleeved, socket and spigot
Asbestos cement	Prefabricated circular and rectangular for flues and chemical exhausts	Socket and spigot
Flexible glass fibre, proofed fabric reinforced with galvanized spring wire helix	Short connections from a duct to a terminal unit	Jubilee clip, waterproof tape
Aluminium, copper, wired glass, stainless steel	Kitchen extract hoods, ornamental use	Flanges
Brick, concrete, timber, fibre- or plasterboard	Recirculation airways within suspended ceilings and floors, surfaces sealed against dust release	As appropriate

drumming effect on flat duct sides caused by air turbulence. Bare metal is painted with metal oxide or zinc chromate paint. Ducts are thermally insulated with resin-bonded glass fibre boards or expanded polystyrene.

An air pressure consisting of the design operational pressure plus 250 pascal (Pa) (1 Pa = N/m^2) is applied after installation for test purposes. The maximum allowed leakage rate is 1% of the system design air flow rate and leaks must not be audible.

Chlorofluorocarbons

Chlorofluorocarbons (CFCs) are numbered to represent the chemical combination. Those in common use are listed in Table 5.4.

When CFCs are released into the atmosphere as a result of leakage from refrigeration systems, the production of expanded foam, venting to the atmosphere during maintenance of refrigeration compressors, the use of aerosols, chemical cleaning or the destruction of refrigerators, they find their way to the upper atmosphere, where they are broken down by the action of ultraviolet solar radiation and chlorine is released. This degradation will continue for many years. Atmospheric ozone is destroyed by the chlorine, and it is reported that the resulting increase in the levels of ultraviolet radiation reaching the earth's surface will cause ecological damage as well as an increase in skin cancer.

In 1986, 100 000 tonnes of CFCs were manufactured, and the 1987 Montréal Protocol agreement was signed by most countries with the target of reducing production to zero by the year 2000. In the immediate time-scale, R22 has a sufficiently low ozone depletion potential (ODP) to be used until a suitable replacement is found. Current use of R12, which is the most common refrigerant fluid, particularly in small refrigeration plants and domestic refrigerators and freezers, is being replaced by hydrogen fluorine alkaline (HFA134A), which is manufactured in Cheshire and the USA. CFC refrigerants are miscible with the mineral lubricating oil used in the compressor, but HFA134A is not and requires synthetic polyglycol alkaline (PGA) lubricant.

Good practice in the use of CFCs involves the following.

1. Avoid leakage by correct use of pipe materials, engineering design, testing and maintenance procedures.
2. Recover fluid from the system by using a vacuum pump.
3. Return CFCs to the manufacturer.

Table 5.4 Chlorinated fluorocarbons in common use

Number	Use	Ozone depletion potential
R11	Foam insulation and furniture	1.0
R12	Refrigeration systems of all size	1.0
R22	Larger refrigeration plant	0.05
R113	Solvent cleaner	1.0
R502	Refrigeration	0.33
Halon	Fire extinguishers	1.0

4. Employ reliable contractors.
5. Use alternative chemicals.

R12 returned to the manufacturer is either cleaned and recycled with new R12 by bulk mixing or pyrolysed at 2500 K in a furnace, where it is completely destroyed and the flue gas is filtered as necessary.

Any replacement for commonly used CFCs must have the following properties:

1. no chlorine content;
2. ODP = 0;
3. non-flammable;
4. low toxicity;
5. similar boiling temperature and vapour pressure to R12;
6. miscible with compressor lubricant and easily separated;
7. thermodynamic and fluid flow properties compatible with currently installed refrigeration systems;
8. cost-effective;
9. no new environmental risks.

Sick building syndrome

Indoor environments may be made artificially close to the warm spring day that most people would like to inhabit, but it is not the genuine atmosphere and will be polluted with synthetic particles and vapours plus other contributory factors:

tobacco smoke;
body odour;
deodorants;
vapours from cleaning fluids, photocopiers, paints and furnishings;
dust;
bacteria;
noise;
flickering lamps;
glare from artificial illumination and the sun;
carpets;
polyvinyl chloride (PVC);
paper;
formaldehyde;
volatile chemicals;
bacteria grown in stagnant water in humidifiers;
treated water aerosols distributed from showers, washing facilities or fountains;
open-plan office;
too many people.

The air temperature, humidity and air movement, which will seem either stagnant or too draughty, can rarely please more than 95% of the occupants and frequently please a lot fewer. The total environmental loading upon the occupants may rise to an unacceptable level, which can be low for those who are hypersensitive, i.e. physically and psychologically unable to fight off such a bombardment of additional foreign agents to the body.

Sick building syndrome (SBS) is epitomized by the occupants' exhibiting a pattern of lethargy, headaches, dry eyes, eye strain, aching muscles, upper respiratory infections, catarrh and aggravated breathing problems such as asthma, upon returning to their workplace after the weekend. Apparent causes are sealed windows, air conditioning, recirculated air, recirculating water humidifiers, high-density occupation, low negative ion content, smoking, air ductwork corrosion, airborne micro-organisms, dust, and excrement from dust mites in carpets. SBS can be defined as a combination of health malfunctions that noticeably affect more than 5% of the building's population.

This means that there should be sick house syndrome as well. Perhaps there is, or perhaps we are more tolerant at home. Cases of formaldehyde vapour irritation after cavity insulation have been noted. The pattern of house occupation is different, and variation of climatic controls is easily achieved.

Relief can be gained by operating windows, temperature control, sun blinds, air grilles, by a brisk walk or by going outdoors to stimulate the body to sweat toxins out.

It has been easy to blame air conditioning for SBS, but the cause is more complex and has much to do with the standards we demand of our buildings and the total internal environment created. Naturally ventilated buildings often have a higher bacteria and dust count than air-conditioned buildings, which use filters and have sealed windows.

Outbreaks of *Legionella* diseases have been attributed to the growth of bacteria in stagnant water in wet cooling towers. These bacteria are distributed on air currents and breathed in by those susceptible to infection, sometimes with fatal results. Dry heat exchangers are preferred for discharging surplus building heat gains back into the external atmosphere but they are rather large. Adequate cleanliness and biocidal dosing of recirculated cooling-tower water is mandatory.

The cure for SBS requires the following actions.

1. Measure pollutants to identify causes.
2. Remove recirculating water humidifiers from air-conditioning plant and replace only with direct steam or water injection.
3. Allow individuals to have control over local air movement, direction and temperature.
4. Clean recirculating water systems such as wet cooling towers and remaining humidifiers and treat them with biocides.
5. Ensure that fresh-air ventilation ductwork, filters, heating and cooling batteries and grilles are internally clean and fully functional.
6. Inspect and clean air-conditioning systems and potential dust-traps regularly.
7. Appraise the lighting system to maximize natural illumination and reduce glare.

Air temperature profile

The recommended upper limit for the room environmental temperature for normally occupied buildings is 27 °C (CIBSE, 1986, Section A8). The external design air temperature for comfort in offices in London (CIBSE, 1986, Table A2.22) may be chosen as 29 °C d.b., 20 °C w.b. Higher outdoor air temperatures occur. The indoor limit of 27 °C will be exceeded in naturally ventilated buildings in the UK and in warmer locations (Chadderton, 1997a, Chapter 3). The elevation of indoor temperature above that of the outdoor air is caused by a combination of the infiltration of external air, solar radiation and indoor heat gains. In parts of the world where high solar radiation intensity and continuously higher external temperatures are common, for example Sydney with 35 °C d.b. and 24 °C w.b., the necessity for controlled air circulation and refrigeration can be recognized.

Environmental temperature is a combination of mean radiant and air temperatures (Chapter 1). Intense solar radiation through glazing during the summer can lead to the mean radiant temperature being higher than the air temperature. The temperature of the air in an office, factory or residence may need to be kept to an upper limit of, say, 26 °C d.b. in order to limit the environmental temperature to 27 °C. Such conditions are tolerable, but not comfortable, for sedentary work. Considerable discomfort is experienced when strenuous physical activity is conducted.

The indoor air temperature fluctuates through each 24 h period owing to the position of the sun relative to the building. South-facing rooms that have a large area of glazing are likely to be exposed to the greatest indoor air temperatures. Figure 5.20 shows the variation of outdoor air, indoor air, window glass and predicted outdoor air temperatures for a south-facing office in Southampton on Sunday 27 June 1993. The windows and doors remained closed throughout the weekend. The office had only natural ventilation, no mechanical cooling, open

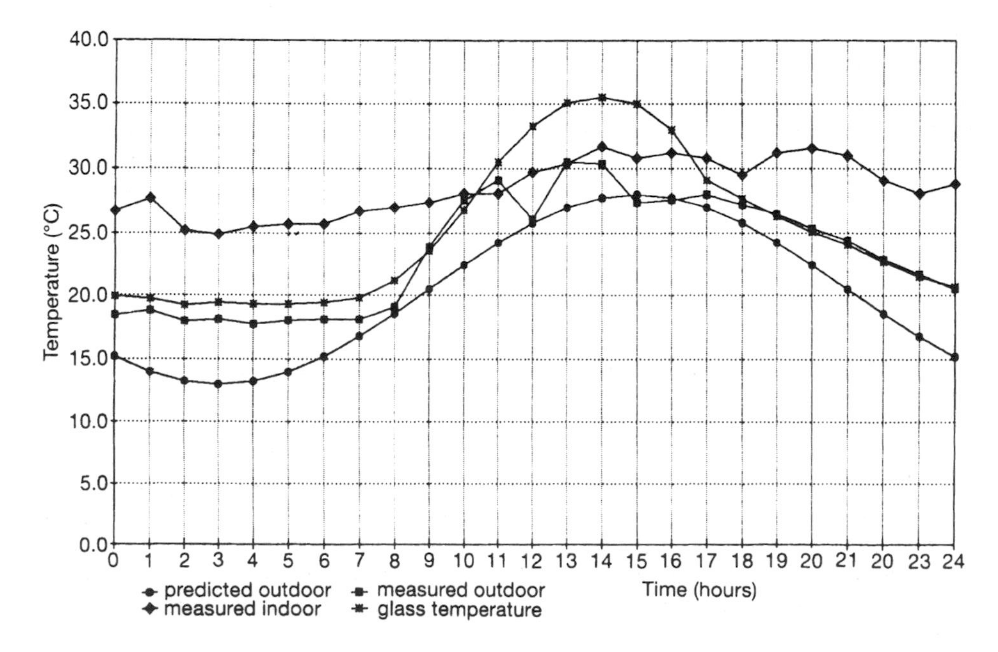

Figure 5.20 Temperature profiles: south-facing office, Sunday 27 June 1993

light grey slatted venetian blinds, and had been used normally for the preceding week. The exterior wall had a 70% glazed area.

The general profile of the outdoor air temperature t_{ao} that is expected can be calculated from a sine wave (Jones, 1985, p. 113):

$$t_{ao} = t_{max} - \frac{t_{max} - t_{min}}{2} \times \left[1 - \sin\frac{\theta\pi - 9\pi}{12}\right]$$

The 24-hour clock time is θ hours: that is, a time between 0 and 24 h. The predicted outdoor air temperature curve has been calculated for maximum and minimum values, t_{max} and t_{min}, of 28 °C d.b. and 13 °C d.b. on an hourly basis for 24 h. This corresponds to the conditions after a week of warm sunny weather in June 1993 in Southampton. This has been a common occurrence since 1990. A thermocouple temperature logger, similar to that in Figure 1.12, measured the outdoor air, indoor air and the internal surface temperature of the outer pane of the double-glazed window, hourly. Figure 5.20 shows that the measured outdoor air temperature follows the general shape of the predicted sine wave. The thermocouple that was adhering to the glass showed a combination of two factors: first, that the air temperature in the cavity between the panes of double glazing rose to 35 °C; second, that the glass absorbed some of the incident solar radiation and was raised in temperature. The internal room air temperature was measured in a shaded location at just above desk height. The room air temperature remained between 25 °C d.b. and 31 °C d.b. During normal use, the same office produced an internal air temperature of 25 °C d.b. when the outdoor air temperature peaked at 27 °C d.b.

While such an example does not prove conclusively that all south-facing rooms in the UK need to be air conditioned for human thermal comfort, it does give some evidence to strengthen the argument in favour of mechanical cooling. Working in air temperatures that move above 24 °C d.b. in naturally ventilated spaces that have significant solar radiation can be noticeably uncomfortable. Whether the performance of human productivity or effectiveness becomes impaired is arguable. Low-cost cooling systems can be designed that make use of cool parts of a building to lower the temperature of the areas that are exposed to solar radiation. Heat pump systems, mechanical ventilation and evaporative water-cooling towers can be used to limit room air temperatures, without the need to involve high-cost refrigeration equipment.

EXAMPLE 5.11

A south-facing hotel lounge in Bournemouth has natural ventilation and a large area of glazing. The maximum and minimum outdoor air temperatures are expected to be 25 °C d.b. and 11 °C d.b. Calculate the outdoor air temperature that is expected at 1700 h. Find the indoor air temperature that will be generated if solar radiation heat gains raise the indoor air by 2.5 °C from that of the outdoor air. Comment upon the thermal comfort conditions that are provided for the hotel guests.

$$\theta = 17 \text{ h for the time of 1700 h}$$

$$t_{ao} = t_{max} - \frac{t_{max} - t_{min}}{2} \times \left(1 - \sin\frac{\theta\pi - 9\pi}{12}\right)$$

$$= 25 - \frac{25 - 11}{2} \times \left(1 - \sin\frac{17\pi - 9\pi}{12}\right)$$

$$= 25 - 7 \times (1 - \sin 2.094)$$

The 2.094 is in radians and not degrees. Switch on the radians mode of the calculator then press the SIN X key to find the answer of $\sin 2.094 = 0.866$. Alternatively, multiply 2.094 by 360 and then divide the answer by (2π). This produces an angle of 119.97 °. Press the SIN key and find sin 119.97 = 0.866. There are 2π radians in 360 °.

$$t_{ao} = 25 - 7 \times (1 - 0.866)$$

$$= 24.1 \text{ °C d.b.}$$

$$\text{indoor air temperature } t_{ai} = 24.1 + 2.5 \text{ °C d.b.}$$

$$= 26.6 \text{ °C d.b.}$$

While this indoor air temperature may be acceptable for reasonable thermal comfort, there is solar radiation and possible glare during the early evening. Guests, and the hotel staff, would benefit from measures to increase the throughput of cooler air from the north side of the building or by means of a mechanical cooling system.

Questions

1. A banking hall is cooled in summer by an air-conditioning system that provides an air flow rate of 5 m^3/s to remove sensible heat gains of 50 kW Room air temperature is maintained at 23 °C. Derive the formula for calculating the supply air temperature and find its value.

2. A room has a sensible heat gain of 10 kW and a supply air temperature of 10 °C d.b. Find the supply air rate required to keep the room air down to 20 °C d.b.

3. Ten people occupy an office and each produces 50 W of latent heat. The supply air flow rate is 0.5 m^3/s and its temperature is 12 °C d.b. If the room is to be maintained at 21 °C d.b. and 50% saturation, calculate the supply air moisture content.

4. The cooling coil of a packaged air conditioner in a hotel bedroom has refrigerant in it at a temperature of 16 °C. Room air enters the coil at 31 °C d.b. and 40% saturation and leaves at 20 °C d.b. at a rate of 0.5 m^3/s.

   (a) Is the room air dehumidified by the conditioner?
   (b) Find the room air wet-bulb temperature and specific volume.
   (c) Calculate the total cooling load in the room.

5. A department store has 340 people in an area of 35 m × 25 m that is 4 m high. Smoking is permitted.

   (a) Calculate the fresh air quantity required to provide 12.5 l/s per person.

   If the air change rate is not to be less than 5 changes per hour, find the following.

   (b) supply air quantity;
   (c) percentage fresh air in the supply duct;
   (d) extract air quantity if 85% of the supply air is to be mechanically withdrawn;
   (e) recirculated air quantity;
   (f) ducted exhaust air quantity.

6. Air enters an office through a 250 mm × 200 mm duct at a velocity of 5 m/s. The room dimensions are 5 m × 3 m × 3 m. Calculate the room air change rate.

7. Show two methods of allowing fresh air to enter a room where extract ventilation is by mechanical means and the incoming air is not to cause any draughts.

8. Discuss the relative merits of centrifugal and axial flow fans used in ventilation systems for occupied buildings.

9. Sketch and describe the arrangements for natural and mechanical ventilation of buildings. State two applications for each system.

10. Describe the operating principles of four different systems of air conditioning. State a suitable application for each.

11. State, with reasons, the appropriate combinations of natural and mechanical ventilation for the following:

    (a) residence;
    (b) city office block;
    (c) basement boiler room;
    (d) industrial kitchen;
    (e) internal toilet accommodation;
    (f) hospital operating theatre;
    (g) entertainment theatre.

12. Explain, with the aid of sketches, how the external wind environment affects the internal thermal environment of a building.

13. A four-storey commercial building is to be mechanically ventilated. Air-handling plant is to be sited on the roof. Each floor has dimensions 20 m × 10 m × 3 m and is to have 6 air changes per hour. Of the air supplied, 10% is allowed to exfiltrate naturally and the remainder is extracted to roof level. The supply and extract air ducts run vertically within a concrete service shaft and the limiting air velocity is 10 m/s. Estimate the dimensions required for the service shaft. Square ducts are to be used and there is to be at least 150 mm between the duct and any other surface.

14. List the procedure for the design of an air-conditioning system for an office block.

15. A lecture theatre has dimensions 15 m × 15 m × 4 m and at peak occupancy in summer has sensible heat gains of 30 kW and latent heat gains of 3 kW. Room and supply air temperatures are to be 23 °C d.b. and 14 °C d.b. respectively. Room air moisture content is to be maintained at 0.008 kg H_2O/kg air. Calculate the supply air volume flow rate, the room air change rate and the supply air moisture content.

16. To avoid draughts, a minimum supply air temperature of 30 °C d.b. is needed for the heating and ventilation system serving a public room. The room has an air temperature of 21 °C d.b. and a sensible heat loss of 18 kW. It is proposed to supply 2 m^3/s of air to the room. Calculate the supply air temperature that is required. If it is not suitable, recommend an alteration to meet the requirements.

17. Describe the operation of the vapour compression refrigeration cycle and sketch a complete system employing chilled-water distribution to cooling coils in an air-conditioning system.

18. Discuss the uses of the absorption refrigeration cycle for refrigerators and air-conditioning systems.

19. Show how refrigeration systems can be used to pump heat from low-temperature sources, such as waste water, outdoor air arid solar collectors, to produce a usable heat transfer medium for heating or air-conditioning systems.

20. Measurements in a mechanically ventilated computer room showed that tracer gas concentration

fell from 10% to 3% in 5 min. Calculate the air change rate.

21. A gymnasium of dimensions 20 m × 12 m × 4 m is to be mechanically ventilated. The maximum occupancy will be 100 people. The supply air for each person is to comprise 20 l/s of fresh air and 20 l/s of recirculated air. Allowing 10% natural exfiltration, calculate the room air change rate, the air flow rate in each duct and the dimensions of the square supply duct if the limiting air velocity is 8 m/s.

6 Hot- and cold-water supplies

Learning objectives

Study of this chapter will enable the reader to:

1. recognize the quality of water supplied to buildings;
2. explain pH value;
3. explain water hardness;
4. identify and apply appropriate water treatment methods;
5. decide appropriate applications for mains pressure and storage tank cold- and hot-water systems;
6. understand pressure-boosted systems for tall buildings;
7. apply economic instantaneous and storage techniques for hot-water provision;
8. understand primary and secondary pipe circulation systems;
9. calculate the heater power for hot-water devices;
10. understand demand units;
11. understand how cold- and hot-water pipe systems are designed;
12. be capable of carrying out basic cold- and hot-water pipe-sizing calculations;
13. understand pipe equivalent length;
14. understand pipe pressure loss calculations;
15. use CIBSE pipe-sizing data;
16. allocate sanitary appliances appropriate to building usage;
17. discuss the use of pipe materials and their methods of jointing;
18. have a basic understanding of how solar energy can be utilized in the provision of hot water in buildings.

Key terms and concepts

Introduction

The convenience of piped water systems is likely to be taken for granted by those who have not been camping or caravanning. The provision of safe and hygienic water supplies is of paramount importance, and a considerable amount of engineering is involved in such provision.

The basics of water treatment are outlined, and then the ways in which water is distributed throughout buildings are discussed. The flows of water to sanitary appliances depend upon their frequency of use, which is not accurately predictable. The concept of pipe sizing with demand units leads to design calculations of water networks.

Calculation of the likely number of sanitary fittings to be installed is demonstrated. Water system materials and the application of solar heating are explained.

Water treatment

About 10% of the rainfall in the UK is used in piped services. Storage in reservoirs allows sedimentation of particulate matter, and then the water is filtered through sand and injected with chlorine for sterilization. A slow sand filter consists of a large horizontal bed of sand or a sand and granulated activated carbon sandwich (Thames Water). The carbon comes from coal and acts as a very efficient filter that traps microscopic traces of pesticides and herbicides. Water percolates down through the bed by gravity. Rapid sand filters have the raw water pumped through a pressurized cylinder that contains the filter medium. This filter material is either crushed silica, quartz or anthracite coal. Filtering removes metallic salts, bacteria and turbidity (muddiness). It also removes colouring effects, odours and particles, which affect the taste of the water (*Encyclopaedia Britannica*, 1980). The naturally occurring pH value and the total dissolved salt concentration are virtually unaltered by the water supply authority.

Water quality varies with the local geology and can be classified as hard, soft, acidic or alkaline. Mineral salts of calcium and magnesium have soap-destroying properties and are considered in the evaluation of water hardness.

Temporary hardness is due to the presence of calcium carbonate, calcium bicarbonate and magnesium bicarbonate, which dissolve in water as it passes through chalky soil. These salts are deposited as scale on heat transfer surfaces during boiling, causing serious reduction in plant efficiency. They are known as carbonate hardness.

Permanent hardness is due to the presence of the non-carbonate salts calcium sulphate, calcium chloride, magnesium chloride and other sulphates and chlorides. Neutralization of these is achieved by means of chemical reactions.

Soft water contains up to 100 mg/l of hardness salts, as in Cornwall, and hard water contains as much as 600 mg/l, as in parts of Leicestershire. Acidic water is produced by contact with decomposing organic matter in peaty localities and normally occurs in soft-water regions. This water is very corrosive to steel, is plumbo-solvent and can cause dezincification of gunmetal pipe fittings.

The pH value denotes acidity or alkalinity due to the presence of free hydrogen ions in the water: acidic water, $pH < 7$; neutral water, $pH = 7$; alkaline water, $pH > 7$. Copper and plastic pipes and fittings can be used in acidic water regions. Hard-water areas are generally alkaline. Water treatment for large boiler plants includes chemical injection to reduce corrosion from dissolved oxygen, and the pH value is raised to 11. Galvanized metal can be used where the pH value is 7.4 if the carbonate hardness is greater than 150 mg/l.

Users of large amounts of water may have treatment plant that removes or converts hardness salts to less harmful salts.

Base exchange

Raw water from the mains passes through a tank of zeolite chemicals where a base exchange takes place:

calcium carbonate + sodium zeolite → sodium carbonate + calcium zeolite
(in water) (in tank) (in water) (left in tank)

Similar base exchanges occur between the zeolites and other hardness salts in the raw water, turning them into non-scale-forming salts. On complete conversion of all the sodium zeolite, the filter bed is backwashed with brine (sodium chloride solution) which undergoes exchange with the calcium zeolite. The normal flow direction can then be resumed. The running cost of the system is limited to consumption of common salt, a small pump, periodic replacement of the zeolites and a small amount of maintenance work.

Steam boilers accumulate the salts passing through the treatment plant, and if they were allowed to become too numerous they would be carried over into the steam pipes and clog safety valves and pressure controllers. Either continuous or intermittent blow-down of boiler water to the drain is designed to control salt concentration. The high-pressure blow-down water is cooled before being discharged into the drains, and the heat is recycled.

Demineralization

Complete removal of mineral salts is very expensive, but it is essential for power station steam boilers, high-performance marine boilers and some manufacturing processes, where the presence of impurities is unacceptable. Raw water is passed through chemical filters in several stages to complete the cycle.

Reverse osmosis

Reverse osmosis is a filtration technique in which untreated water is pumped alongside a semipermeable membrane in a pipe system. Clean water passes through the membrane. This method is used to produce drinking water in desert regions.

Self-cleaning depth filter

Hard and deformable solid particles, which are suspended in potable water systems, are removed by pumped filtration through a bed of sand at the treatment plant prior to chlorination. A compact filter element of crimped fibrous nylon yarn (Kalsep Limited) provides an alternative. This filter element is backwashed with a low flow rate at intervals of typically 4 to 6 hours. Figure 6.1 depicts the fibre bundle. The fibres are rotated and stretched by pneumatic motors to displace the particles by a mechanical wringing action. Figure 6.2 shows the filtering and cleaning water flows. Up to 98% of particles down to 2 μm diameter (2×10^{-6} m) are removed. The filter plant operation, backwashing

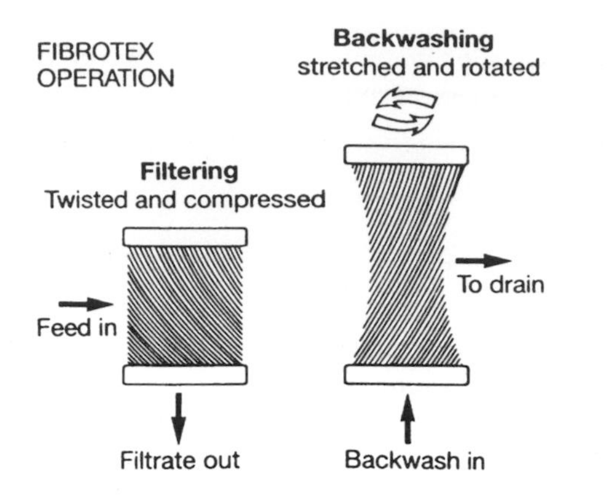

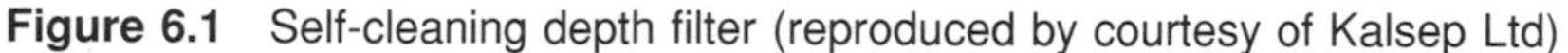

Figure 6.1 Self-cleaning depth filter (reproduced by courtesy of Kalsep Ltd)

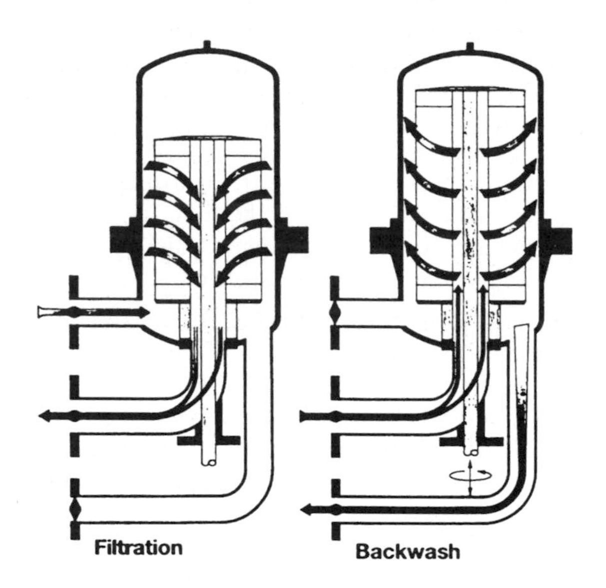

Figure 6.2 Fibrotex filtration and backwash operations (reproduced by courtesy of Kalsep Ltd)

cycle and pressure drop are monitored and controlled by a programmed logic controller. The plant may be used as a mobile unit. It can pre-filter borehole or sea water prior to the use of reverse osmosis treatment. This is how it was used for the provision of water for drinking, concrete mixing and cooling of the boring machines for the Channel Tunnel. Desalinated sea water was provided for this project at up to 11 400 tonnes per day.

Solar distillation

Solar stills consist of glass-covered water troughs in which solar radiation evaporates the water, which then condenses on the cooler sloping glass roof and is collected in channels. This method can be used in hot locations.

Cold-water services

Mains water is used in two ways: direct from the main and as low-pressure supplies from cold-water storage tanks.

Mains supplies

At least one tap per dwelling and taps at suitable locations throughout large buildings are connected to the main for drinking water. The main also supplies ball-valves on cold-water storage tanks and machines requiring a high-pressure inlet.

The economical use of water is important for safety, environmental and cost control reasons. The manual flush control of WCs and the tap operation of other appliances allows responsible usage. Urinals present a particular hygiene and water consumption contradiction. The user has no control over the flushing of water through the trough or bowl. The absence of flushing water leaves the urinal unpleasantly odorous and discoloured. Cleaning staff may counteract this by the excess dumping of deodorant blocks into the urinal. Perfumed toilet blocks are up to 100% paradichlorobenzene. Toilet-cleaning fluid contains phosphoric acid. These toxic chemicals are passed to the sewage treatment plant through the drain system. Uncontrolled flushing when the urinals remain unused, particularly overnight, results in wasteful water consumption and no benefit to the user. In the UK the supply of potable water plus the removal of waste water from consumers costs £1.17 per m^3 (Southern Water Services Limited, 1993/94) from a meter on the supply inlet pipe. An uncontrolled urinal cistern of 9 l would flush, say, four times per hour, 24 h per day for 365 days in a year and consume 315 m^3 of water costing up to £369.

The installation of a water inlet flow control valve to a range of urinals will only allow flushing when appliances have been used, saving consumption. The valve may be operated from a passive infrared presence detector, discharge water temperature sensor or a variation in the water pressure within the same room. Figure 6.3 shows a valve that is controlled from a water pressure accumulator. A short-term water flow to a WC or basin causes the stored water pressure within the bellows (2) to exceed the pressure in the pipeline. The diaphragm (1) opens and allows water to flow to the urinal cistern until the accumulator pressure again equals the pipeline pressure. The water quantity that is passed can be adjusted (3) to avoid wastage.

Low-pressure supplies

Static water pressures in tall buildings are reduced by storing water at various levels. Sealed storage tanks are used for drinking water. Open water tanks become contaminated with airborne bacteria and are only used for sanitary purposes. Cold-water services are taken to taps, WC ball valves, hot-water storage cylinders and equipment needing low-pressure supplies. A separate cold feed is taken to a shower or group of showers to avoid the possibility of scalding. Tanks are sized to store the total cold-water requirement for a 24 h period.

The minimum mains water pressure available in the street is 100 kPa (1 bar), which is 1 atmosphere gauge or 10 m height of water. The water supplier may be

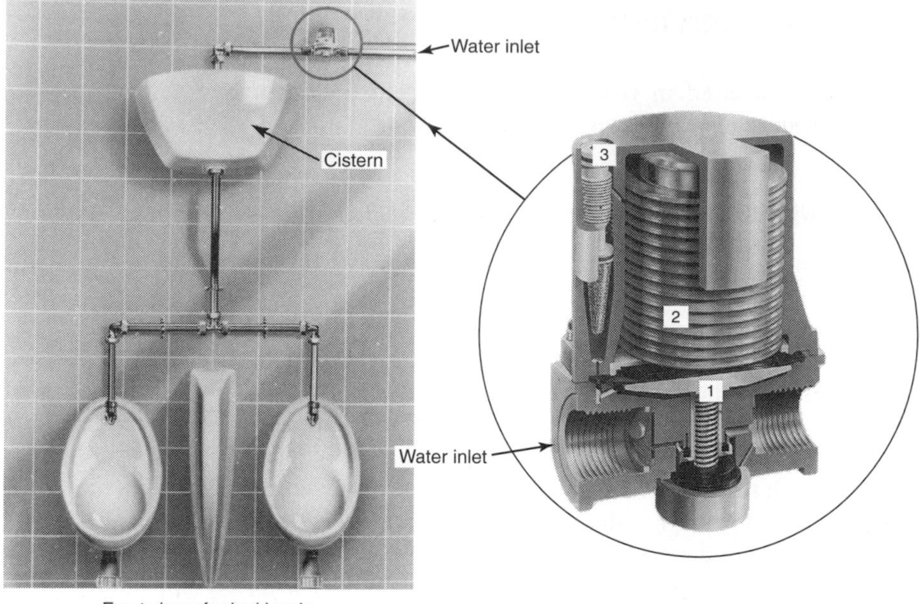

Figure 6.3 Urinal flush control valve (reproduced by courtesy of Cistermiser Ltd)

able to provide 300 kPa, or enough pressure to lift water to the top of a building 30 m high; however, allowance has to be made for friction losses in pipelines and discharge velocity, which effectively limits the vertical distance to between two and six storeys.

Separation of the contaminated water being used within the building for washing, flushing sanitary appliances, circulating within heating and air-conditioning cooling systems, evaporative cooling towers, ornamental fountains, agricultural irrigation or manufacturing processes from potable mains water is achieved by using the following:

1. a storage tank with ball valve (break tank);
2. a permanent air gap between the tap discharge and the contaminated water level (e.g. wash basin);
3. a single-seat non-return valve (check valve);
4. a double-seat check valve.

The Water Byelaws 1989 classify the risk of contamination from the building reaching upstream into the water main in three groups, each having its own protection (Table 6.1).

Cold-water storage tanks are expected to contain water of similar quality to that supplied from the main and so must be covered to exclude foreign matter, insects and light as well as being thermally insulated and not contaminating the stored water themselves. Tanks are generally not larger than 2 m long by 1 m wide by 1 m high, and pipe connections must ensure that water flushes through all of them to eliminate stagnation.

Table 6.1 Classification of contamination risks

Class	Risk	Example of risk class	Type of protection
1	Serious danger to life	WC, bidet, urinal, agricultural or industrial process	Permanent air gap or break tank
2	May cause minor illness	Clothes, dishwashing and drinks vending machines, commercial water softeners	Permanent air gap, break tank or double check valve
3	May cause an unpleasant taste, odour or discoloration	Single-outlet mixer taps and domestic water softeners	Any class 1 or 2 protection, or a single check valve

Servicing or isolating valves are located on the inlet to all ball valves on storage tanks and WC cisterns to facilitate maintenance without unnecessary water loss or inconvenience to the occupier. A servicing valve is required on all outlets from tanks of more than 15 l, i.e. larger than a WC cistern.

The drinking and food-rinsing water tap at a kitchen sink must be connected to the water main before any water softener enters and a check valve is required between this tap and the softener.

Service entry into a building is via an underground pipe passing through a drain pipe sleeve through the foundations and rising in a location away from possible frost damage. An external stop tap near the boundary of the property is accessible from a brick or concrete pit. A ground cover of 760 mm is maintained over the pipe. A stop valve and drain tap are fitted to the main on entry to the building to enable the system to be emptied if the building is to be unoccupied during cold weather.

A water meter is the next pipe fitting. This has a rotary flow sensor, which is used to integrate the quantity of water that has passed. The cubic metres of water that are supplied, and charged for, are assumed to be discharged into the sewer. A separate charge is levied for the supply of potable water and for the acceptance of the contaminated discharge foul water. The consumer normally has no choice but to pay both the charges.

In tall buildings the pressure required to reach the upper floors can be greater than the available head, or pressure, in the mains. A pneumatic water pressure boosting system is used, as shown in Fig. 6.4. Float switches in the storage tanks operate the pump to refill the system and minimize running times to reduce power consumption. A delayed action ball valve on the cold-water storage tanks can be used. This delays the opening of the ball valve until the stored water has fallen to its low-level limit. System pressure is maintained by a small air compressor and pneumatic cylinder. The controller relieves excess pressure and switches on the compressor when the air pressure falls. During much of the day, water is lifted pneumatically at much lower cost than if it were pumped.

Cold-water storage to cover a 24 h interruption of supply (CIBSE, 1986) ranges from 45 l/person for offices to 90 l/person for dwellings and 135 l/person for hotels.

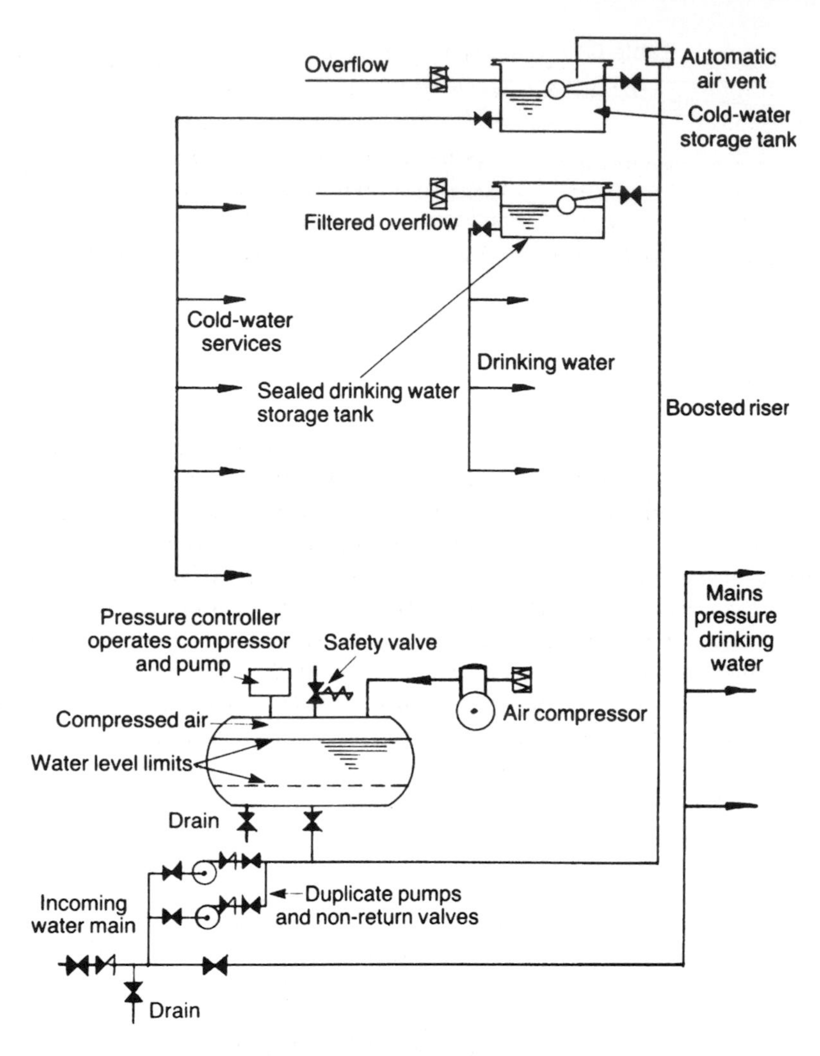

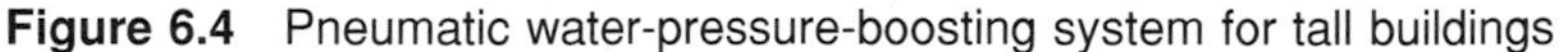

Figure 6.4 Pneumatic water-pressure-boosting system for tall buildings

Hot-water services

Hot water can either be generated by the central boiler plant and stored, or produced close to the point of use by a more expensive fuel.

Central hot-water storage

The low-cost fuel used for the central heating plant is also used for the hot-water services boiler. This is located within the main boilerhouse and a large-volume storage cylinder is employed. A small power input boiler is run almost continuously, winter and summer, under thermostatic control from the stored hot water. Primary circulation pipes are kept short and well insulated.

This system can meet sudden large demands for hot water. Secondary circulation pipes distribute hot water to sanitary appliances. A pump is fitted in the secondary return; its function is to circulate hot water when the taps are shut and it does not appreciably assist draw-off rates from taps. Connections from the secondary flow to the tap are known as dead-legs and are limited to 5 m of 15 mm diameter pipe. This minimizes wastage of cold water in the non-circulating pipework when running a tap and waiting for hot water to arrive.

Decentralized system

The decentralized system is mainly for small hot-water service loads distributed over a large building or site where it would be uneconomic to use a central storage cylinder and extensive secondary pipework. Electricity or gas can be used in small storage or instantaneous water heaters located at the point of use. They are connected directly to the water main. Figures 6.5 and 6.6 show the operational features of gas-fuelled instantaneous and storage water heaters.

Electric instantaneous heaters have power consumptions of up to 6 kW and produce water at 40 °C and up to 3 l/min at 100% efficiency. Small electric storage heaters of 7 l are fitted over basins or sinks. Capacities of up to 540 l operate on the off-peak storage principle. Immersion heaters are controlled by time switches and thermostats and are connected in 3 kW stages.

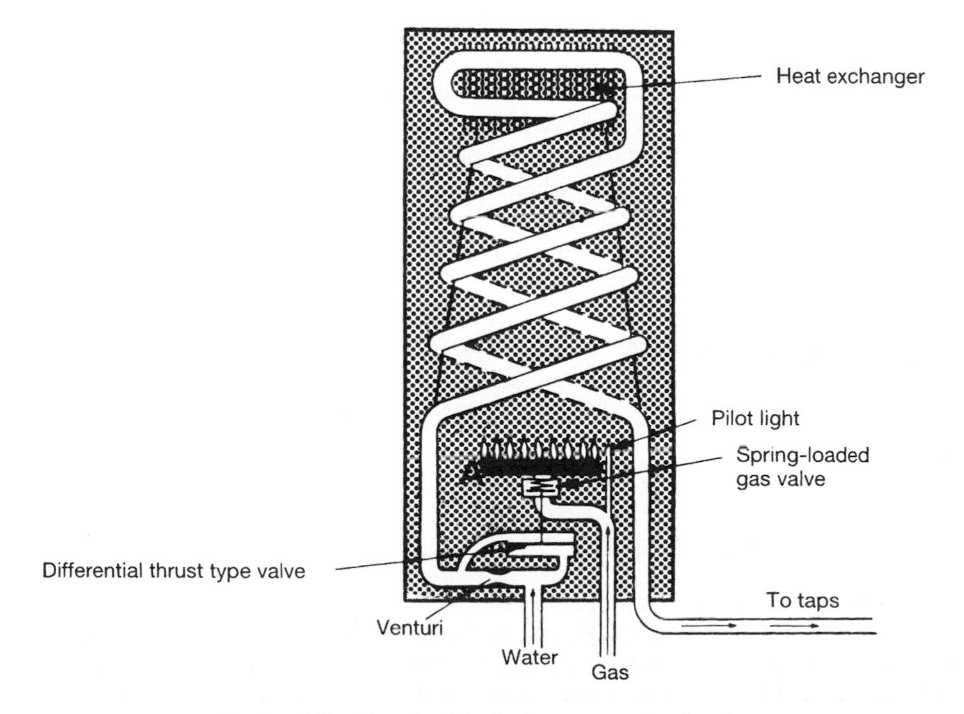

Figure 6.5 Gas-fired instantaneous water heater

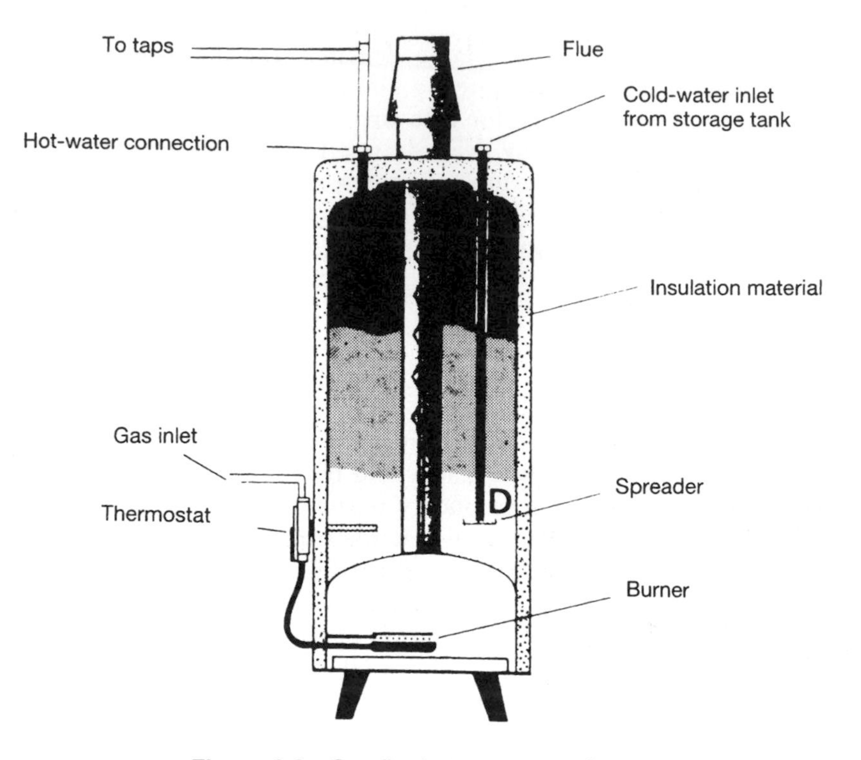

Figure 6.6 Gas-fired storage water heater

The indirect hot-water system

The basic layout of the combined central heating and indirect hot-water service system is shown in Fig. 6.7. The cylinder is insulated with 75 mm fibre glass and should have a thermostat attached to its surface at the level of the primary return. Water is stored at 65 °C, and when fully charged the thermostat closes the motorized valve on the primary return. This 'off' signal may also be linked into the pump and boiler control scheme to complete the shut-down when the central heating controls are satisfied.

Hot-water pipes are insulated with a minimum of 25 mm of insulation, as are tanks exposed to frost. The primary system feed and expansion tank has a minimum capacity of 50 l, and the cold-water storage tank has a capacity of at least 230 l.

Hot-water storage requirements at 65 °C are as follows: office, 5 l per person; dwelling, 30 l per person; hotel and sports pavilion, 35 l per person (CIBSE, 1986).

EXAMPLE 6.1

A dwelling has a 135 l hot-water storage indirect cylinder. The stored cold-water temperature is 10 °C and the hot water is to be at 65 °C. Calculate the necessary heat input rate to provide a 3 h recovery period from cold.

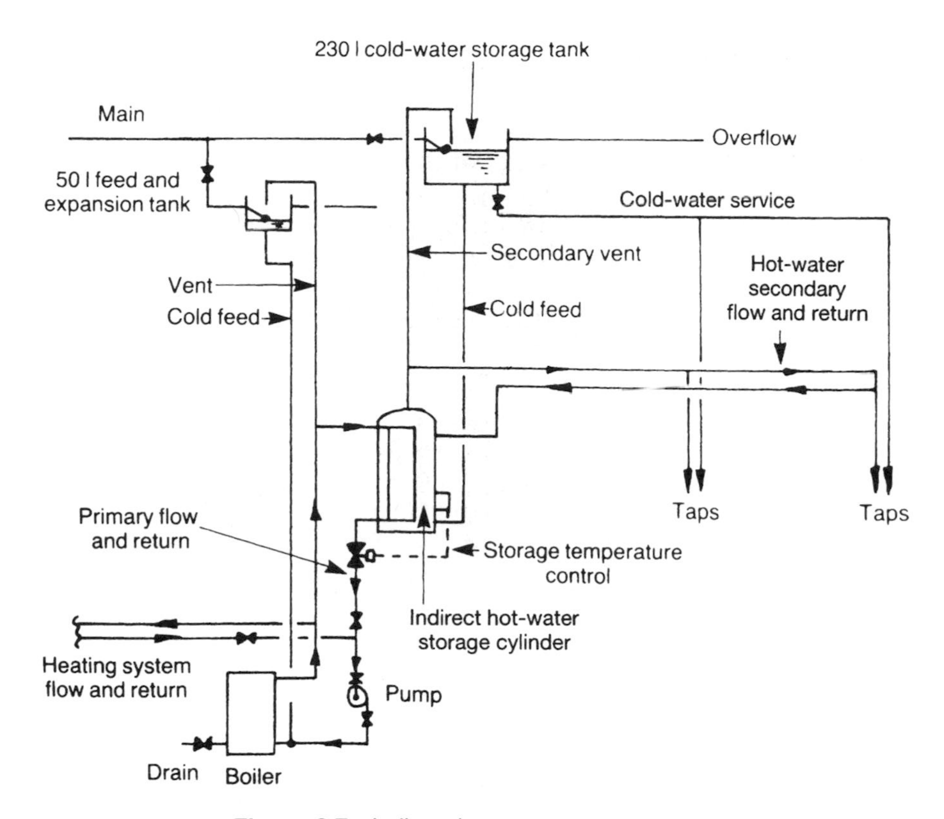

Figure 6.7 Indirect hot-water storage system

$$\text{heat input rate} = \frac{\text{mass kg}}{\text{time h}} \times \text{SHC}\ \frac{\text{kJ}}{\text{kg K}} \times \text{temperature rise K}$$

$$= \frac{135\ \text{l}}{3\ \text{h}} \times \frac{1\ \text{kg}}{1\ \text{l}} \times 4.19\ \frac{\text{kJ}}{\text{kg K}} \times (65 - 10)\ \text{K}$$

$$= \frac{135}{3} \times 4.19 \times (65 - 10)\ \frac{\text{kJ}}{\text{h}} \times \frac{1\ \text{h}}{3600\ \text{s}} \times \frac{\text{kWs}}{\text{kJ}}$$

$$= 2.881\ \text{kW}$$

This can be rounded to 3 kW to allow for heat losses and is adequate for most dwellings with two to four occupants.

Pipe sizing

The demand for water at sanitary appliances is intermittent and mainly random but has distinct peaks at fairly regular times. The pipe sizes for maximum possible peak flows would be uneconomic. Few appliances are filled or flushed simultaneously. To enable designers to produce pipe systems that adequately match likely simultaneous water flows, demand units (DUs) are used.

DUs are dimensionless numbers relating to fluid flow rate, tap discharge time and the time interval between usage.

They are based on a domestic basin cold tap water flow rate of 0.15 l/s for a duration of 30 s and an interval between use of 300 s. This application is given a theoretical DU of 1 and other appliances are given relative values. Table 6.2 lists practical DUs.

The use of spray taps and small shower nozzles greatly reduces water consumption. Design water flow rates of 0.05 l/s for a spray tap, 0.1 l/s for a shower spray nozzle over a bath and 0.003 l/s per urinal stall can be used in place of DUs. Figure 6.8 (CIBSE, 1986) is used to convert DUs into water flow rates.

The design procedure for pipe sizing is as follows.

1. Draw the pipework layout on the building plans.
2. Mark the DUs appropriate to each sanitary fitting on the drawing.

Table 6.2 Practical demand units

Fitting	Application		
	Congested	Public	Private
Basin	10	5	3
Bath	47	25	12
Sink	43	22	11
WC (13.5 l cistern)	35	15	8

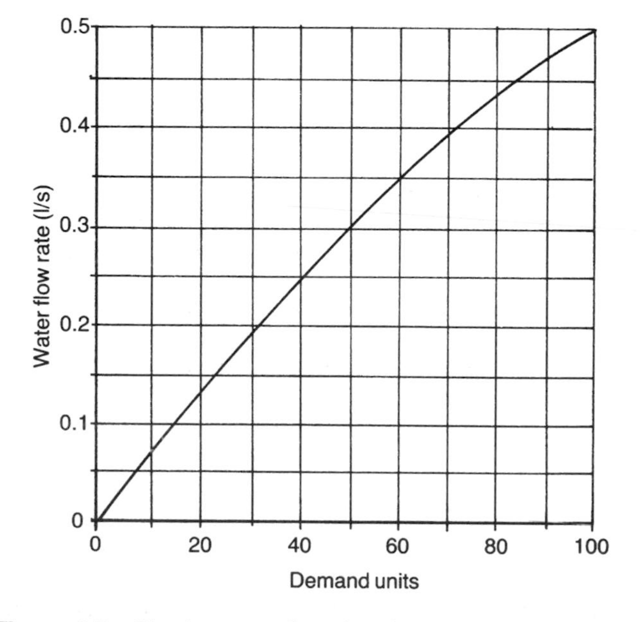

Figure 6.8 Simultaneous flow data for water draw-off points

3. Sum all the DUs along the pipework to the water source, which will be the storage tank or incoming water main.
4. Convert DUs to water flow rates using Fig. 6.8.
5. Find the head of water H (in metres) causing the flow to each floor level.
6. Estimate the equivalent length (EL) of the pipe run to each floor in metres. This can be taken as the measured length plus 30% for the frictional resistance of bends, tees and the tap.
7. Find the index circuit. This is the circuit with the lowest ratio of H to EL.
8. Choose pipe sizes from Table 6.3 for the index circuit.
9. Determine the other pipe sizes from the H/EL figure appropriate to each branch of the index circuit.
10. Determine the water flow rate and head for a bronze-body hot-water service secondary pump, if one is required.

A notation system can be adopted to gather pipe-sizing data together on drawings, as shown in Fig. 6.9.

EXAMPLE 6.2

A domestic cold-water service system is shown in Fig. 6.10. The water main in the street is 50 m from the entry point shown and the supply authority provides a minimum static pressure of 20 m water gauge. The velocity energy and the frictional resistance at the ball valve amount to 2 m water gauge. Determine the pipe sizes.

The pipe-sizing data are shown in Fig. 6.11. Taps and ball valves will be 15 mm on domestic sanitary appliances and 22 mm on baths. DU values are taken from Table 6.2: WC, 8; basin, 3; sink, 11; bath, 12. These are entered into the appropriate boxes on the working drawing and then totalled back to the water main. Great accuracy is not needed in reading the water flow rates appropriate to each DU.

The heads of water causing flow are 4 m for the upper floor and 8 m for the lower floor. The measured length of pipe from the storage tank to the furthest

Table 6.3 Flow of water in copper tube of various diameters

		Water flow rate q (kg/s)					
$\Delta p/EL$(N/m^3)		15 mm	22 mm	28 mm	35 mm	42 mm	
1000		0.160	0.429	0.933	1.60	2.58	
1500	$v = 1.5$	0.201	0.537	1.170	2.00	3.22	$v = 3$
2000		0.236	0.630	1.370	2.34	3.77	
2500	$v = 2$	0.268	0.712	1.540	2.65	4.26	$v = 4$
3000		0.296	0.787	1.710	2.92	4.70	

v = water velocity (m/s).
Source: Reproduced from CIBSE *Guide* (CIBSE, 1986) by permission of the Chartered Institution of Building Services Engineers.

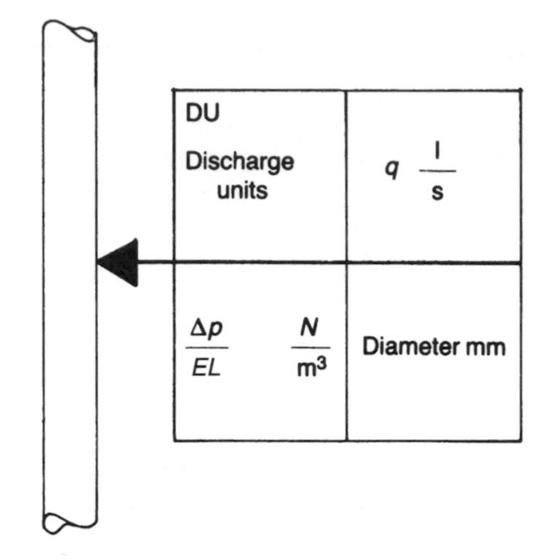

Figure 6.9 Notation for pipe-sizing data on drawings

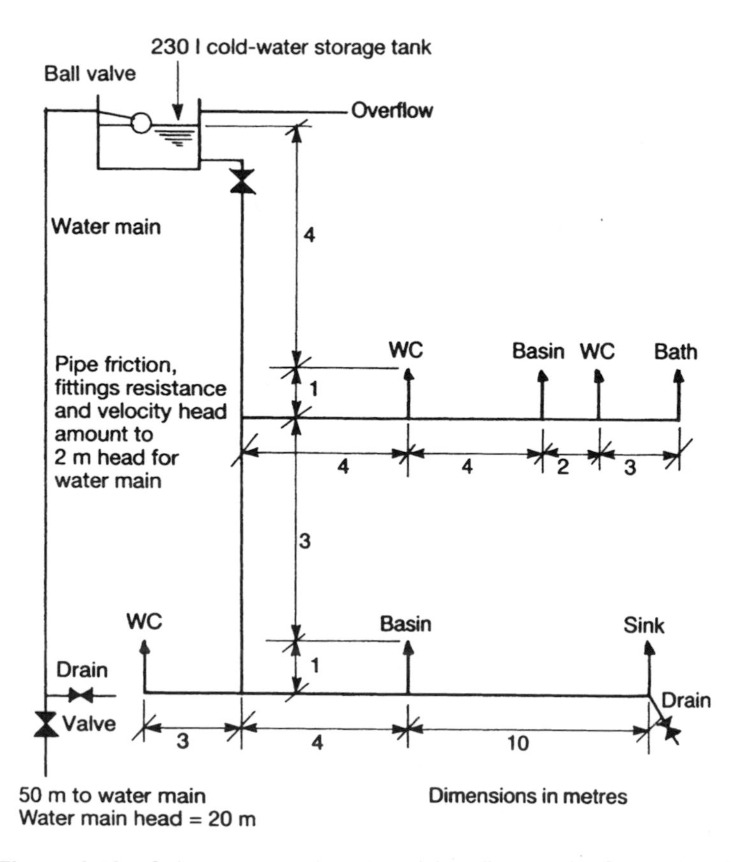

Figure 6.10 Cold-water service pipe-sizing diagram for Example 6.2

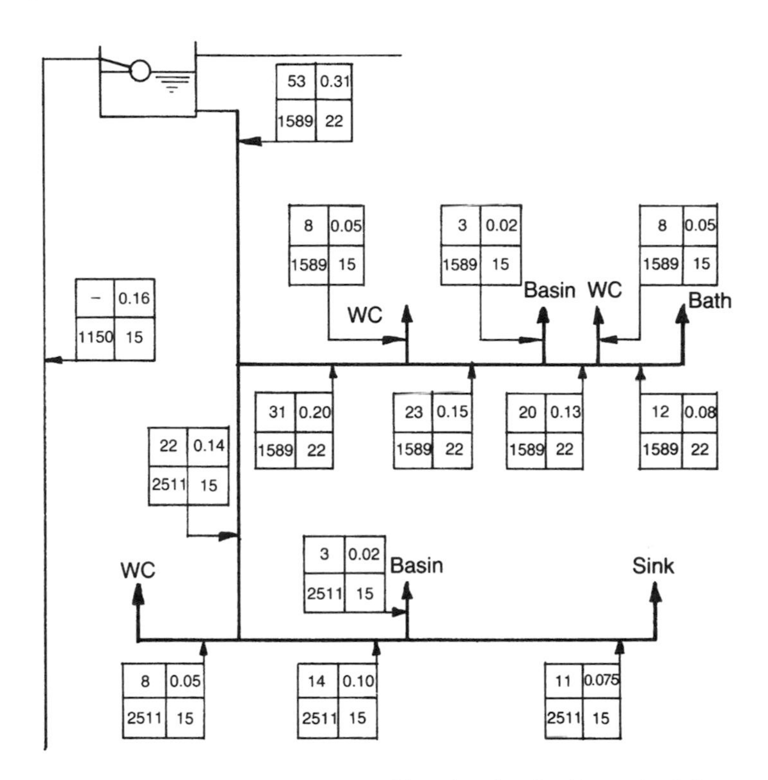

Figure 6.11 Pipe-sizing working drawing for Example 6.2

fitting on the upper floor, the bath, is

$$L_1 = (4 + 1 + 4 + 4 + 2 + 3 + 1)\ \text{m} = 19\ \text{m}$$

and the equivalent length of the circuit to the bath is

$$EL_1 = 1.3 \times 19\ \text{m}$$
$$= 24.7\ \text{m}$$

Similarly, the equivalent length for the lower floor circuit to the sink is

$$EL_2 = 1.3 \times (4 + 1 + 3 + 1 + 4 + 10 + \text{l})\ \text{m} = 31.2\ \text{m}$$

The head loss rate to the bath is

$$\frac{H_1}{EL_1} = \frac{4\ \text{m}}{24.7\ \text{m}}$$
$$= 0.162\ \text{m head/m run}$$

The head loss rate to the sink is

$$\frac{H_2}{EL_2} = \frac{8\ \text{m}}{31.2\ \text{m}}$$
$$= 0.2564\ \text{m head/m run}$$

The pipe circuit to the bath has the lowest H/EL figure; this is the index circuit. H_1/EL_1 is the available pressure loss rate, which drives water through the upper part of the system. Branches to other fittings on the same floor level can be sized from the same figure. All pipes below the upper floor are sized using the value of H_2/EL_2 that is appropriate to that circuit. Now,

$$H = \frac{\Delta p}{9807} \text{ m } H_2O$$

where Δp is the pressure exerted by a water column of height H m. Therefore

$$\frac{\Delta p_1}{EL_1} = \frac{H_1}{EL_1} \times 9807 \text{ N/m}^3$$

$$= 0.162 \times 9807 \text{ N/m}^3$$

$$= 1589 \text{ N/m}^3$$

and

$$\frac{\Delta p_2}{EL_2} = 0.2564 \times 9807 \text{ N/m}^3$$

$$= 2514.5 \text{ N/m}^3$$

These pressure loss rates are rounded and entered on Fig. 6.11. A different pressure loss rate could be calculated for each pipe but sizing them all on the index values is sufficiently accurate at this stage. Suitable pipe diameters are chosen from either Table 4.3 or Table 6.3, or the full data tables in CIBSE (1986). Notice that the bath connection size has been used rather than the 15 mm pipe, which would satisfy the design data for much of the upper floor branch. The lower water velocities produced will minimize noise generation in the pipework. The pressure loss rate causing flow in the water main is

$$\frac{H_3}{EL_3} = \frac{\text{head available for overcoming pipeline friction}}{\text{equivalent pipe length}}$$

Now

$$H_3 = \text{water main pressure} - \text{vertical lift} - \text{ball-valve resistance}$$

$$- \text{water velocity pressure head at ball valve}$$

$$= 20 - (4 + 1 + 3 + 1) - 2 \text{ m}$$

$$= 9 \text{ m water gauge}$$

and

$$EL_3 = 1.3 \times (50 + 4 + 1 + 3 + 1) \text{ m}$$

$$= 76.7 \text{ m}$$

Therefore

$$\frac{H_3}{EL_3} = \frac{9 \text{ m water}}{76.7 \text{ m run}}$$

$$= 0.1173 \text{ m water/m run}$$

and

$$\frac{\Delta p}{EL_3} = 0.1173 \times 9807 \text{ N/m}^3$$

$$= 1150.4 \text{ N/m}^3$$

While this pressure loss rate is available, a water main 15 mm in diameter would provide a flow of a little over 0.16 kg/s. Then, while the taps are shut, the storage tank would refill in

$$230 \text{ kg} \times \frac{\text{s}}{0.16 \text{ kg}} \times \frac{1 \text{ h}}{3600 \text{ s}} = 0.4 \text{ h}$$

EXAMPLE 6.3

A hot-water service secondary system is shown in Fig. 6.12. Estimate the sizes of the pipes and specify the pump size.

Taps on the lower floor are within the limit for dead-leg non-circulating pipes and a secondary return is shown for the group of appliances on the upper floor. Figure 6.13 is the working drawing. Water flow through the cold-feed pipe into the indirect cylinder is at the same rate as the expected outflow. The pressure loss rate to X from the cold-water storage tank is

$$\frac{H_1}{EL_1} = \frac{3 \text{ m head}}{1.3 \times 24 \text{ m run}}$$

$$= 0.0962 \text{ m head/m run}$$

Then

$$\frac{\Delta p_1}{EL_1} = 0.0962 \times 9807 \text{ N/m}^3$$

$$= 943.4 \text{ N/m}^3$$

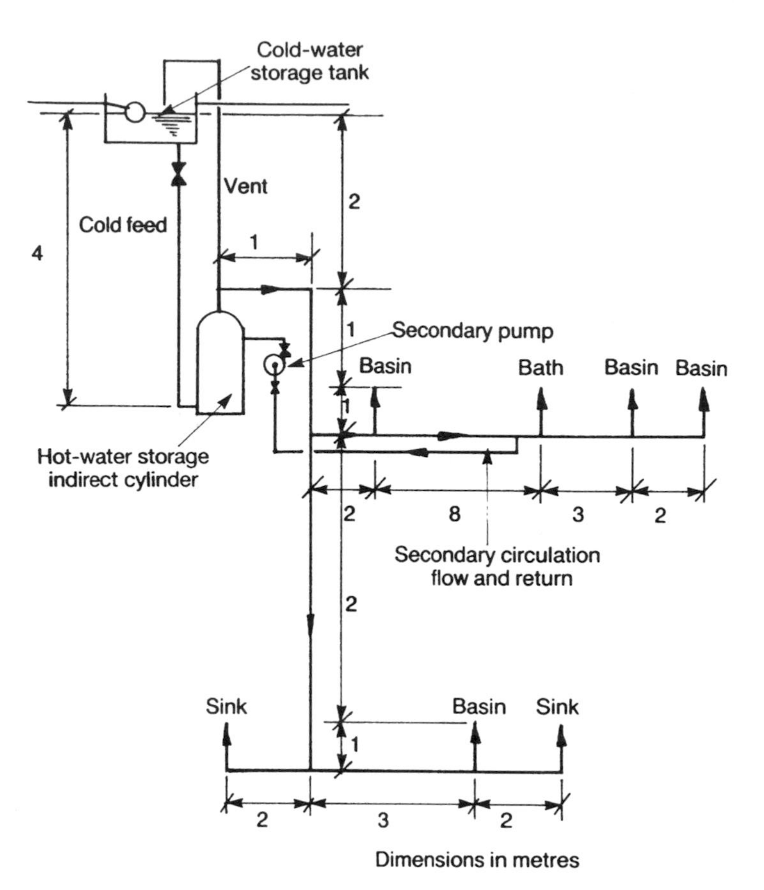

Figure 6.12 Secondary hot-water service system for Example 6.3

The pressure loss rate to Y is

$$\frac{\Delta p_2}{EL_2} = \frac{6}{1.3 \times 18} \times 9807 \text{ N/m}^3$$

$$= 2514.6 \text{ N/m}^3$$

The secondary return pipe is not intended to play an active part in water discharge from the taps but only to pass an amount of water relating to circulation pipe heat losses. Pipes insulated with 25 mm glass fibre have heat emissions as shown in Table 6.4.

The heat loss from the secondary circulation is

$$\text{heat loss} = 13 \text{ m} \times 0.23 \frac{\text{W}}{\text{m K}} \times (65 - 20) \text{ K}$$

$$+ 13 \text{ m} \times 0.19 \frac{\text{W}}{\text{m K}} \times (65 - 20) \text{ K}$$

$$= 245.7 \text{ W} = 0.2457 \text{ kW}$$

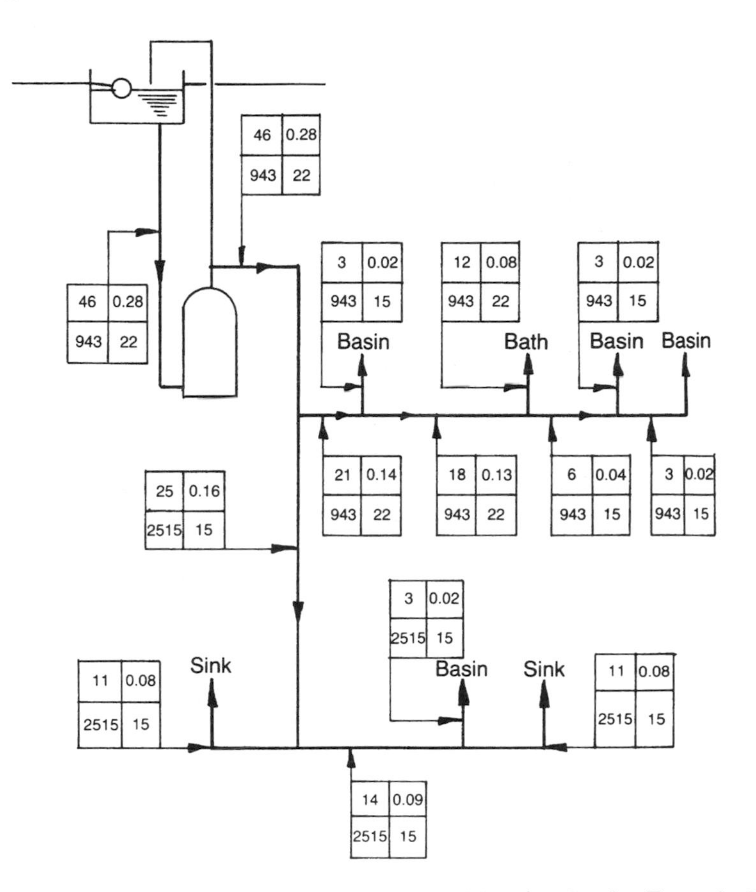

Figure 6.13 Hot-water service pipe-sizing working drawing for Example 6.3

Table 6.4 Insulated pipe heat emission

Pipe diameter (mm)	15	22	28	35	42
Heat emission (W/m K)	0.19	0.23	0.25	0.29	0.32

Source: Reproduced from *CIBSE Guide* (CIBSE, 1986) by permission of the Chartered Institution of Building Services Engineers.

The water flow rate necessary to offset this pipe heat loss while losing say 5 °C between the outlet and inlet connections at the cylinder is

$$q = \frac{0.2457}{4.19 \times 5} \text{ kg/s}$$

$$= 0.0117 \text{ kg/s}$$

By reference to Table 4.3, a pipe 15 mm or 22 mm in diameter carrying 0.0117 kg/s has a pressure loss rate of much less than 200 N/m^3. A gross overestimate of the pump head for this circuit is

$$H = 26\ \text{m} \times 1.3 \times 200 \times \frac{1}{9807}$$

$$= 0.69\ \text{m water}$$

A pump that delivers 0.0117 kg/s at a head of 0.69 m would meet the requirement. Pump C in Fig. 4.16 would provide a far higher flow rate than this, and either a smaller pump would be used or the control valves would be partially closed to avoid the production of noise due to high water velocities.

Allocation of sanitary appliances

The recommended numbers of sanitary fittings are given in Table 6.5.

EXAMPLE 6.4

An adult college is to have 160 male and 40 female staff. The student population is considered to be 1100 males and 770 females. Recommend a suitable allocation of sanitary accommodation.

Table 6.5 Recommended allocation of sanitary fittings

Building accommodation	No. of occupants	Water closets		Urinals	Basins	
		Male	Female		Male	Female
Staff	1 – 100	1 + 1 per 25	1 + 1 per 14	1 + 1 per 25	1 + 1 per 25	1 + 1 per 14
	Over 100	+1 per 30	+1 per 20	+1 per 30	+1 per 30	+1 per 20
Transient public	1 – 200	1 per 100	2 per 100	1 per 50	1 per WC	1 per WC
	200 – 400	1 per 100	+1 per 100	1 per 50	1 per WC	1 per WC
	Over 400	+1 per 250	+1 per 100	1 per 50	1 per WC	1 per WC

Source: Reproduced from *IHVE Guide* (CIBSE, 1986 [IHVE, 1970]) by permission of the Chartered Institution of Building Services Engineers.

Table 6.6 Allocation of sanitary accommodation for Example 6.4

Staff	Students
7 male WCs	7 male WCs
4 female WCs	10 female WCs
7 urinals	22 urinals
7 male basins	7 male basins
4 female basins	10 female basins

Using Table 6.5, we obtain the results given in Table 6.6. The accommodation is to be distributed around the site to ensure uncongested access and reasonable walking distances. A tall building would ideally have toilets on each floor, close to the stairways and lifts, so that all the pipework can run vertically in a service duct. Two male and two female toilets for disabled people should be included at suitable locations, with appropriate access, in the above schedule.

Materials for water services

The materials used in hot- and cold-water systems are listed in Table 6.7. Corrosion protection is provided by ensuring that incompatible materials are not mixed in the same pipework system, by recirculating the water in central heating systems to reduce fresh oxygen intake, and by adding inhibiting chemicals to the water. Hot- and cold-water service systems are continually flushed with fresh water, making it necessary to use galvanized metal, copper or stainless steel.

Copper and galvanized steel should not be used in the same system because electrolytic action will remove the internal zinc coating and lead to failure. A galvanized metal cold-water storage tank can be successfully used with copper pipework as the low temperature in this region limits electrolytic action. Heat accelerates all corrosion activity.

Black mild steel is used in recirculatory heating systems, and an initial layer of mill scale, which is metal oxide scale formed during the high-temperature working of the steel during its manufacture, helps to slow further corrosion. Discoloration of the central heating water from rust to black during use shows steady corrosion. A black metallic sludge forms at low points after some years. Large hot-water and steam systems have the mill scale chemically removed during commissioning and corrosion-inhibiting chemicals are mixed with the water to maintain cleanliness and avoid further deterioration.

The formation of methane gas in heating systems during the first year of use is due to early rapid corrosion, and radiators need frequent venting to maintain water levels. Proprietary inhibitors should be added to all central heating systems. These control methods of corrosion are anti-bacterial. Without them, steel boilers and radiators can rust through in 10 years.

Solar heating

Solar heating can be employed to assist the generation of hot water in secondary storage systems with a consequent reduction in energy costs. The highly variable nature of solar radiation in the UK produces a financial return on the capital invested in equipment of around 10% per annum when electrical water heating is used. Further information can be obtained from Building Research Establishment Digest 205: 1977 and Courtney (1976a, b), McVeigh (1977), Szokolay (1978) and Palz and Steemers (1981). Solar energy is used to provide:

1. comfort heating through architectural design in a 'passive' system;
2. comfort heating using collectors, with air as the heat transfer fluid, in an 'active' system;

Table 6.7 Materials for hot- and cold-water systems

Material	Application	Jointing
Lead	Elderly hot- and cold-water pipes up to 22 mm; water becomes contaminated during storage in the pipework and must not be consumed	Hot molten lead wiped joints and swaged (flared) connections to copper pipe
Copper	All water services, gas and oil pipelines; 6–10 mm soft copper tube supplied in rolls for oil lines and microbore heating; semi-hard tube in various thicknesses in diameters of 15 mm upwards; hand- or machine-formed large radius bends are popular; aluminium finned pipe is used in convector heaters	Compression, manipulative, silver solder, bronze weld, flanged or push-fit ring seal using polybutylene fittings
Black mild steel	Indirect hot-water heating systems, pipework and radiators	British Standard pipe thread (BSPT), screwed and socketed, flanged or welded
Galvanized mild steel	Hot- and cold-water pipework on open systems and water mains; cold-water storage tanks and indirect cylinders	BSPT screwed, socketed, flanged or welded
Stainless steel	Hot- and cold-water pipework 15–28 mm; thickness and diameter correspond to those of semi-hard copper and can be bent in the same way; larger diameters are used for chemicals or for sterile fluids in hospital services	Compression, silver solder, flanged or push-fit ring seal using polybutylene fittings
Cast iron	Central heating radiators, boilers and centrifugal pump bodies	BSPT screwed fittings and gasket joints
Brass	Pipe fittings and valves	BSPT screwed
Gunmetal	Pipe fittings and valves; pump impeller; also body of a secondary hot-water service centrifugal pump	BSPT screwed
Polybutylene	Cold- and hot-water pipes and fittings in 15 m and 22 mm diameters, withstanding 90 °C at 3 bar (atm) pressure	Push-fit ring seal, compression fusion weld
Polyethylene	Underground mains cold water	Compression
Unplasticized poly (vinyl chloride) (UPVC)	Mainly cold-water services	Compression, solvent weld
Cast aluminium	Boiler body, central heating radiators with convection fins	BSPT screwed

3. comfort heating and hot-water using collectors, with water as the heat transfer fluid, in an active system.

Thermal storage of the collected energy is needed to balance the supply of solar heat with its times of use, usually when the supply has greatly diminished. The following methods of thermal storage are used:

1. thick concrete or brick construction for passive systems, combined with large areas of south-oriented glazing;
2. rock heat storage, heated by air recirculation;
3. water tanks;
4. phase change salts.

Flat-plate solar collectors are the most popular as they can heat water to 30–40 °C without the danger of boiling. They can be incorporated into the architectural design of a sloping roof and are reasonably cheap. They are used to preheat the water supplied to the hot-water service storage cylinder, as shown in Fig. 6.14. Only the simplest arrangement is indicated, where the cold-water storage tank supplies only the hot-water cylinder. A wide variety of system designs are in use, depending upon requirements. The solar pump switches off when the temperature sensors find no measurable water temperature rise. All the

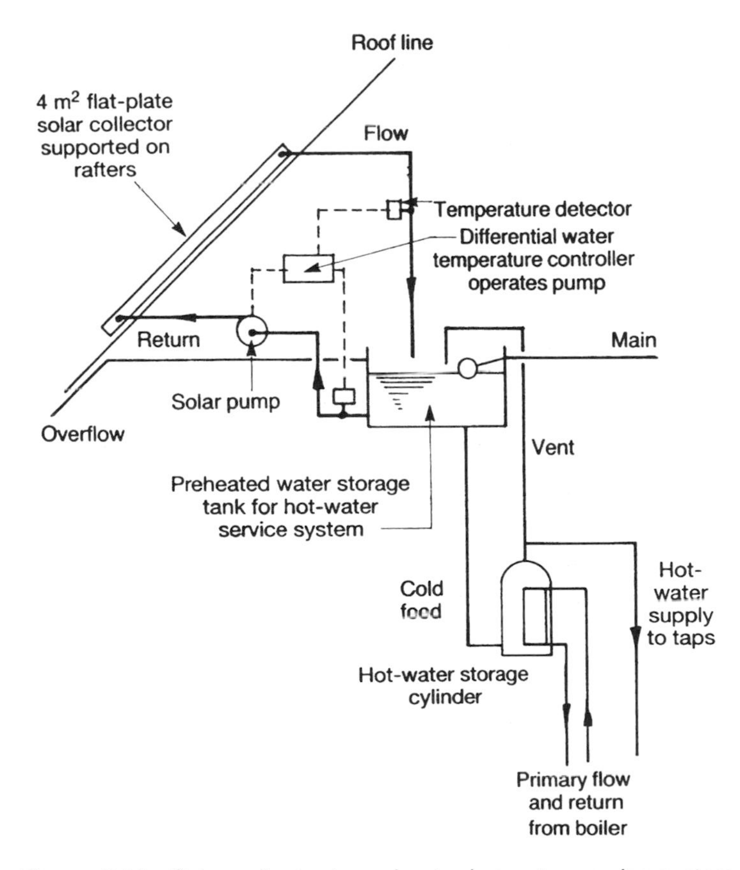

Figure 6.14 Solar collector to preheat a hot water-service system

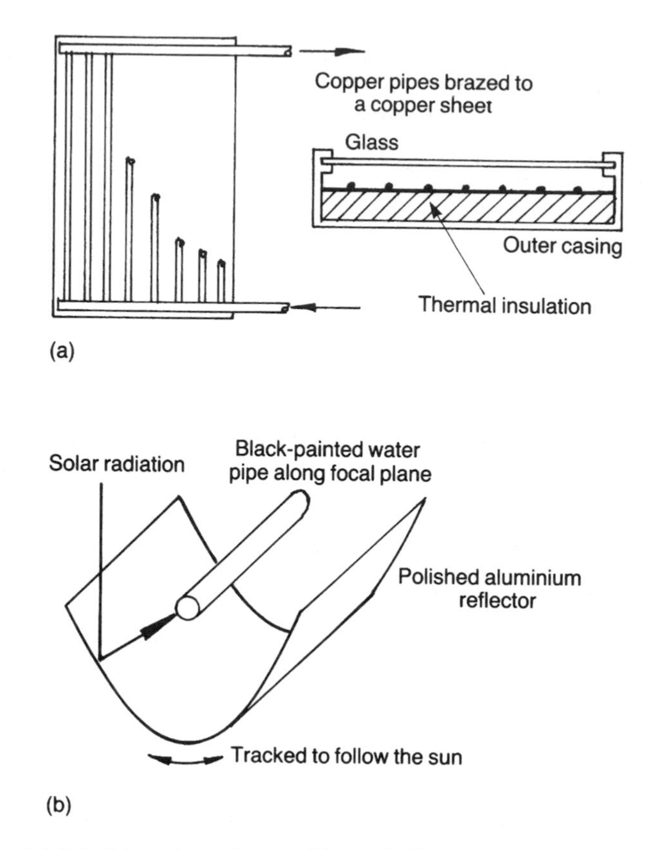

Figure 6.15 (a) flat-plate solar collector; (b) parabolic trough concentrating solar collector

water then drains back into the storage tank to avoid frost damage. The system will be shut down during the coldest weather.

Concentrating solar collectors are used to generate water temperatures of 80 °C and over for hot-water service heating or air-conditioning systems. They are usually driven by electric motors, through reduction gears, to track the sun's movement during each day so that the collecting pipe is kept in the focal plane of the parabolic mirror or polished aluminium reflector. Typical flat-plate and concentrating solar collectors are shown in Fig. 6.15.

Questions

1. Sketch and describe the earth's natural water cycle.

2. List and describe the sources of water and the methods used for its storage and treatment.

3. What pollutants are present in naturally occurring water and where do they come from?

4. Explain the terms 'temporary' and 'permanent' hardness, and list their characteristics.

5. State the use of service reservoirs and describe mains water distribution methods.

6. State the design parameters for cold-water mains and storage systems within buildings, giving parti-

cular information on protection from frost damage, suitability for drinking and protection of the mains against contamination from the building.

7. List the design parameters for hot-water service systems, giving typical data.

8. Sketch the layout of a water services system in a house, showing typical sizes of equipment and methods of control. Show how wastage of water is minimized.

9. Sketch and describe the methods used to generate hot water, noting their applications, economy and thermal performance.

10. Sketch and describe a suitable cold-water services installation for a 20-storey hotel where the mains water pressure is only sufficient to reach the fifth floor.

11. Draw the layout of an indirect hot-water system employing a central-heating boiler and secondary circulation. Show all the pipework and control arrangements.

12. A small hotel is designed for 20 residents and 4 staff. Hot water to be stored at 65 °C is taken from a cold-water supply at 10 °C and heated during a 4 h period. Calculate the heat input rate required.

13. Explain the meaning and use of 'demand units'.

14. The cold-water service system in Fig. 6.9 has three identical sanitary fittings in the locations shown for each type. The pipe lengths shown are to be multiplied by 1.5 but vertical dimensions remain the same. Calculate the pipe sizes.

15. A cold-water storage tank in a house with five occupants is to have a capacity of 100 l/per person and be fed from a water main able to pass 0.25 l/s. How long will it take to fill the tank?

16. A secondary hot-water service system has 55 m of 28 mm circulation pipework and 40 m of 15 mm circulation pipework. Water leaves the cylinder at 65 °C and returns at 60 °C. Air temperature around the pipes is 15 °C. Pressure loss rate in the pipework is 260 N/m^3. The frictional resistance of the pipe fittings is equivalent to 25% of measured pipe length. Specify the head and flow rate required for the secondary circulation pump and choose a suitable pump. Use Table 4.3 and Fig. 4.16.

17. A bank building is to house 115 male and 190 female staff. Recommend a suitable allocation of sanitary accommodation.

18. Draw cross-sections through four different types of pipe joint used for water services, showing the method of producing a water seal in each case.

19. List the factors involved in the provision of a pipework system for the conveyance of drinking water within a curtilage. Comment on the suitability, or otherwise, of jointing materials, lead pipes and storage tanks in this context.

20. Describe the corrosion processes that take place within water systems and the measures taken to protect equipment.

21. Sketch and describe the ways in which solar energy can be used within buildings for the benefit of the thermal environment and to reduce primary energy use for hot-water production. Comment on the economic balance between capital cost and expected benefits in assessing the viability of such equipment.

22. List the types of sanitary appliance available and describe their operating principles, using appropriate illustrations. Comment on their maintenance requirements, water consumption, long-term reliability and materials used for manufacture.

7 Soil and waste systems

Key terms and concepts

Learning objectives

Study of this chapter will enable the reader to:

1. define the parts of waste and drain systems;
2. understand the type of fluid flow in a waste pipe;
3. explain the ventilation requirements of drainage pipes;
4. list and explain the ways in which the water seal can be lost from traps;
5. know the permitted suction pressure in a waste system;
6. understand air pressure distribution in a stack;
7. know how to connect waste pipes into a stack;
8. know the diameters, slopes and maximum permitted lengths of waste and drain pipes for above-ground systems;
9. know how to design waste and drain pipes for ranges of sanitary appliances;
10. design domestic, high-rise and commercial building waste installations;
11. understand and use discharge units for pipe sizing;
12. explain the uses of different materials and jointing methods;
13. understand how drain systems are tested;
14. explain how drain systems are maintained.

Introduction

The terminology of drainage systems is outlined and then the characteristic flow within the pipework is explained. Understanding how fluid flows through waste and drain pipework is fundamental to correct design.

The potential for water seal loss in traps beneath sanitary fittings and the deposition of solids in long sloping drains is examined.

Various standard pipework layouts for above-ground systems are shown. The fluid flow through drain pipes is subject to diversity in timing and duration, as are the hot- and cold-water supplies to the same appliances. However, the characteristic flows into and out of the appliance are not the same, and the use of discharge units for drains is explained.

The materials and jointing methods used for pipework are demonstrated, as are the testing and maintenance procedures.

Definitions

The following terms are used (CIBSE, 1986).

Bedding:	material around a buried pipeline assisting in resisting imposed loads from ground and traffic
Benching:	curved smooth surfaces at the base of manholes, which assist the smooth flow of fluids
Combined system:	a drainage system in which foul and surface-water are conveyed in the same pipe
Crown:	the highest point on the internal surface of a pipe
Discharge stack:	vertical pipe conveying foul fluid/solid
Foul drain:	a pipe conveying water-borne waste from a building
Foul sewer:	the pipework system provided by the local drainage authority
Invert:	the lowest point on the internal surface of a pipe
Manhole:	an access chamber to a drain or sewer
Separate system:	a drainage system in which foul and surface-water are discharged into separate sewers or places of disposal
Subsoil drains:	a system of underground porous or unjointed pipes to collect groundwater and convey it to its discharge point
Stack:	vertical pipe
Surface-water drain:	a pipe conveying rain water away from roofs or paved areas within a single curtilage
Surface-water sewer:	the local authority pipework system
Waste pipe:	pipe from a sanitary appliance to a stack

Fluid flow in waste pipes

The discharge of fluid from a sanitary appliance into a waste, soil or drain pipe is a random occurrence of short duration exhibiting a characteristic curve similar to that shown in Fig. 7.1.

Flows in waste pipes occur as surges, or plugs of fluid, which last for a short time. The pipe flows full at some time and a partially evacuated space appears towards the end of discharge, as shown in Fig. 7.2. Separation between the water attempting to remain in the P-trap and the plug falling into the soil stack causes an air pocket to form. The static pressure of this air will be subatmospheric. Air from the room and the ventilated soil stack bubbles through the water to equalize the pressures and a noisy appliance operation results. The inertia of the discharge may be sufficient to syphon most of the water away from the trap, leaving an inadequate or non-existent seal. The problem is avoided (BS 5572: 1978) by using

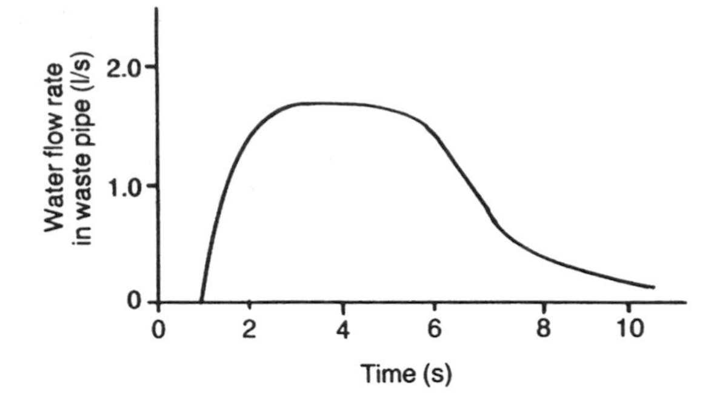

Figure 7.1 Discharge of water from a sanitary appliance

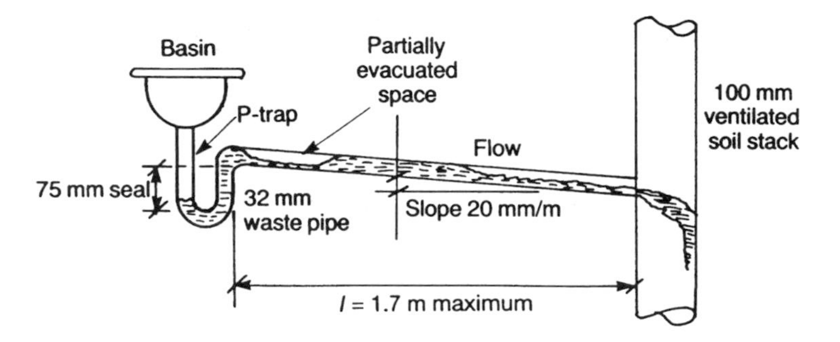

Figure 7.2 Design of a basin waste pipe to avoid self-syphonage

32 mm basin waste pipes when the length is restricted to 1.7 m at a slope of 20 mm/m run (about 1°).

The sloping waste pipe can be up to 3 m long if its diameter is raised to 40 mm after the first 50 mm of run. This allows aeration from the stack along the top of the sloping section. Longer waste pipes with bends and steeper or even vertical parts have a 25 mm open vent pipe as shown in Fig. 7.3.

Vertical soil and vent stacks are open to the atmosphere 900 mm above the top of any window or roof-light within 3 m. Underground foul sewers are thus atmospherically ventilated. Water discharged into the stack from an appliance entrains air downwards and establishes air flow rates of up to a hundred times the water volume flow rate. Air flow rates of 10–150 l/s have been measured. The action of water sucking air into the pipe lowers the air static pressure, which is further reduced by friction losses.

Water enters the stack as a full-bore jet, shooting across to the opposite wall, falling and establishing a downward helical layer attached to the pipe surface. Restricted air passageways at such junctions further lower the air static pressure by their resistance to flow. Atmospheric pressure will be re-established at the base of the stack because of the flow of air into the low-pressure region. The falling fluid tends to fill the pipe near the base and positive air static pressures can be generated. Appliances connected to such a region may have their water seals intermittently forced out. Figure 7.4 shows the probable air static pressures during the simultaneous discharge of three appliances.

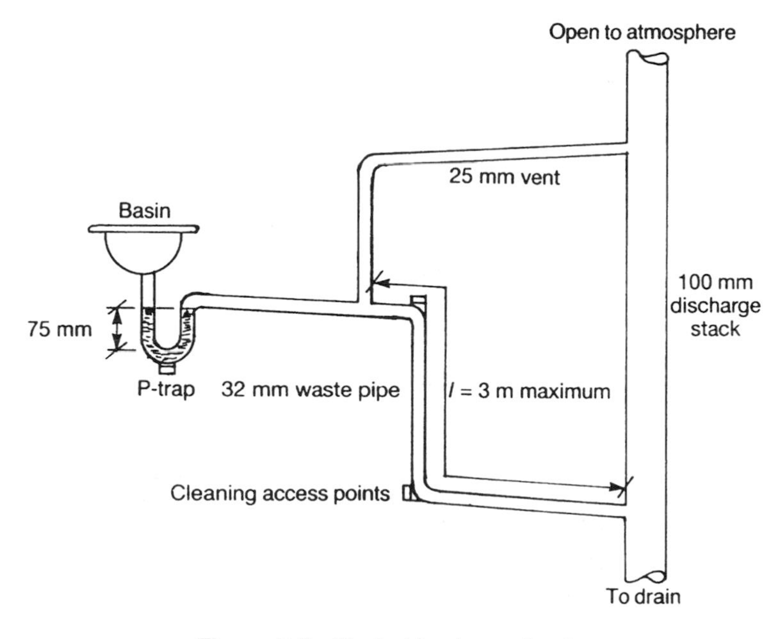

Figure 7.3 Vented basin waste pipe

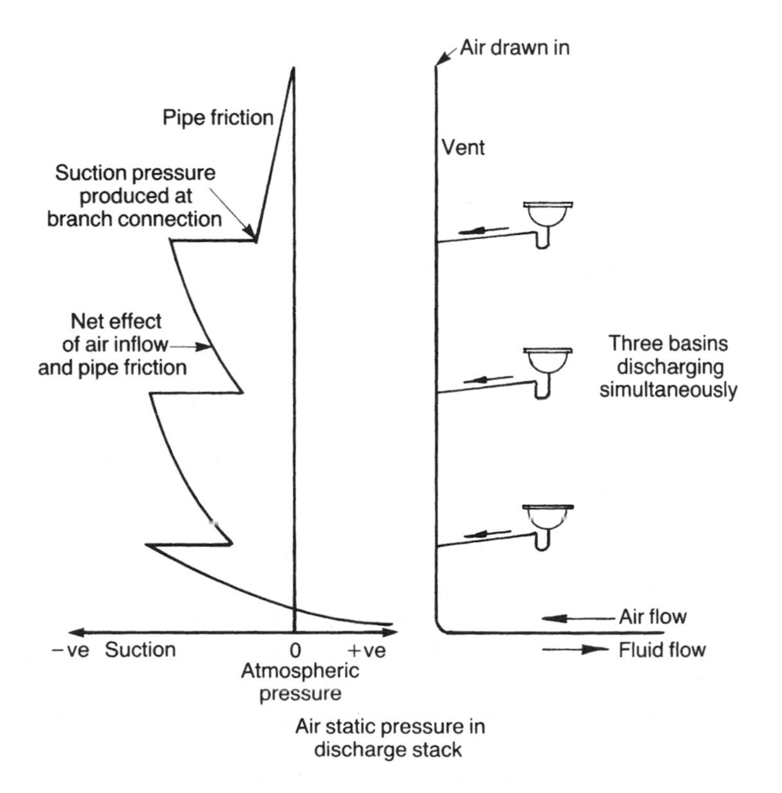

Figure 7.4 Air static pressure distribution in soil and vent pipes

The pressure gradient shown in Fig. 7.4 can be drawn with the aid of data from experimental work (Wise, 1979). The maximum permitted pressure is −375 Pa as this is equivalent to the recommended trap depth of 75 mm water gauge for single-stack drain installations. When suction of this magnitude is applied to a 75 mm water seal, some of the water is sucked from the trap, leaving about 56 mm. This is sufficient to stop fumes entering the building.

Loss of water seal from a trap can occur through the following mechanisms.

1. Self-syphonage: this can be avoided by placing restrictions on lengths and gradients and venting long or steep gradients.
2. Induced syphonage: water flow past a waste pipe junction in a stack or along a sloping horizontal range of appliances can suck out the water seal. This is overcome by suitable design of the pipe diameters, junction layouts and venting arrangements.
3. Blow-out: a positive pressure surge near the base of a stack could push out water seals of traps connected in that region. Waste pipes are not connected to the lower 450 mm of vertical stacks, measured from the bottom of the horizontal drain.
4. Cross-flow: flow across the vertical stack from one appliance to another. Waste pipes are not connected to soil and vent pipes where cross-flow, particularly from WC branches, could be caused, as shown in Fig. 7.5.
5. Evaporation: this amounts to about 2.5 mm of seal loss per week while appliances are unused.
6. Wind effects: wind-induced pressure fluctuations in the stack may cause the water seal to waver out. The vent terminal should be sited away from areas subject to troublesome effects. Wind-tunnel tests using smoke as a tracer are performed for large developments.
7. Bends and offsets: sharp bends in a stack can cause partial or complete filling of the pipe, leading to large pressure fluctuations. Foaming of detergents through highly turbulent fluid flow will aggravate pressure fluctuations. Connections to the vent stack before and after an offset equalize air pressures. A bend of minimum radius 200 mm is used at the base of a soil stack to ensure constant ventilation.
8. Surcharging: an underground drain that is allowed to run full causes large pressure fluctuations. Additional stack ventilation is required.
9. Intercepting traps: where a single-stack system is connected into a drain with an interceptor trap nearby, fluid flow is restricted. Additional stack ventilation is used.
10. Admission of rainwater into soil stacks: when a combined foul and surface-water sewer is available, it is possible to admit rainwater into the discharge stack. Continuous small rainwater flows can cause excessive pressure fluctuations in buildings of about 30 storeys. Flooding of the stack during a blockage would cause severe damage.
11. Pumped or pneumatically ejected sewage lifting: the discharge stack is gravity-drained into a sump, from where it is pumped into a street sewer at a higher level. A separate vent is used for the sump chamber and pumped sewer pipe to avoid causing pressure surges.
12. Capillary: lint or hair remaining in a trap may either block the capillary or empty it. Additional maintenance is carried out in high-risk locations.

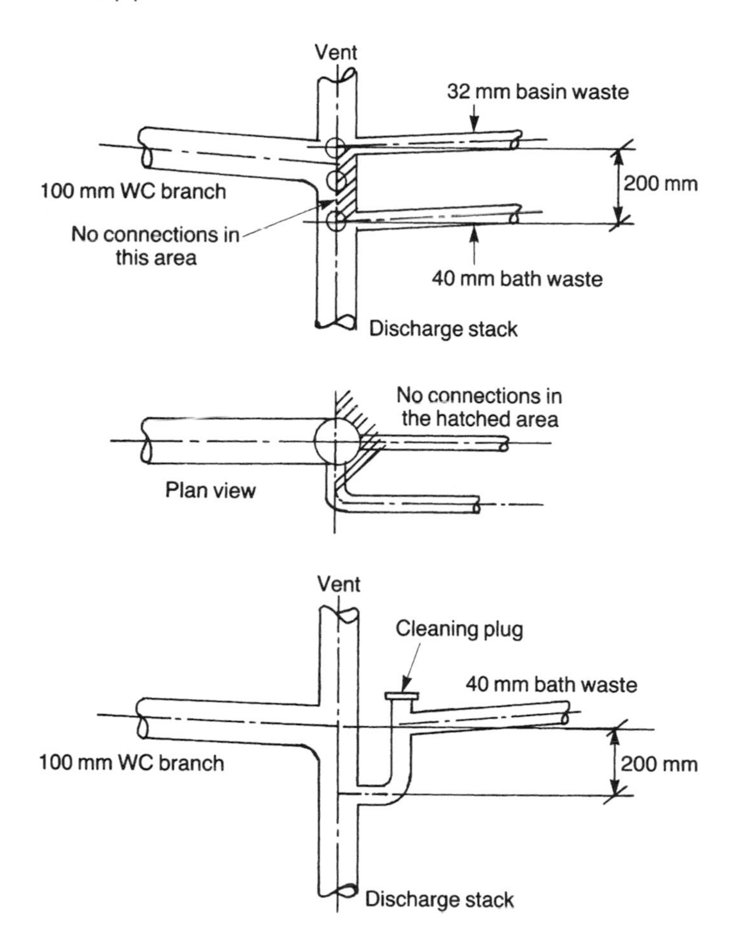

Figure 7.5 Permitted stack connections avoiding cross-flow

13. Leakage: leakage can occur through mechanical failure of the joints or the use of a material not suited to the water conditions.

Figure 7.6 shows the principle of operation of an anti-syphon trap (Marley Extrusions Limited). When excessive suction pressure occurs in the waste pipe, some of the water in the trap is syphoned out. When the central ventilation passage becomes uncovered, it connects the inlet and outlet static air pressures. This returns the waste pipe to atmospheric pressure and the syphonage ceases. Sufficient water remains in the trap to maintain a hygienic seal.

Drainage installations should remove effluent quickly and quietly, be free from blockage, and be durable and economic. They are normally expected to last as long as the building and be replaced only because of changed requirements or new technology. Blockages occur when the system is overloaded with solids, becomes frozen, suffers restricted flow at poorly constructed bends or joints, or has building material left inside pipe runs. Each section of discharge pipework must be accessible for inspection and internal cleaning.

Transport of solids from WCs is the controlling problem in the design, installation and maintenance of sloping drains. Swaffield and Wakelin (1976)

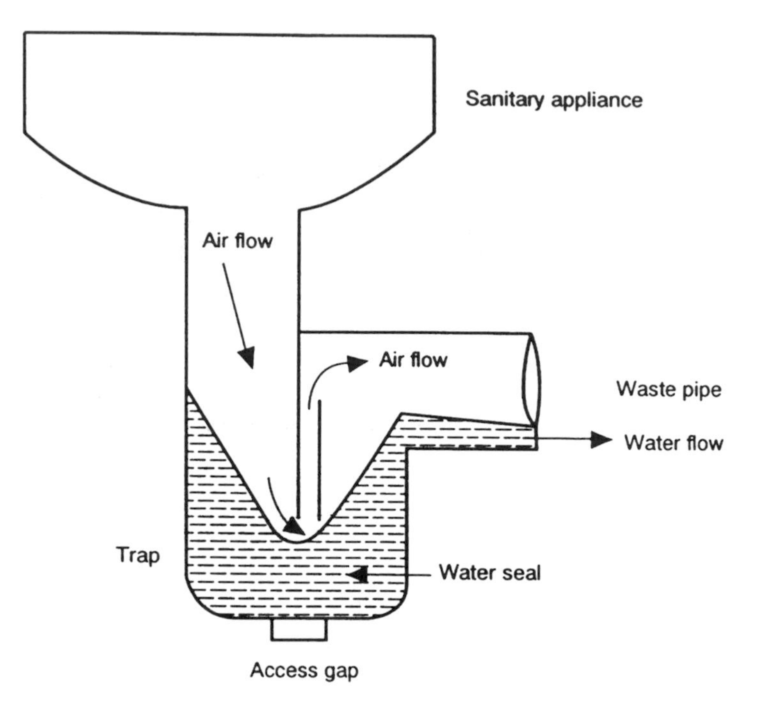

Figure 7.6 Anti-syphon trap

showed that, to maintain the flow from a WC and avoid deposition of solids in the drain, the ratio of the length L of sloping drain (metres) to the gradient G must be

$$L/G = 35^2$$

Pipe bends produce rapid deceleration of solids downstream, followed by velocity regain as the remaining flush water catches up with and accelerates the solids with minimal loss of inertia. When minimum gradients are used, solid deposition could occur at a bend. To avoid this, the equivalent length of a bend can be taken as 5 m of straight pipe in design calculations. Solid deposition can also occur at a top entry into a sewer. Branch connections should be at 45 °C to the horizontal.

EXAMPLE 7.1

A WC is to be connected to a 100 mm soil stack, which runs in a 250 mm deep service duct formed from a false ceiling. It is intended that the WC be 16 m from the stack. There is a 90° bend just before the drain enters the stack. Determine whether the proposed layout will be satisfactory. Figure 7.7 shows the intended arrangement.

The allowed fall will be approximately 150 mm, subject to free available passage. The equivalent length of the sloping drain is $(L + 5)$ m. Then the

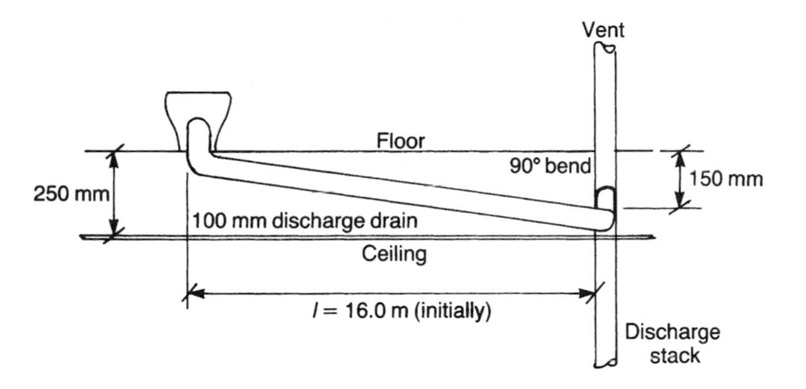

Figure 7.7 Sloping drain in a false ceiling for Example 7.1

gradient is

$$G = \frac{0.15}{L \text{ m}}$$

The intended conditions are

$$G = 0.15/16 = 0.009\,375$$

and

$$L = 16 + 5 = 21 \text{ m}$$

Using $L/G = 35^2$ for an equivalent length of 21 m the gradient cannot be less than

$$G = \frac{L}{35^2}$$

$$= \frac{21}{35^2}$$

$$= 0.01714$$

The intended gradient is less than the minimum allowable gradient and so the design must be modified.

Assuming that the WC can be brought nearer to the stack, how far away can it be? There are two conditions to be met:

$$G = \frac{0.15}{L} \tag{7.1}$$

$$\frac{L + 5}{G} = 35^2 \tag{7.2}$$

Substitute equation (7.1) into equation (7.2) to eliminate G:

$$(L+5)\times\frac{L}{0.15}=35^2$$

This can be rearranged to

$$L^2+5L=183.75$$

$$L^2+5L-183.75=0$$

This is a standard quadratic equation of the form,

$$ax^2+bx+c=0$$

where $x=L$, $a=1$, $b=5$ and $c=-183.75$. The solution of the quadratic is

$$x=\frac{-b\pm\sqrt{(b^2-4ac)}}{2a}$$

Therefore

$$L=\frac{-5\pm\sqrt{[5^2-4(-183.75)]}}{2}$$

$$=\frac{-5\pm\sqrt{760}}{2}$$

$$=11.284 \text{ m}$$

Thus, if the WC is situated 11 m from the stack, the installed gradient of $0.15/11=0.014$ is steeper than that required by the design guide ($L/35^2$):

$$G=\frac{11+5}{35^2}=0.013$$

and hence solid deposition should not occur.

Pipework design

The arrangement of pipework (Building Research Establishment Digest 205, 1977) for various sanitary appliances is shown in Figs 7.8–7.12.

Groups of appliances for dwellings are depicted in Figs 7.13 and 7.14. A pumped WC discharge unit, as shown in Fig. 7.15, enables the use of a 22 mm diameter copper pipe to run long distances, and upwards, to connect into the soil and vent stack at a convenient location.

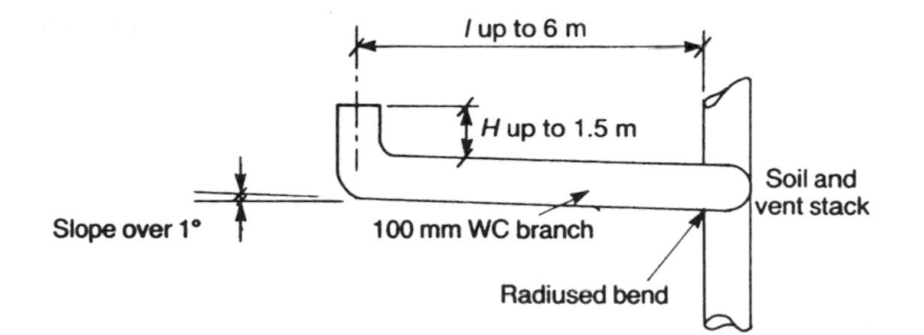

Figure 7.8 Branch pipe to a WC

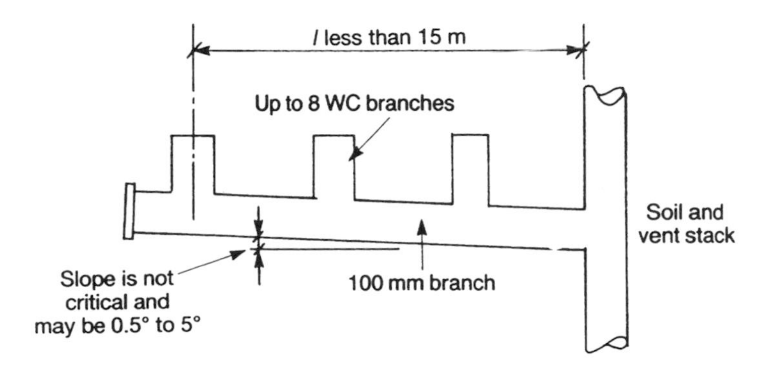

Figure 7.9 Branch for a range of WCs

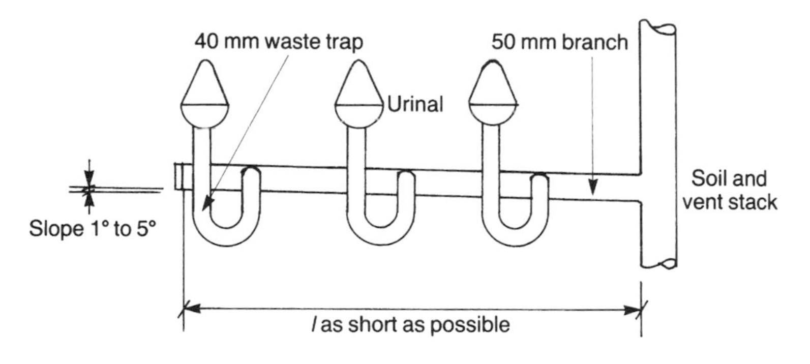

Figure 7.10 Branch for a range of urinals

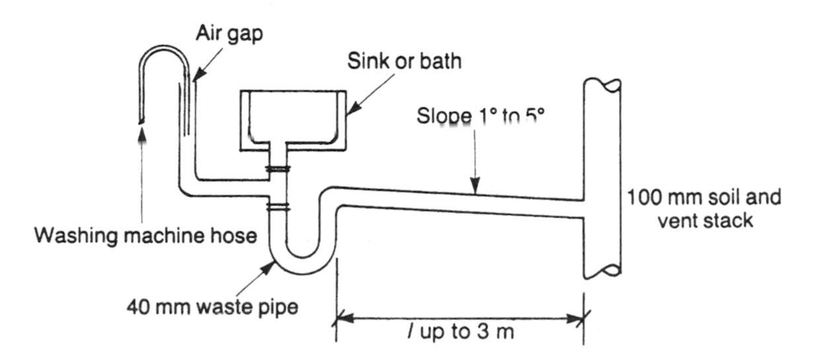

Figure 7.11 Branch from a sink or bath

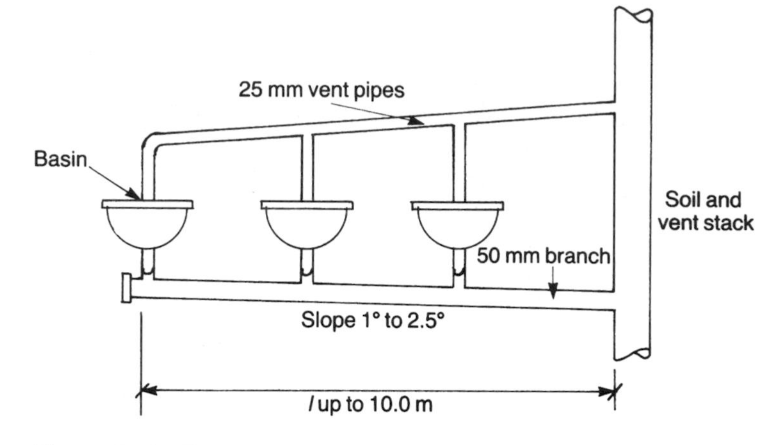

Figure 7.12 Branch discharge pipe for a range of up to ten basins

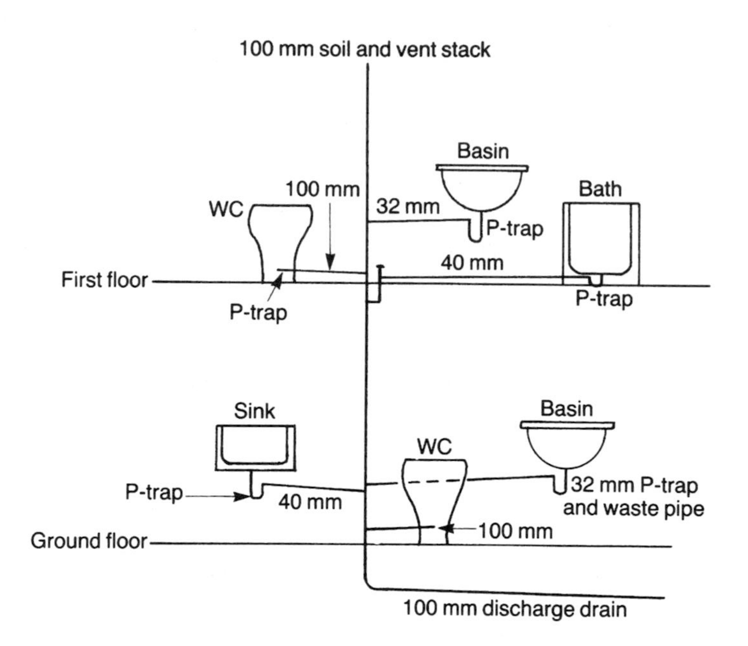

Figure 7.13 Soil and vent stack in housing

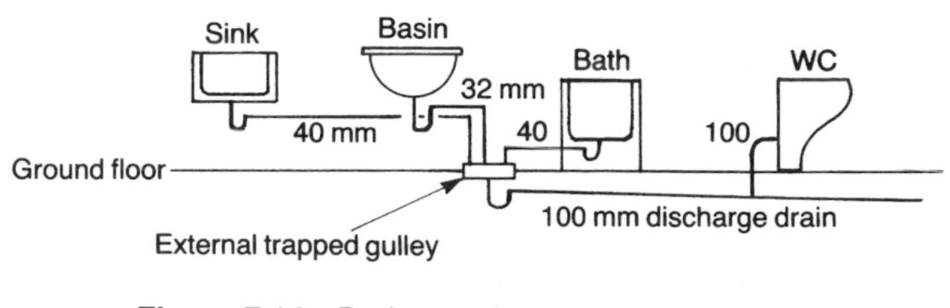

Figure 7.14 Drainage pipework for a bungalow

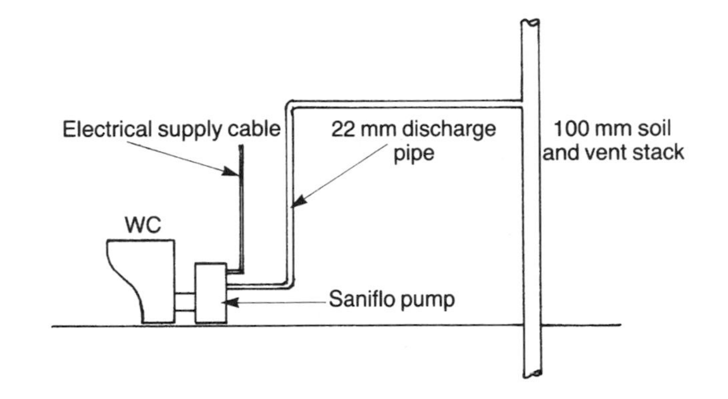

Figure 7.15 Saniflo pumped WC discharge system

Discharge unit pipe sizing

The intermittency of discharge from appliances necessitates the use of discharge units that relate to the flow volume, flow time and interval between flows from sanitary fittings in a similar way to the demand units for water supplies to such fittings. Typical discharge units are as follows (domestic use): WC, 14; basin, 3; bath, 7; urinal, 0.3; washing machine, 4; sink, 6. A group consisting of WC, bath, sink and two basins has a value of 14 discharge units.

A 100 mm diameter stack can carry 750 discharge units, a 125 mm diameter stack can carry 2500 discharge units and a 150 mm diameter stack can carry 5500 discharge units.

Materials used for waste and discharge systems

The materials available for waste pipes and soil and vent stacks are listed in Table 7.1.

A clearance of 30 mm should be left between external pipes and the structure to allow free access and for painting. Secure bracketing to the structure is essential and allowance for thermal expansion should be made. Pipes passing through walls or floors should be sleeved with a layer of inert material to prevent the ingress of moisture into the building and provide the elasticity required for thermal movement. This is particularly important with plastics.

Testing

Inspection and commissioning tests on drainage installations are carried out as follows.

Inspection

During installation, regular inspections are made to check compliance with specifications and codes. Particular attention is given to quality of jointing,

Table 7.1 Materials for waste and discharge pipework

Material	Application	Jointing
Cast iron	50 mm and above vent and discharge stacks	Lead caulking with molten or fibrous lead; cold compound caulking
Galvanized steel	Waste pipes	BSPT screwed
Copper	Waste pipes and traps	Compression, capillary, silver solder, bronze weld or push-fit ring seal
Lead	Waste pipes and discharge stacks	Soldered or lead welded
ABS	Up to 50 mm waste and vent pipes	Solvent cement and push-fit ring seal
High-density polyethylene	Up to 50 mm waste and ventilating pipes and traps	Push-fit ring seal and compression fittings
Polypropylene	Up to 50 mm waste and ventilating pipes and traps	Push-fit ring seal and compression couplings
Modified PVC	Up to 50 mm waste and vent pipes	Solvent cement and push-fit ring seal
Unplasticized PVC	Over 50 mm soil and vent stacks; vent pipes under 50 mm	Solvent cement and push-fit ring seal
Pitch fibre	Over 50 mm discharge and vent stacks	Driven taper or polypropylene fitting with a push-fit ring seal

ABS, acrylonitrile butadiene styrene; PVC, polyvinyl chloride.

security of brackets and removal of swarf, cement or rubble from inside pipe runs.

Air pressure

Prefabricated waste pipe systems will be factory tested before delivery to the site. The complete system will be tested on completion by filling the water seals and inserting air bags (expandable bungs) into the ends of stack pipes. A rubber hose is inserted into the vent stack through a WC water trap. The air pressure in the stack is hand-pumped up to 38 mm water gauge, measured on a U-tube manometer. This pressure must remain constant for 3 min without further pumping. Soap solution wiped onto joints will reveal leak locations.

Smoke

Existing stacks can be tested by the injection of smoke from an oil-burning generator or a smoke cartridge, provided that it will not cause damage to the drain materials. This is less severe than the air test, as smoke pressure remains low and damage from the test itself is less likely. Suitable warnings must be given to the occupants of the building.

Syphonage

The simultaneous discharge of several appliances should reveal a minimum remaining water seal of 25 mm in all traps. Discharge should take place quietly and smoothly.

Maintenance

Periodic inspection, testing, trap clearance, removal of rust and repainting should be a feature of an overall service maintenance schedule. Washers on access covers require occasional replacement. The use of chemical descaling agents, hand or machine-operated rodding and high-pressure blockage removal must be carefully related to the drainage materials and the skill of the operator.

Lime scale removal

Lime scale is found in hard-water areas. A dilute corrosion-inhibited acid-based descaling fluid is applied directly to scale visible on sanitary appliances and is then thoroughly flushed with clean water. The fluid is a mixture of 15% inhibited hydrochloric acid and 20% orthophosphoric acid.

Removal of grease and soap residues

A strong solution of 1 kg of soda crystals and 9 l of hot water is flushed through the system. The soda crystals are mixed with the hot water in a basin. When the soda is fully dissolved, the plug is released. This may be necessary frequently in commercially used appliances.

Blockage

A hand plunger may be sufficient but repeated blockage should be investigated. Hand rodding from the nearest access point can be performed using various tools as appropriate. A kinetic ram gun can be used for blockage in branch pipes. The impact of compressed air from the gun creates a shock wave in the water, which dislodges the solids. However, a blow-back from a stubborn blockage may injure the operator and damage the pipework and therefore the ram gun must be limited to the removal of soft materials. Coring and scraping mechanical tools can be used to remove hard lime scale in 100 mm pipes, provided that the materials will withstand the maintenance operation. A steel cutter is revolved by a flexible drive fed through the drain pipework.

> **EXAMPLE 7.2**
>
> Draw a suitable arrangement of a sanitary pipe system for a 15-storey block of flats with two groups of appliances on each floor. Each group consists of a WC, a bath, a sink and a basin, all sited close to the stack.

$$\text{discharge units carried by the stack} = (15 \times 2)\text{ flats} \times 14\text{ DU}$$

$$= 420\text{ DU}$$

Therefore a 100 mm soil and vent stack can be used. A typical floor layout is shown in Fig. 7.16.

> **EXAMPLE 7.3**
>
> A 10-storey office block is to have ranges of three WCs, three basins and three urinals on each floor. The WCs and urinals are situated on each side of the stack, but the common waste pipe from the basins is to be 5 m long and have four bends. The following discharge units can be assumed: urinal, 0.3; WC, 14; basin, 3. Draw a suitable discharge pipework arrangement and state the pipe sizes to be used.

$$\text{discharge units carried by the stack} = 10 \times (3 \times 0.3 + 3 \times 14 + 3 \times 3)$$

$$= 519\text{ DU}$$

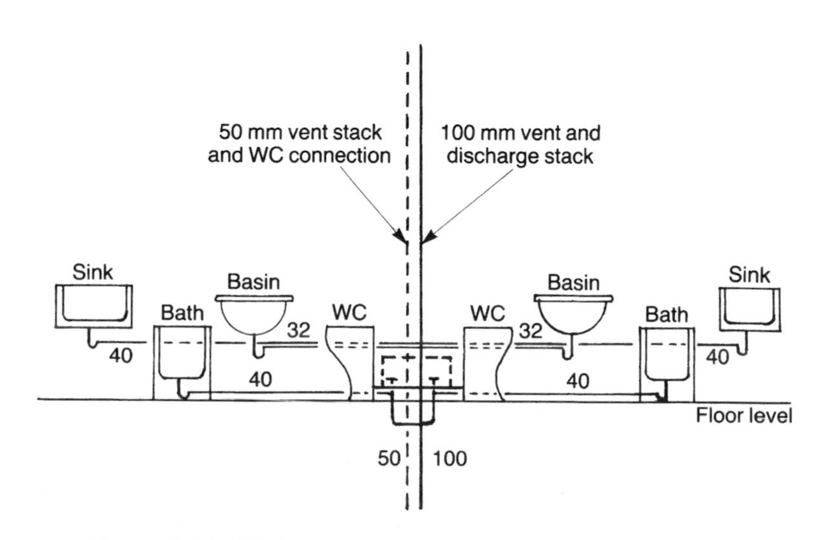

Figure 7.16 Typical floor layout for two flats in Example 7.2

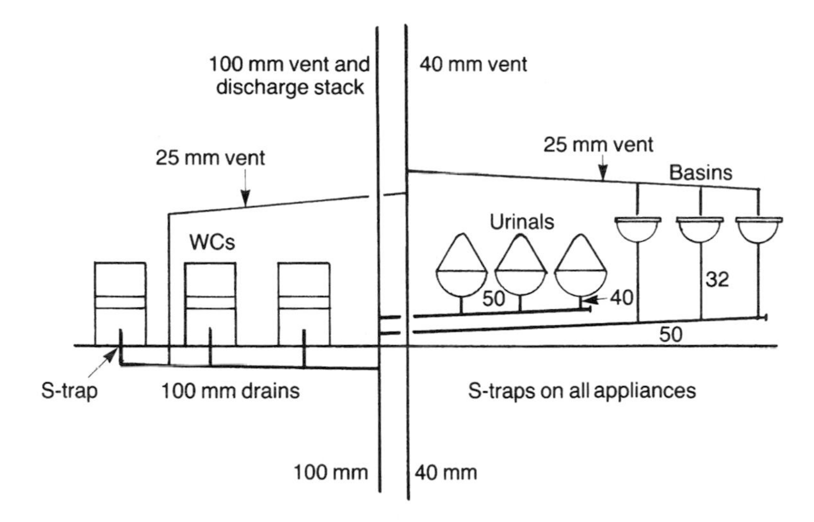

Figure 7.17 Typical floor layout for Example 7.3

A 100 mm discharge and vent stack can be used in the arrangement shown in Fig. 7.17. An additional ventilating stack of diameter 40 mm is recommended in BS 5572: 1978, as shown.

Questions

1. List the ways in which an above-ground drainage installation satisfies its functional requirements.

2. Describe, with the aid of sketches, the ways in which the water seal can be lost from a trap and the precautions taken to avoid this happening.

3. Describe the types of fluid flow encountered in drainage pipework.

4. Outline the development of current drainage practice from the early Roman occupation of Britain.

5. State the meaning of the following terms: bedding; combined system; drain; sewer; manhole; separate system; stack; discharge pipe; vent.

6. Sketch the pipework layout for a typical group of sanitary appliances in a dwelling, where they are all connected into a stack. Show suitable pipe sizes, slopes and details of the connections at the stack.

7. A range of WCs is to be connected into a common branch pipe of outside diameter 125 mm fitted within a false ceiling 300 mm deep. It is intended that the furthest WC should be 18 m from the stack. The branch has a 90° bend between the last WC and the stack. Determine whether the proposed arrangement would be satisfactory. If it is not, calculate the maximum distance that could be allowed between the furthest WC and the stack.

8. State the meaning of the term 'discharge unit'. How many WCs can be connected into a discharge stack of diameter 100 mm?

9. Sketch sections through four types of joint used in drainage pipework to show their constructional features, method of providing a seal, and thermal expansion facility.

10. Sketch the installation of a vertical discharge stack within a plasterboard service duct in a house, clearly showing suitable dimensions and support.

11. Sketch and describe the methods of testing above-ground drainage installations.

12. Describe the maintenance work needed to support the efficient operation of drainage installations in residences, laundries, canteens and hotels.

13. Draw a suitable sanitary pipework installation for a 10-storey block of flats with two groups of appliances on each floor connected to one stack. Show pipe sizes and routes.

14. A 20-storey office block is to have ranges of five basins, five urinals and five WCs on alternate floors. Draw a suitable pipework installation, stating pipe sizes and slopes.

8 Surface-water drainage

Learning objectives

Study of this chapter will enable the reader to:

1. calculate rainfall run-off into surface-water drains;
2. calculate gutter water flow capacity;
3. choose appropriate gutter and rainwater downpipe combinations to create an economical design;
4. calculate and find gutter sizes;
5. assess methods for the disposal of surface-water;
6. calculate soakaway pit design.

Key terms and concepts

Introduction

Design calculations for roofs, gutters and ground drainage are presented along with practical exercises in suitable arrangements. It is important for the designer to maintain the closest contact with the architect during this process because of the required integration.

Flow load

In the UK, surface-water systems are designed on the basis of a rainfall intensity of 50 mm/h (75 mm/h for roofs). The quantity of water entering a drain depends on the amount of evaporation into the atmosphere and natural drainage into the ground. The drain flow load is represented by the impermeability factor, and typical figures are shown in Table 8.1.

The drain water flow rate Q is given by

$$Q = \text{area drained m}^2 \times \text{rainfall intensity } \frac{\text{mm}}{\text{h}} \times \text{impermeability factor}$$

Table 8.1 Ground impermeability factors

Nature of surface	Impermeability factor
Road or pavement	0.90
Roof	0.95
Path	0.75
Parks or gardens	0.25
Woodland	0.20

EXAMPLE 8.1

Footpaths, roadways and gardens on a housing estate cover an area of 10 000 m^2, of which 40% is garden and grassed areas. Estimate the surface-water drain flow load in litres per second.

From Table 8.1, impermeability factors are 0.9 for the roads and paths and 0.25 for gardens and grass. Therefore

$$\text{average impermeability} = 0.4 \times 0.25 + 0.6 \times 0.9$$
$$= 0.64$$

Hence

$$Q = 10\ 000\ \text{m}^2 \times 50\ \frac{\text{mm}}{\text{h}} \times \frac{1\ \text{h}}{3600\ \text{s}} \times \frac{1\ \text{m}}{10^3\ \text{mm}} \times 0.64 \times \frac{10^3\ \text{l}}{1\ \text{m}^3}$$
$$= 88.9\ \text{l/s}$$

Roof drainage

A rainfall intensity of 75 mm/h occurs for about 5 min once in 4 years. An intensity of 150 mm/h may occur for 3 min once in 50 years, and where overflow cannot be tolerated this value is used for design. The flow load Q for a roof is calculated from

$$Q = A_r\ \text{m}^2 \times 75\ \frac{\text{mm}}{\text{h}} \times \frac{1\ \text{h}}{3600\ \text{s}} \times \frac{1\ \text{m}}{10^3\ \text{mm}} \times \frac{10^3\ \text{l}}{1\ \text{m}^3}$$
$$= A_r\ \text{m}^2 \times 0.021\ \text{l/s}$$

where A_r is the surface area of a sloping roof of pitch up to 50° and no evaporation takes place, which is characteristic of a cold saturated atmosphere.

For a roof pitch of greater than 50°

$$Q = 0.021 \times (1 + 0.462 \tan\theta) \times A_r\ \text{l/s}$$

where θ is the roof pitch in degrees.

The flow capacity of a level half-round gutter is given by

$$Q = 2.67 \times 10^{-5} \times A_g^{1.25} \text{ l/s}$$

where A_g is the cross-sectional area of the gutter (mm^2). For level gutters other than half-round the flow capacity can be found from

$$Q = \frac{9.67}{10^5} \times \sqrt{\left(\frac{A_0^3}{W}\right)} \text{ l/s}$$

where A_0 is the area of flow at the outlet (mm^2) and W is the width of the water surface (mm). The depth of flow in a level gutter discharging freely increases from the outlet up to a maximum at the still end. It can be assumed that the depth of flow at the outlet is about half that at the still end. Thus the depth at the outlet is half the gutter depth. A_0 is found from the gutter cross-sectional area at half its depth. W will normally be the width across the top of the gutter.

A fall of 1 in 600 increases flow capacity by 40%. The frictional resistance of a sloping gutter may reduce water flow by 10%, and each bend can reduce this further by 25%. Water flow in downpipes is much faster than in the gutter and they will never flow full. Their diameter is usually taken as being 66% of the gutter width. Some typical gutter flow capacities are given in Table 8.2.

For calculation purposes, a roof is divided into areas served by a gutter with an end-outlet rainwater downpipe. If the whole roof is drained into a gutter with an outlet at one end, then the gutter carries water from the entire roof area. However, when a centre outlet is used, a smaller gutter size might be possible as it only carries half the flow load. The number and disposition of the downpipes is considered in relation to the gutter size, architectural appearance, cost and complexity of the underground drainage system.

Rectangular gutter sizes can be found from

$$A_0 = \frac{WD}{2}$$

where D is the gutter depth (mm).

Table 8.2 Typical flow capacities for a PVC half-round gutter at a 1 in 600 fall

Nominal gutter width (mm)	Q (l/s)	
	End outlet	Centre outlet
75	0.46	0.76
100	1.07	2.10
125	1.58	2.95
150	3.32	6.64

Source: Reproduced from *IHVE Guide* (CIBSE, 1986 [IHVE, 1970]) by permission of the Chartered Institution of Building Services Engineers.

EXAMPLE 8.2

A flat roof of dimensions 20 m × 10 m is laid to fall to a PVC half-round gutter along each long side. Find an appropriate gutter size when the gutter is to slope at 1 in 600.

The flow load is

$$Q = 0.021\ A_r\ \text{l/s}$$
$$= 0.021 \times 20 \times 10\ \text{l/s}$$
$$= 4.2\ \text{l/s}$$

Each gutter will carry half of Q, i.e. 2.1 l/s. For one gutter, the fall will increase the carrying capacity by 40% and friction will reduce it by 10%. The required gutter area can then be found from

$$Q = 1.4 \times 0.9 \times 2.67 \times 10^{-5} \times A_g^{1.25}\ \text{l/s}$$

Hence

$$A_g^{1.25} = \frac{10^5 \times Q}{2.67 \times 1.4 \times 0.9}$$
$$= \frac{10^5 \times 2.1}{3.3642}$$
$$= 0.624 \times 10^5$$

Therefore

$$A_g = (0.624 \times 10^5)^{1/1.25}\ \text{mm}^2$$
$$= (0.624 \times 10^5)^{0.8}\ \text{mm}^2$$
$$= 6857\ \text{mm}^2$$

For a half-round gutter

$$A_g = \frac{\pi W^2}{8}$$

and hence

$$W = \sqrt{\frac{8A_g}{\pi}}$$
$$= \sqrt{\left(\frac{8 \times 6857}{\pi}\right)}\ \text{mm}$$
$$= 132\ \text{mm}$$

Thus a 150 mm half-round gutter with an end outlet would be used along each side. This can be checked with the data in Table 8.2. A smaller gutter would be possible if a centre outlet was appropriate to the appearance of the building and the underground drain layout.

EXAMPLE 8.3

A roof sloping at 60° has a level box gutter 100 mm wide and 75 mm deep. The roof is 10 m long and 5 m up the slope. Calculate whether the gutter will adequately convey rainwater when the rainfall intensity is 75 mm/h.

The flow load is given by

$$Q = 0.021 \times (1 + 0.462 \times \tan 60°) \times 10 \times 5 \text{ l/s}$$

$$= 1.89 \text{ l/s}$$

If an end outlet is used, the water depth at the outlet will be half the gutter depth, i.e. 37.5 mm. Thus

$$A_0 = 100 \times 37.5 \text{ mm}^2$$

$$= 3750 \text{ mm}^2$$

The gutter flow capacity is expected to be

$$Q = \frac{9.67}{10^5} \times \sqrt{\frac{3750^3}{100}} \text{ l/s}$$

$$= 2.221 \text{ l/s}$$

This is greater than the imposed flow load and will produce the required service. If the gutter flow capacity was inadequate, a centre outlet would have the effect of halving the flow load on the gutter.

Cast iron covers over drainage gulleys in roadways pass 20 l/s or more, depending upon surface-water speed, degree of flooding and blockage from debris.

Disposal of surface-water

Surface-water can be removed from a site by one or more of the following methods.

Sewer

Where the local authority agrees that there is adequate capacity, surface-water is drained into either a combined sewer or a separate surface-water sewer. Surface-water from garage forecourts and car parks is run in open gullies to an interceptor chamber. Ventilation of explosive and poisonous petrol vapour is essential, as a concentration of 2.4% in air is fatal. It is illegal to discharge petrol, oil or explosive vapour into public sewers. The interceptor chamber is an underground storage tank of concrete and engineering bricks, which allows separation of the clean water from the oily scum remaining on its surface. It is intermittently pumped out and cleaned. The discharge drain to the sewer is turned downwards to near the bottom of the interceptor and three separate chambers are used in series.

Soakaway

Ground permeability is established using borehole tests to measure the rate of natural drainage within a curtilage. If running underground water is found, a simple rock-filled pit can be used. Slow absorption is overcome by constructing a perforated precast concrete, dry stone or brick pit, which stores the rainfall quantity. The stored volume is found from an assumed steady rainfall of 15 mm/h over a period of 2 h. This is exceeded around once in 10 years, so there may be occasional flooding for short periods. A soakaway pit is circular with its depth equal to its diameter.

Storage

An artificial pond or lake, or even an underground storage tank, will be necessary if the expected run-off from a curtilage is at a greater rate than could be accommodated by a sewer or watercourse.

Watercourse

The relevant local authorities may allow the disposal of surface-water into watercourses. Expected flow rates at both normal and flood water levels must be established.

EXAMPLE 8.4

Storage soakaway pits 2 m deep are to be employed for a tarmac-covered car park of dimensions 30 m × 18 m. Determine the number and size of the pits needed. Draw a suitable drainage layout for the car park.

$$\text{volume to be stored} = 15\ \frac{\text{mm}}{\text{h}} \times 2\ \text{h} \times \frac{1\ \text{m}}{10^3\ \text{mm}} \times 30\ \text{m} \times 18\ \text{m} \times 0.9$$

$$= 14.58\ \text{m}^3$$

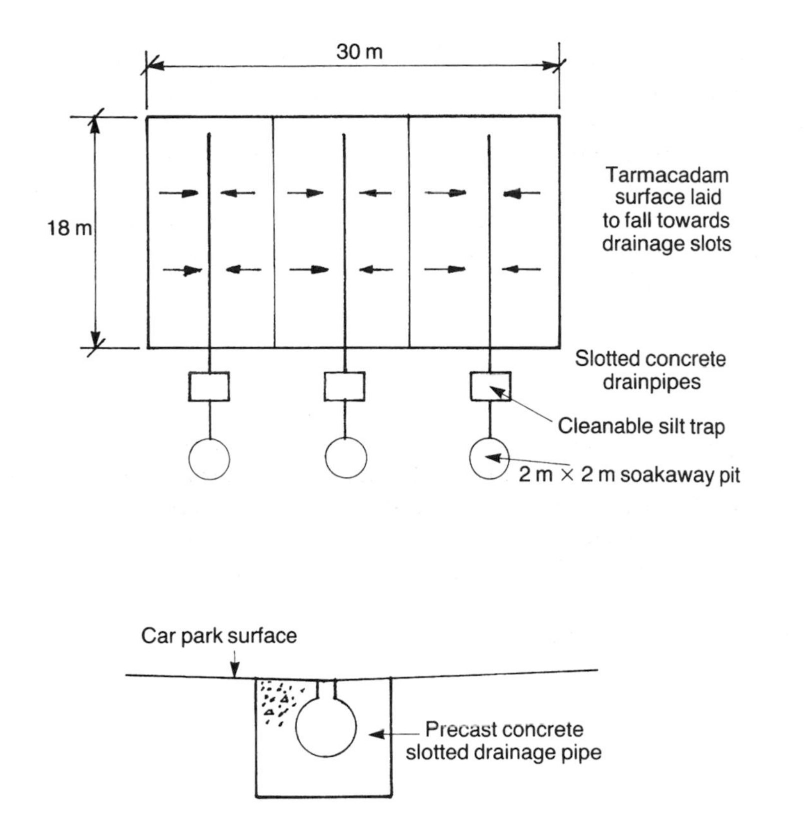

Figure 8.1 Surface-water drainage for the car park in Example 8.4

Pits 2 m deep will have diameters of 2 m. Therefore

$$\text{pit volume} = \frac{\pi \times 2^2}{4} \times 2 \text{ m}^3 = 6.283 \text{ m}^3$$

and hence

$$\text{number of pits} = \frac{14.58}{6.283} = 3$$

Figure 8.1 shows a suitable arrangement of precast concrete slot drainage channels and soakaway pits.

Questions

1. Explain how the rainwater flow load is calculated for roof and groundwater drainage system design.
2. A housing estate has footpaths and roads covering an area of 4000 m^2. Calculate the rainwater flow load and the number of drain gullies required.

3. Find the flow load for a flat roof of dimensions 20 m × 12 m.

4. Find the flow load for a roof of dimensions 10 m × 4 m with a pitch of 52°.

5. Calculate the flow capacity of a level half-round gutter 125 mm wide. Ignore friction.

6. Calculate the flow capacity of a level box gutter 200 mm wide and 150 mm deep when running full. Ignore friction.

7. A PVC half-round gutter 150 mm wide slopes at 1 in 600 and has an end outlet to a rainwater pipe. The water depth at the outlet is half the gutter depth. Assume that A_0 is half the gutter cross-sectional area. Take W as the gutter width. Calculate the gutter flow capacity.

8. A house has two sloping roofs, each side of a ridge, of dimensions 15 m × 8 m. Calculate the flow load and determine the gutter and rainwater pipe design from Table 8.2.

9. The flat roof of a school is to be of dimensions 30 m × 20 m with a rectangular gutter on each long side and sloping at 1 in 600 to an outlet at each end. Calculate suitable dimensions for the gutter.

10. A pitched roof of dimensions 20 m × 5 m drains into a level box gutter 120 mm wide and 80 mm deep on one long side. The gutter has one end outlet. Calculate whether this is a satisfactory arrangement.

11. Describe the actions taken during the design and construction of a surface-water disposal system, stating what options are investigated.

12. Storage soakaway pits 2 m deep are to be used to dispose of rainwater from a roof of dimensions 10 m × 8 m. Determine a suitable size and number of pits.

13. List the techniques used for subsoil drainage systems.

14. Describe the features and maintenance requirements of surface-water drainage systems for car parking, garage forecourts and large paved areas in shopping centres.

9 Below-ground drainage

Learning objectives

Study of this chapter will enable the reader to:

1. understand the design principles for underground drainage pipework;
2. use discharge units;
3. use a pipe-sizing chart;
4. calculate flow capacity;
5. calculate loads on buried pipelines;
6. identify materials and jointing methods;
7. understand sewage-lifting requirements;
8. know testing procedures;
9. carry out a design assignment;
10. explain the principles of below-ground drain layout;
11. know the location and types of access fitting.

Key terms and concepts

Introduction

Below-ground drainage systems are designed to operate without the input of energy, wherever possible, to be reliable and to require little, if any, maintenance. Their layout has to be such that drains are not subject to undue stress from foundations or traffic and are fully accessible for occasional clearance.

Design calculations can be made on the basis of flow rates, utilizing discharge units, gradients, pipe material and pipe diameter. Stress loads, pipe materials and jointing, sewage lifting and testing are discussed, and a design layout assignment is given.

Design principles

Sanitary discharge services operate by gravity flow and require no energy input. Parts of buildings or sites that are below the sewer invert require a pump to raise

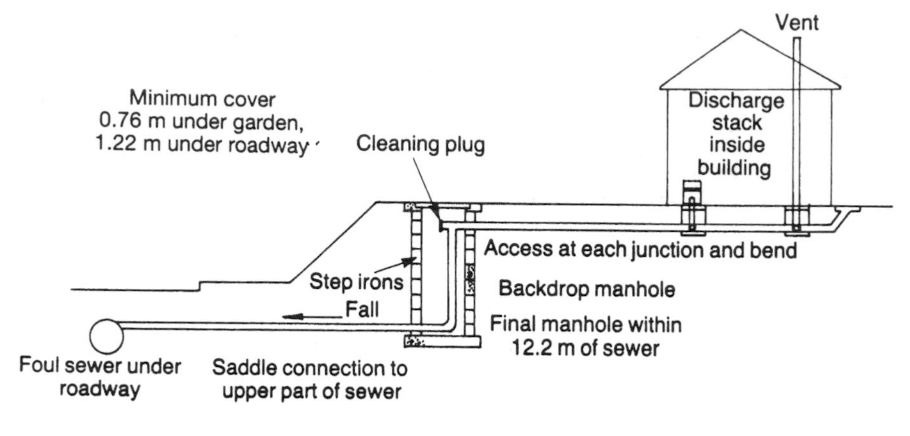

Figure 9.1 Foul drainage installation

the fluid. These operate intermittently to minimize electrical power consumption. Drains are laid to fall at an even gradient, which produces a self-cleaning water velocity so that potential deposits are accelerated and floated downstream. Large drops in drain level are accommodated in a back-drop manhole, rather than a lengthy steep slope, in order to minimize excavation.

Pipes are laid in a series of straight lines between access points used for inspection, testing and cleaning. Branch connections are made obliquely in the direction of flow. There is a preference against running drains beneath buildings owing to possible settlement and the potential cost of later excavation to make repairs, and because the drain invert is lower than the floor damp-proof membrane and rising drains need a waterproof seal against groundwater.

Selected trench bedding and backfill material is used to provide continuous pipe support, to spread imposed ground loads due to the weight of soil and passing traffic, to protect drains from sharp objects and other services, as well as to divert stresses imposed by building foundations. Temporary boards are used to protect exposed drain trenches during construction work. Figure 9.1 shows a typical foul drain system from a single stack.

Access provision

Blockages may happen, as drain systems are likely to be in place for a hundred years or more and demands upon them continue to increase. Cleanability is an essential feature of good design. Good health depends upon satisfactory drainage. Domestic drains are likely to be located less than 1.5 m below ground level, at a maximum gradient of 1:40 and 100 mm in diameter, possibly increasing to 150 mm in diameter at the downstream end of an estate or large building.

Vitrified clay or PVC pipe and fittings are often used, with flexible joints to accommodate ground movement and thermal expansion due to variations in fluid temperature. Brick, concrete, PVC or glass-reinforced plastic (GRP) access chambers are used. All changes in direction are either through 135° or large radius swept bends.

Access points are provided for removing compacted material and for using rigid rods to clear blockages in the direction of flow, even though flexible water-jetting techniques are currently available and it is possible to clear obstructions from either direction. Airtight covers are desirable to avoid access points allowing a health hazard or flooding.

Figure 9.2 shows the types of access.

Rodding eye: a 100 mm diameter drain pipe extended from any depth to ground level to allow rodding in the downstream direction

Shallow access chamber: a removable threaded cap on a branch fitting to allow access in either direction located such that the distance from ground level to drain invert is less than 600 mm to facilitate reaching into the drain

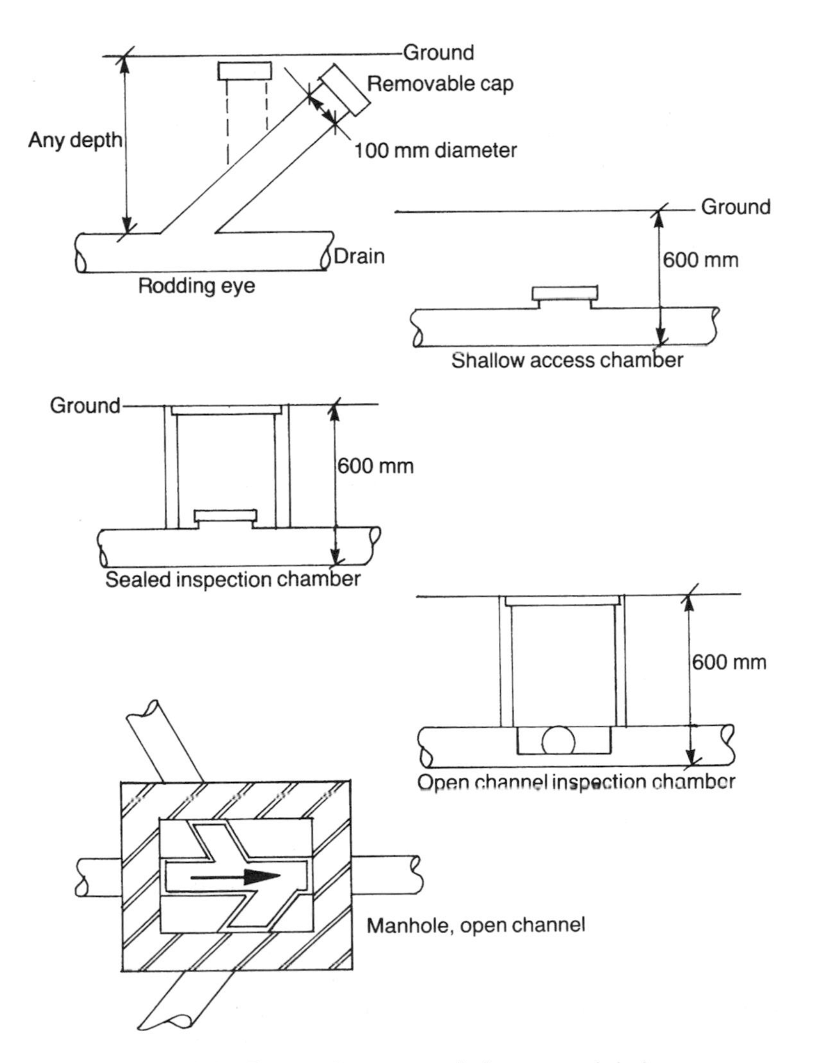

Figure 9.2 Types of access to below-ground drainage

Sealed inspection chamber: a 600 mm deep, 500 mm diameter chamber for access to screwed caps on drain junctions

Open-channel inspection chamber: a 600 mm deep, 500 mm diameter access chamber with benched smooth surfaces for drain junctions

Manhole: the main access point for an operative wearing breathing apparatus to climb down steps to any depth; a 1 m deep manhole is 450 mm square, and a 1.5 m deep manhole has dimensions of 1200 mm × 750 mm or 1050 mm diameter, and a cover 600 mm square

Gulley: ground-level connection point for various waste pipes and the below-ground 100 mm diameter drain providing a water trap against sewer gas and allowing debris removal and rodding access; it may have a sealed lid or open grating

The first access point close to the building is either a gulley, a removable WC or a shallow access chamber just after the base of the internal drainage stack. It is not necessary to fit access points at every change in drain direction, but pipe junctions are made with access chambers. The maximum spacings between access points are 12 m from the start of the drain to the first access, 22 m from a rodding eye to a shallow access chamber, 45 m from a rodding eye to an access chamber or manhole and 90 m between manholes. Figure 9.3 demonstrates a typical housing estate drain layout. Careful integration with the surface-water

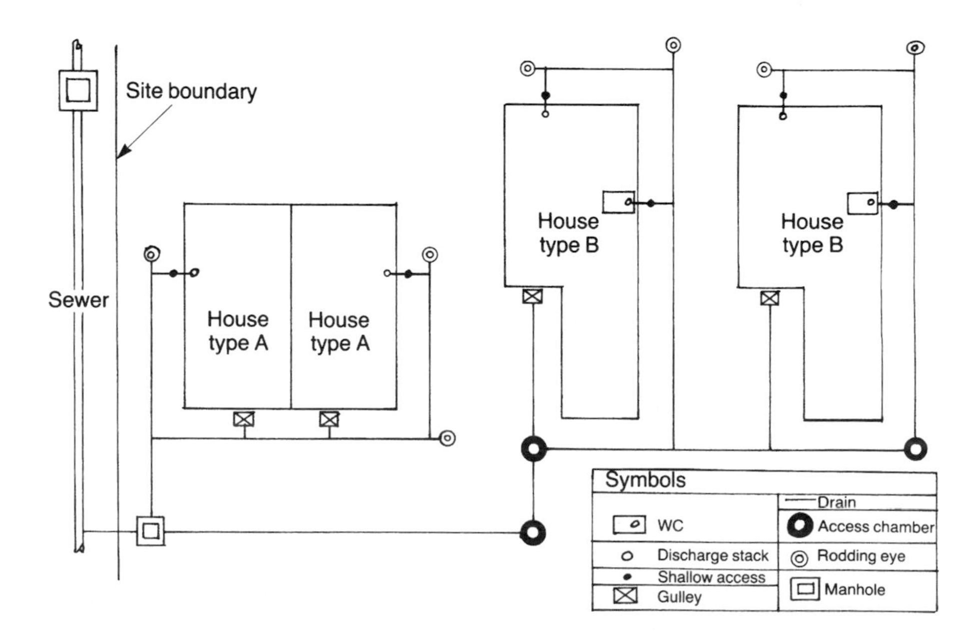

Figure 9.3 Typical site layout showing access

drainage system is necessary as falls to the sewers are preconditioned by the sewer inverts, and the two drains may run within the same trenches and cross each other where branch connections are made.

Pipe diameters for surface-water pipework are based on the flow loads discharging from each downpipe. Those for foul drains are found from the discharge units in each stack. Flows in underground drains are found by totalling calculated flow rates along the route of the collecting drain run. Discharge units are converted into flow rate using Fig. 9.4. Pipe sizes and fluid velocities can be read on Fig. 9.5 from the calculated flow rate and desired gradient as appropriate to the maximum allowable fall available on the site.

EXAMPLE 9.1

The flow from a 100 mm stack is equivalent to 750 discharge units and is to run underground for 30 m before entering the foul sewer at a depth 375 mm lower than at the building end. Find a suitable diameter for a spun precast concrete drain so that it will not be more than two-thirds full at maximum flow rate.

From Fig. 9.4, 750 discharge units are equivalent to a flow rate of 7 l/s. The gradient of the drain is given by

$$\text{gradient} = \frac{0.375}{30} = 1 \text{ in } 80$$

From Fig. 9.5, a flow rate of 7 l/s at a gradient of 1 in 80 requires a 150 mm diameter drain. The fluid velocity will be 1.05 m/s.

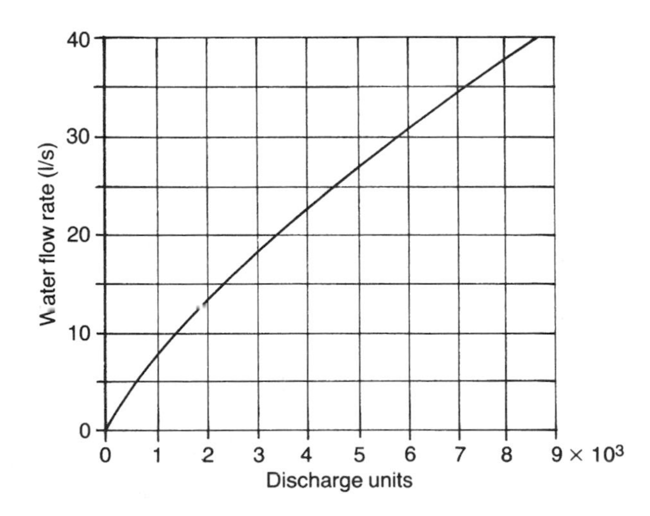

Figure 9.4 Simultaneous flow data for foul- and surface-water drains (reproduced from data in *CIBSE Guide* (CIBSE, 1986) by permission of the Chartered Institution of Building Services Engineers)

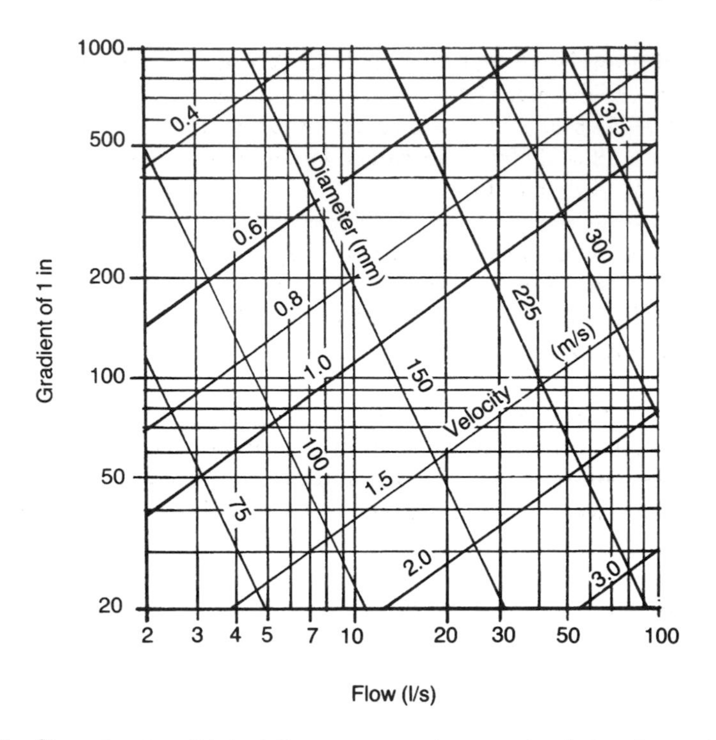

Figure 9.5 Sizes for two-thirds full spun precast concrete drains (reproduced from *IHVE Guide* [IHVE, 1970], now superseded, by permission of the Chartered Institution of Building Services Engineers [CIBSE, 1986]).

It is worth noting that, at this gradient, a 100 mm drain will carry only 5.3 l/s. A 100 mm drain would carry 7 l/s if its gradient was increased to 1 in 48, but this may be impractical. A 150 mm drain is the next available size up from the design point on the chart but it is grossly oversized for the present duty. At 1 in 80 it can carry a flow rate of 15 l/s, which corresponds to 2200 discharge units. This represents a possible future increase in discharge unit capacity of

$$\frac{2200 - 750}{750} \times 100\% = 193\%$$

Where drain use may be increased by additional site or area development, this is a useful advantage.

External loads on buried pipelines

Ground and traffic-induced stresses in buried pipelines are related to bedding type by

$$W_e \leqslant \frac{W_t F_m}{F_s}$$

where W_e is the total external vertical load (kN/m), W_t is the standard test crushing strength of the pipe (kN/m), F_m is the Marston bedding factor and F_s is the design safety factor. The crushing strength W_t for concrete pipes, *IHVE Guide* (1970), is 19.8 kN/m. The safety factor F_s is taken as 1.25. Marston bedding factors F_m are as follows.

1. $F_m = 3.4$ where the drain is supported along its length by a 120° reinforced concrete cradle and then well-compacted fill to the top of the pipe, and is covered with lightly compacted material (class A).
2. $F_m = 2.6$ when supported as in 1 but with plain concrete (class A) used for the pipe.
3. $F_m = 1.9$ when the drain is supported along its length by a 180° well-compacted granular bedding, filled to the top of the pipe and well compacted, and then covered with lightly compacted material (class B). This is most suitable for very wet unstable soils where the possibility of settlement exists, as found in the UK.

Table 9.1 shows some values for the total external vertical loads W_e imposed upon buried pipelines.

EXAMPLE 9.2

A 150 mm glazed vitrified clay drain pipe is to be used under gardens, paths and roads on a housing estate. The test crushing strength of the pipe is 19.6 kN/m. State whether this material will resist the probable imposed loads, and which bedding class is to be used.

For class B bedding $F_m = 1.9$. Using $F_s = 1.25$ and $W_t = 19.6$ kN/m, we obtain

$$W_e = 19.6 \frac{\text{kN}}{\text{m}} \times \frac{1.9}{1.25} = 29.8 \text{ kN/m}$$

This installation will withstand a total vertical external load W_e of 29.8 kN/m. The imposed loads will be 13.5 kN/m under gardens and 16 kN/m under the estate roads. The installation will be satisfactory.

Table 9.1 Total external vertical loads applied to buried pipelines

Trench details		W_e(kN/m) for diameters of			
Traffic	Depth of cover	100 mm	150 mm	225 mm	300 mm
Garden	0.9	8.9	13.5	19.5	26.0
Estate roads	1.2	11.0	16.0	24.0	32.0

Source: Reproduced from *IHVE Guide* (CIBSE, 1986 [IHVE, 1970]) by permission of the Chartered Institution of Building Services Engineers.

Materials for drainage pipework

Traditionally, glazed vitrified clay (GVC) pipes have been used because they represent an efficient use of UK national resources. The finished internal surface of GVC pipes offers less frictional resistance to flow than that of concrete pipes and is resistant to chemical attack and abrasion. Rigid joints consist of a socket and spigot cemented together. The brittle nature of such pipe runs has led to the introduction of flexible joints, which can withstand ground movement due to thermal and moisture variations and settlement of buildings. Plastic and rubber sealing ring joints allow up to 5° of bending and longitudinal expansion and contraction. Pipe sizes range from 75 mm to 750 mm in diameter.

Spun concrete drain pipes of diameter up to 1.83 m with oval cross-sections, which maintain flow velocity at periods of low discharge, are used. Plastic sleeves with rubber sealing rings give joints flexibility and a telescopic action.

Asbestos cement pressure pipes in lengths of up to 4 m have been used because of their lower weight. Flexible sleeve joints with rubber ring seals are used. Diameters from 100 mm to 600 mm are produced.

Pitch fibre pipes are formed by impregnating wood fibre with pitch. They are lightweight and can be used for some drainage applications. Lengths of 2.5 m are easily handled and can be hand sawn. Push taper joints are made using a hand-operated chamfering tool. Pipelines have flexibility and require well-selected backfill and careful protection during site work. Hot fluid or chemical discharges may lead to the early collapse of the pipe from ground pressure. Plastics are used for bends and other pipe fittings. Diameters are in the range 75–200 mm.

Cast iron drain runs are used for overground sections and where the ground movement might otherwise cause fracture. Pipework beneath buildings can either be cast iron encased in concrete or short lengths with flexible joints. Rigid socket and spigot joints are caulked with tarred yarn and then filled with hot lead or lead wool.

Plastics have increasingly replaced naturally occurring materials owing to their low weight and high degree of prefabrication. Complete systems from the sanitary appliance to the sewer, using one supplier and material, are common. Such materials are derived from crude oil and their higher cost needs to be compensated by reduced site time. Smooth bore drain systems can be assembled with minimum skill and they are highly resistant to corrosion. Thermal expansion is greater, and telescopic joints are used. Short-term discharges from some appliances (for example some types of washing machine) can be at temperatures of 80 °C or higher. Polypropylene and acrylonitrile butadiene styrene (ABS) pipes are suitable for the high-temperature applications.

Sewage-lifting pump

Where sanitary appliances discharge into drain pipework that is below the foul sewer invert, a collecting sump, pump and fluid level controller are used in the manner shown in Fig. 9.6. Either a large clearance centrifugal pump, driven by a 440 V three-phase electric motor, or a pneumatic ejector, is used.

The storage chamber is sized to accommodate several hours of normal discharge so that the pump only runs for short periods and total electrical power

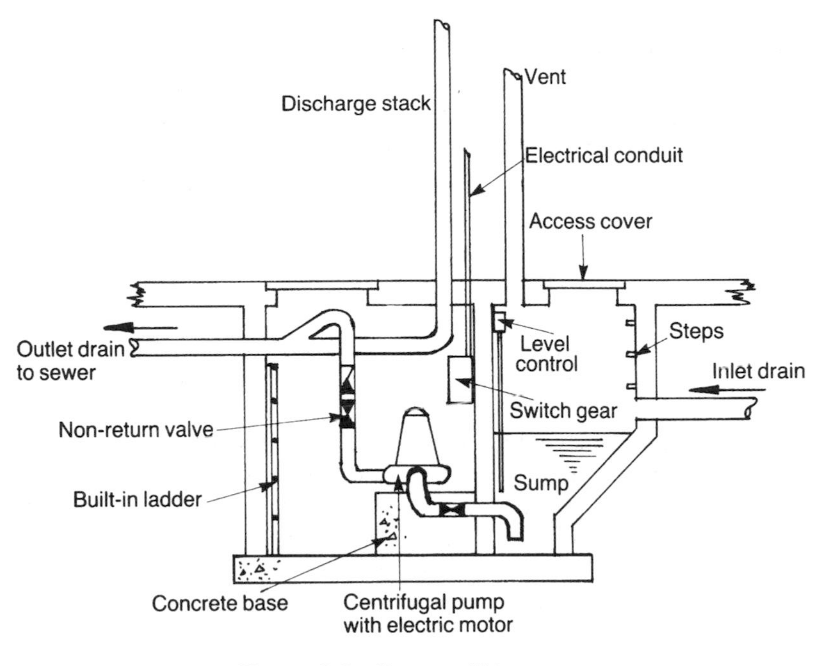

Figure 9.6 Sewage-lifting pump

consumption need not be high. Duplicate pump sets ensure a continuity of service during breakdowns and maintenance.

The pneumatic ejector collects the discharge in a steel tank containing a float. At the upper fluid level, the float operates a change-over valve, which admits air from a compressor and storage vessel. The incoming compressed air drives the sewage into the outlet drain at the higher level. Non-return valves are fitted to the inlet and outlet pipes to stop the possibility of reverse flow.

Both types of sewage-lifting equipment have open vent pipes to ensure that back pressures are not imposed upon the soil stack.

Testing

In addition to the inspection and smoke tests previously described, a water test can be applied to underground drainage pipelines. A test is carried out before, and sometimes after, backfilling. Drain runs are tested between manholes. The lower end is sealed with an expandable plug. A temporary upstand plastic or aluminium pipe is connected at the higher manhole. Water is admitted to produce a static head of 1.4–2.4 m and maintained for an hour or more. Some drainage materials will absorb water, and the initial water level is replenished from a measuring cylinder or jug. The maximum allowable water loss is 1 l/h for 10 m of 100 mm diameter pipe or *pro rata* for other diameters and lengths.

An air test can be conducted in a similar manner: a static pressure of 75 mm water gauge on a U-tube manometer should be maintained for a period of 5 min without further pumping.

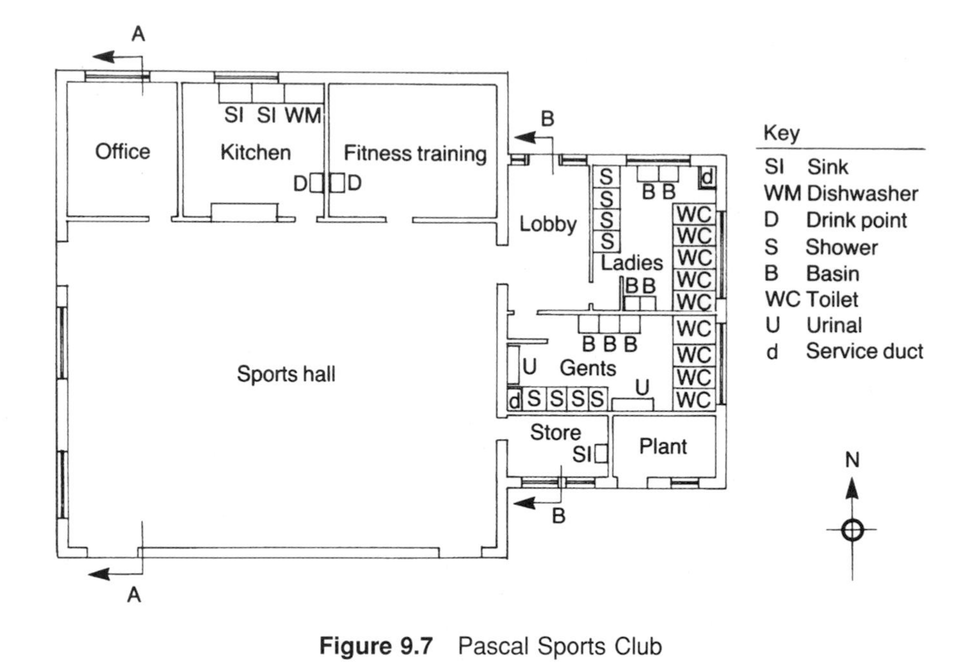

Figure 9.7 Pascal Sports Club

Questions

1. List the principal requirements for an underground drainage installation.

2. Explain, with the aid of sketches, the differences between the following types of drain and sewer system: separate, combined and partially separate.

3. A 100 mm discharge stack connects to a drain laid to a 1 in 80 gradient and is expected to carry 500 discharge units. Find a suitable diameter for the drain.

4. A building has four 100 mm discharge stacks, which connect into a common underground drain. The stacks have discharge unit values of 400, 500, 600 and 700. Find the diameters of each part of the collecting drain if it has a gradient of 1 in 100.

5. A 100 mm PVC drain runs for 30 m to connect between a discharge stack and the sewer. If its gradient is 1 in 80 and it commences its route with minimum ground cover under a garden, what will its invert be at the sewer connection if the ground is level?

6. There are 236 houses on a new development. Each has a group of sanitary appliances with a discharge unit value of 14. The common drain is laid at a gradient of 1 in 100. Find the diameter of the common drain and the maximum possible number of houses that it could serve.

7. Sketch and describe the types of bedding used for drains, giving an application for each.

8. A 100 mm clayware drain with a test crushing strength of 19.6 kN/m is to be laid under gardens and estate roads. Find whether class B bedding will prove to be satisfactory.

9. A 225 mm concrete sewer is to be laid under housing estate roads in class B bedding. Determine whether the imposed load can be withstood.

10. Under what circumstances may drains and sewers become damaged during their construction and service periods?

11. Describe, with the aid of sketches, the materials and jointing techniques used for below-ground drain systems.

12. Sketch and describe the operation of a sewage pumping installation. Draw the details of the construction of the below ground chambers.

13. State the performance criteria for tests on below-ground drain systems.

14. Design a below-ground drainage system for the Pascal Sports Club shown in Fig. 9.7. The foul sewer is at an invert of 2 m and 25 m to the right of the east wall of the club. Only one connection is allowed to be made to the 300 mm diameter sewer, and this is to its upper half.

    It will be necessary to design the above-ground waste pipework from all sanitary appliances in order to optimize the gulley positions, the 100 mm diameter pipe routes and the location of the one ventilation stack at the high point of the whole system. Minimize the use of underfloor pipework, all of which must be 100 mm in diameter and fully accessible.

    Modifications can be made to the building to construct above-ground service ducts to accommodate hot- and cold-water pipes as well as wastes and drains.

    A 100 mm diameter rainwater downpipe is located 500 mm from each external corner of the building on the north and south sides. These connect to 100 mm diameter below-ground drains, which run to the surface-water sewer alongside the foul sewer. Ensure that both drain systems are fully integrated and separated by a bedding of at least 100 mm thick shingle or broken stones of maximum size 5–10 mm. Access to the surface-water pipework is of the same standard as that to the foul pipework.

    The last access prior to the sewer for both drains should be a manhole. There is no manhole at the junction of the drain and sewer. The shower rooms will have trapped floor gulleys that connect to the foul drain.

    No model solution is provided as the design should be discussed with tutor and colleagues, and reference should be made to manufacturers' guides.

10 Condensation in buildings

Key terms and concepts

Learning objectives

Study of this chapter will enable the reader to:

1. identify the moisture content of humid air by its vapour pressure;
2. understand dew-point temperature;
3. identify the sources of moisture within a building;
4. understand the flow and storage characteristics of moisture flows found in habitable buildings;
5. explain the causes of condensation;
6. discuss the damage which can be caused by condensation;
7. calculate vapour diffusion resistance;
8. calculate vapour flow rate;
9. calculate air vapour pressure;
10. calculate air dew-point temperature;
11. understand atmospheric pressure terms;
12. use the e^x calculator function;
13. use the $\log_e$ calculator function;
14. be able to convert from e^x to $\log_e$ forms;
15. calculate and draw thermal temperature gradients through structures;
16. calculate and draw dew-point temperature gradients through structures;
17. identify condensation zones within structures;
18. discuss surface and interstitial condensation;
19. understand where to install thermal insulation and vapour barriers in relation to condensation risk and thermal and structural integrity requirements.

Introduction

Condensation risk is analysed during design of a building, when retrofit measures such as additional thermal insulation, double glazing or ventilation control are being considered or where damage from condensation has been discovered. Anti-condensation measures are linked to temperature control systems and ventilation provision in that they determine the size of plant and resulting operating costs.

The fundamentals of air and water vapour mixtures are introduced and then the moisture diffusion properties of building materials are analysed. A convenient form of equation to enable the air dew-point temperature to be found using a student's scientific calculator is derived and this saves the need to refer to charts or tables.

Thermal and dew-point temperature gradients are calculated, allowing moisture flow rate to be found. The rate of moisture deposition within the structure can then be assessed for its damage potential.

Sources of moisture

Air is a mixture of dry gases and water vapour. The water vapour exists in the form of finely divided particles of superheated steam at the air dry-bulb temperature. Total atmospheric pressure consists of the sum of the partial pressures of the two main constituents.

Typically, one standard atmosphere exerts a pressure of 1013.25 mb at sea level. When the air conditions are 25 °C d.b. and 20 °C w.b., the standard atmosphere is made up of 993.08 mb dry gas and 20.17 mb water vapour pressure. If this air is allowed to come into contact with a surface at a temperature of 17.6 °C, the air becomes saturated with moisture and can no longer support all the water in its vapour state. This temperature is known as the air dew-point t_{dp}, and is shown on a sketch of the CIBSE psychrometric chart in Fig. 5.3. Further data on the properties of humid air can be obtained from the *CIBSE Guide* (CIBSE, 1986). Figure 10.1 is a reduced psychrometric chart that may be used to find data. It is reproduced by permission of the Chartered Institution of Building Services Engineers.

The sources of water vapour in an occupied building are as follows:

1. people, upwards of 0.7 kg per 24 h;
2. cooking;
3. washing, bathrooms, drying clothes;
4. humidifiers and open water surfaces;
5. animals (dogs exhale more moisture than people produce overall);
6. combustion of paraffin (the complete combustion of 1 kg of C_9H_{20} produces 1.41 kg of water vapour).

Porous structural surfaces, furniture and fabrics within the building absorb moisture and then release it into the internal atmosphere when the temperature and humidity allow this. Some moisture travels through the structure and evaporates externally unless it is prevented from doing so by an impervious layer or vapour barrier. The majority of internally produced humidity is removed by ventilation.

The warm internal atmosphere is able to hold more moisture than the cooler external air; thus the partial pressure of the vapour, i.e. the vapour pressure p_s is higher inside than outside. This vapour pressure difference causes mass transfer

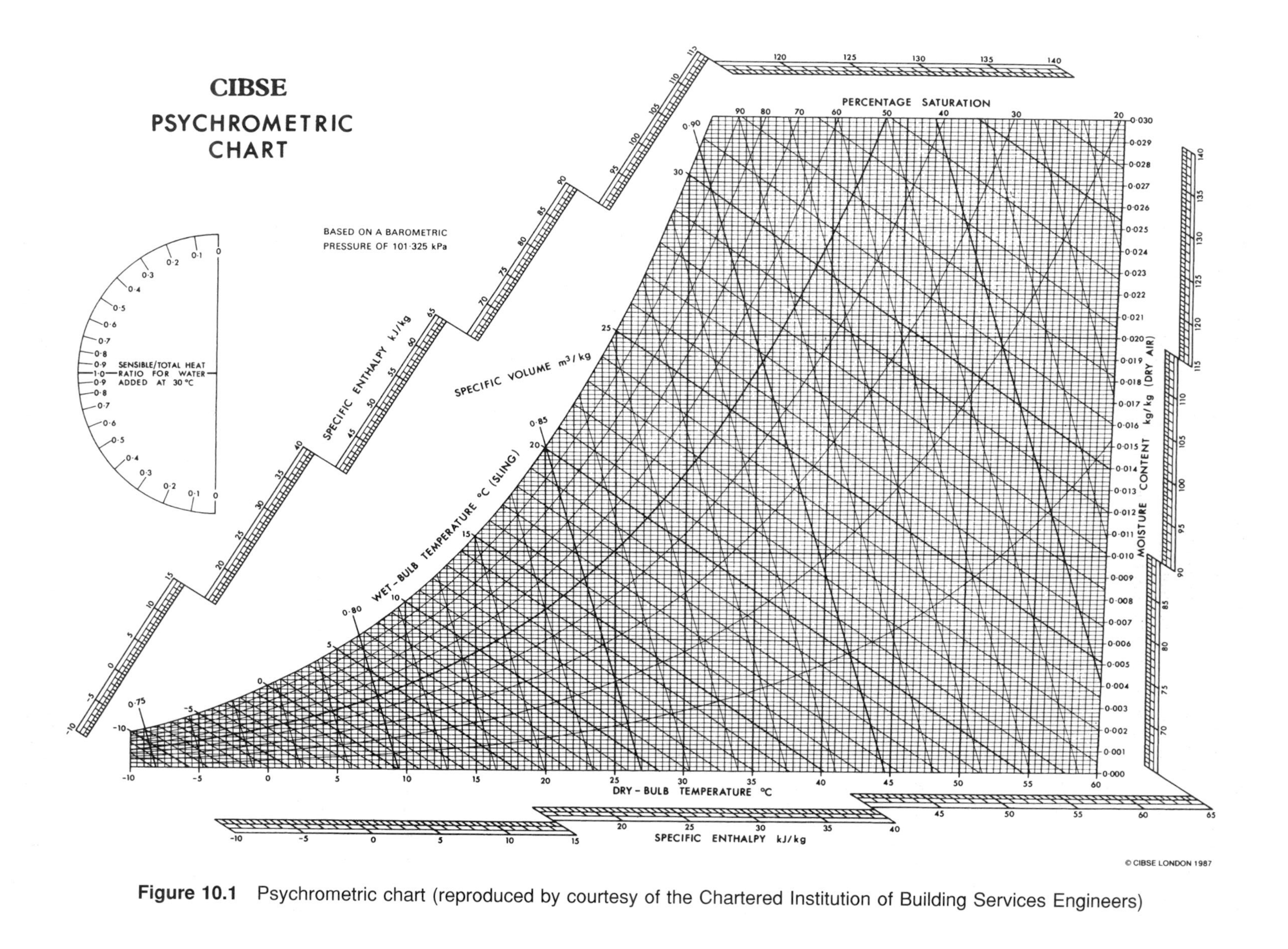

Figure 10.1 Psychrometric chart (reproduced by courtesy of the Chartered Institution of Building Services Engineers)

of moisture out of the building through the porous structure and by the ventilation air flow. When dense materials such as precast concrete or impervious barriers form a major part of the construction, moisture removal from the normal habitation is slow and condensation occurs on cold surfaces. Brickwork or masonry that is already saturated has the same effect.

Condensation and mould growth

Condensation and mould may readily form on window sills and in the corners of rooms, where the surface temperatures are lower compared with large areas. Gloss paint stops the absorption of moisture into an otherwise porous plaster or wooden component and water droplets form on the surface. The production of surface condensation in heated rooms is generally avoided in structures with a U value lower than 1.4 W/m^2 K.

Dampness in the timber of roofs, not caused by the ingress of rain, may be due to condensation on low-temperature surfaces. A well-insulated flat plaster ceiling may produce these low-temperature conditions in the roof construction. Natural ventilation through gaps between the tiles and roofing felt is normally sufficient to stop rot and damp patches on the plasterboard. Well-sealed roofs, with boarding under the felt, should have their ventilation increased by means of openings in the soffit of the eaves after extra thermal insulation. Humid air enters the roof space through gaps in the flat plaster ceiling around access hatches and pipes. Sealing these substantially reduces condensation risk. Roof insulation should stop at the wall head and not be pushed into the eaves, or ventilation will be restricted (Saunders, 1981).

Condensation forms within a structure or inside a solid material wherever the temperature falls below the dew-point of the moist air at that location. This is known as interstitial condensation. Vapour diffusion through building materials is calculated in a similar manner to the calculation of heat flow.

Vapour diffusion

The flow of water vapour through a porous building material or composite slab is analogous to the flow of heat through the structure. Convection currents transfer heat and moisture at the fluid–solid boundary. Conduction heat transfer is similar to vapour diffusion through a porous material, and its resistance to moisture flow varies with density, as does thermal resistance but in the opposite sense.

The mass flow rate of moisture through a composite structure consisting of a number of plane slabs and surfaces in series is

$$G = \frac{p_{s1} - p_{s2}}{R_v}$$

where G is the mass flow rate of vapour (kg/m^2 s), R_v is the vapour resistance of the structure (N s/kg) and p_{s1} and p_{s2} are the vapour pressures on surfaces 1 and

2 on each side of the slab (N/m^2 or Pa). The total vapour resistance R_v is given by

$$R_v = r_v \times l$$

where R_v is the total vapour resistance of a slab of homogeneous material (GN s/kg), r_v is the vapour resistivity of a material (GN s/kg m), and l is the thickness of material (m). Surface films and air cavities have only slight resistance to the flow of vapour and they are not normally included. Some typical values of vapour resistivity are given in Table 10.1. The complete resistance to the flow of vapour through some typical vapour barrier films is given in Table 10.2.

Further data are available in the reference source and in BRE Digest 369, February 1992, Building Research Establishment. Note that the values quoted in

Table 10.1 Vapour resistivity

Material	Vapour resistivity (GN s/kg m)
Brickwork	40
Dense concrete	200
Aerated concrete	30
Glass fibre wool	10
Foamed urea formaldehyde	30
Foamed polyurethane, open cell	30
Foamed polyurethane, closed cell	1000
Foamed polystyrene	500
Hardboard	520
Insulating fibreboard	20
Mineral fibre wool	6
Plaster	50
Plywood	520
Wood wool/cement slab	15
Wood	50

Source: Reproduced from *CIBSE Guide* (CIBSE, 1986) by permission of the Chartered Institute of Building Services Engineers.

Table 10.2 Vapour resistances of films

Material	Vapour resistance (GN s/kg)
Aluminium foil	over 4000
Double layer Kraft paper	0.35
Glass paint	8
Interior paint	3
Polythene film, 0.1 mm	200
Roofing felt	4 (and up to 100)

Source: Reproduced from *Guide* (CIBSE, 1986) by permission of the Chartered Institute of Building Services Engineers.

BRE Digest 369 are in MN s/g m (meganewton seconds per gram metre) for vapour resistivity and MN s/g for the vapour resistance of films. These units of measurement are the same as GN s/kg m (giganewton seconds per kilogram metre) and GN s/kg as both the numerator and denominator have been reduced by 1000 times.

Values of vapour pressure are available in CIBSE (1986) but can be calculated with sufficient accuracy from the following curve fit to the saturation conditions data:

$$p = 600.245 \exp(0.0684\ t_{dp})\ \text{Pa}$$

where t_{dp} is the air dew-point temperature from the psychrometric chart (°C) and $\exp(x) = e^x$ where $e = 2.718\,28$ is the exponential operator. Tabulated vapour pressures are in millibars (mb), and since 1 bar = 100 000 N/m^2 = 100 000 Pa, 1 mb = 100 N/m^2 = 100 Pa.

EXAMPLE 10.1

State the vapour pressure for air at 25 °C d.b., 20 °C w.b. in pascals.

The vapour pressure for air under these conditions is 20.17 mb. Thus

$$p_s = 20.17\ \text{mb} \times \frac{100\ \text{Pa}}{1\ \text{mb}} = 2017\ \text{Pa}$$

EXAMPLE 10.2

Calculate the vapour pressure for saturated air at 25 °C d.b., 20 °C w.b. from the dew-point.

The dew-point is $t_{dp} = 17.6$ °C. Therefore

$$p_s = 600.245 \times \exp(0.0684 \times 17.6)\ \text{Pa}$$

Scientific calculators have an e^x function, and in this calculation $x = 0.0684 \times 17.6 = 1.203\,84$. Consequently

$$p_s = 600.245 \times e^{1.203\,84}$$

Having left 1.203 84 in the displayed x register, execute the e^x function, producing 3.332 89, and multiply by 600.245 to obtain

$$p_s = 2000.6\ \text{Pa}$$

This is less than 1% different from the tabulated vapour pressure and is of sufficient accuracy bearing in mind the other figures involved in the problem.

The dew-point temperature can be found from the saturation vapour pressure by rearranging the equation

$$p_s = 600.245 \times \exp(0.0684\ t_{dp})\ \text{Pa}$$

to give

$$\frac{p_s}{600.245} = \exp(0.0684\ t_{dp})$$

This is a logarithmic equation of the form

$$y = e^x$$

where y is the number whose logarithm to base e is x. Logarithms to base e are called natural logarithms and are expressed as follows:

$$\log_e y = x \quad \text{or} \quad \ln y = x$$

EXAMPLE 10.3

Compute the natural logarithm of 2 and then raise e to the power of this logarithm.

Enter 2 into the calculator x display and press the ln key; the answer is 0.6931. Therefore

$$\ln 2 = 0.6931 \quad \text{or} \quad \log_e 2 = 0.6931$$

Now, as $x = 0.6931$ is displayed in the calculator, execute e^x. This results in

$$e^{0.6931} = 2$$

Thus it is seen that e^x is the antilogarithm of ln x, and the two expressions $y = e^x$ and ln $y = x$ are interchangeable to suit the problem. Thus for

$$\frac{p_s}{600.245} = \exp(0.0684\ t_{dp})$$

we can write

$$\ln\left(\frac{p_s}{600.245}\right) = 0.0684\ t_{dp}$$

and

$$t_{dp} = \frac{1}{0.0684} \times \ln\left(\frac{p_s}{600.245}\right)\ ^\circ C$$

EXAMPLE 10.4

Calculate the dew-point for saturated air with a vapour pressure of 2000.6 Pa

$$t_{dp} = \frac{1}{0.0684} \times \ln\left(\frac{2000.6}{600.245}\right)\ ^\circ C$$

$$= \frac{1}{600.245} \times \ln 3.33$$

$$= 17.6\ ^\circ C$$

An alternative general equation is stated in BRE Digest 369 for the calculation of vapour pressure:

$$p_s = 0.6105 \times \exp\left(\frac{17.269 \times t_{dp}}{237.3 + t_{dp}}\right)\ \text{kPa}$$

This is less convenient to use when a dew-point temperature is to be calculated from a known vapour pressure. The curve-fit equation that has been demonstrated here is of sufficient accuracy for most manual estimations of condensation risk.

Typical values of vapour pressure are given in Table 10.3. These will accommodate some applications without reference to tables or equations.

EXAMPLE 10.5

Calculate the total vapour resistance of a cavity wall constructed from 13 mm plaster, 100 mm aerated concrete, 40 mm mineral wool, 10 mm air space and 105 mm brickwork.

For each layer

$$R_v = r_v \times l$$

For the whole structure

$$\Sigma(R_v) = \Sigma(r_v \times l)$$

Table 10.3 Vapour pressures

Air condition t_a (°C d.b.)	% saturation	Dew-point t_{dp} (°C)	Vapour pressure (Pa)
−5	100	−5	402
−3	80	−5.6	381
0	80	−2.7	489
1	80	−1.8	526
2	80	−0.9	565
5	80	1.9	699
10	80	6.7	984
14	60	6.5	965
15	60	7.4	1030
20	50	9.4	1182
22	50	11.3	1339

Source: Reproduced from *CIBSE Guide* (CIBSE, 1986) by permission of the Chartered Institute of Building Services Engineers.

From Table 10.1 the vapour resistivities are plaster 50, aerated concrete 30, mineral wool 6, brickwork 40. The surface films and air space have no resistance to the flow of vapour. Material thicknesses are used in metres.

$$\Sigma(R_v) = 50 \times 0.013 + 30 \times 0.1 + 6 \times 0.04 + 40 \times 0.105$$

$$= 8.09 \text{ GN s/kg}$$

EXAMPLE 10.6

The cavity wall in Example 10.5 is to be used for a dwelling exposed to an external environment of −1 °C d.b., 80% saturation, where the heating system is designed to maintain the internal air at 22 °C and 50% saturation. If the wall has a surface area of 110 m^2, find the moisture mass flow rate taking place through the wall.

From the CIBSE psychrometric chart, the internal and external air dew-point temperatures are found to be

$$\text{internal } t_{dp} = 11.5 \text{ °C}$$

$$\text{external } t_{dp} = -3.5 \text{ °C}$$

The internal air vapour pressure is

$$p_{s1} = 600.245 \times \exp(0.0684 \times 11.5) \text{ Pa}$$

$$= 1318.09 \text{ Pa}$$

and the external air pressure is

$$p_{s2} = 600.245 \times \exp[0.0684 \times (-3.5)] \text{ Pa}$$

$$= 472.45 \text{ Pa}$$

Using $R_v = 8.09$ GN s/kg from Example 10.5, we obtain the moisture mass flow rate through the wall as

$$G = \frac{p_{s1} - p_{s2}}{R_v} \text{ kg/m}^2 \text{ s}$$

$$= (1318.09 - 472.45) \frac{\text{N}}{\text{m}^2} \times \frac{\text{kg}}{8.09 \text{ GN s}} \times \frac{1 \text{ GN}}{10^9 \text{ N}}$$

$$= 1.045 \times 10^{-7} \text{ kg/m}^2 \text{ s}$$

For the whole wall

$$G = \frac{1.045}{10^7} \frac{\text{kg}}{\text{m}^2 \text{ s}} \times 110 \text{ m}^2$$

$$= 1.15 \times 10^{-5} \text{ kg/s}$$

Temperature gradient

Heat flows through a structure from an area of high temperature to one of lower temperature. Homogeneous materials have a linear temperature gradient through their thickness, as shown in Fig. 10.2. Temperature drops 1–2 and 3–4 are caused by the internal and external surface film resistances. To determine the surface and intermediate temperatures, the overall rate of heat, flow through the whole structure is equated with the individual heat flows in each slab:

$$Q_f \text{ W} = U \frac{\text{W}}{\text{m}^2 \text{ K}} \times A \text{ m}^2 \times (t_1 - t_5) \text{ K}$$

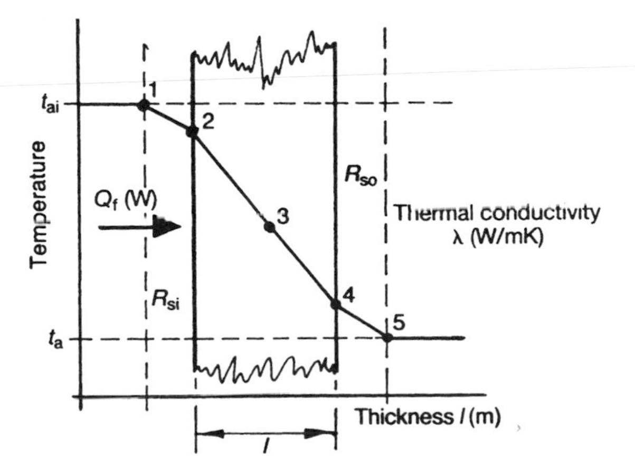

Figure 10.2 Temperature gradient through a solid construction

This same rate of heat flow Q_f also passes through the internal surface film, the concrete and the external surface film. Therefore

$$Q_f = \frac{1}{R_{si}} \frac{\text{W}}{\text{m}^2\ \text{K}} \times A\ \text{m}^2 \times (t_1 - t_2)\ \text{K}$$

and

$$Q_f = \frac{\lambda}{l} \frac{\text{W}}{\text{m}^2\ \text{K}} \times A\ \text{m}^2 \times (t_2 - t_4)\ \text{K}$$

and

$$Q_f = \frac{1}{R_{so}} \frac{\text{W}}{\text{m}^2\ \text{K}} \times A\ \text{m}^2 \times (t_4 - t_5)\ \text{K}$$

The heat flow rate Q_f can easily be evaluated from U, t_{ai} and t_{ao}. The only unknowns in the other equations are the second temperatures t_2 and t_4. An intermediate temperature t_3 can be calculated at half the concrete thickness by using $l/2$:

$$Q_f = \frac{2\lambda}{l} \frac{\text{W}}{\text{m}^2\ \text{K}} \times A\ \text{m}^2 \times (t_2 - t_3)\ \text{K}$$

If the wall area is taken as 1 m^2, then

$$Q_f = \frac{1}{R_{si}} (t_1 - t_2)$$

and

$$t_2 = t_1 - Q_f\, R_{si}$$

Similarly

$$Q_f = \frac{\lambda}{l} (t_2 - t_4)$$

and

$$t_4 = t_2 - \frac{l}{\lambda} Q_f$$

Also

$$Q_f = \frac{1}{R_{so}} (t_4 - t_5)$$

and

$$t_5 = t_4 - Q_f\, R_{so}$$

Calculating the outdoor air temperature t_5 is a check on the accuracy of the calculations and method. It should agree with the original value used in finding Q_f to within ±1%.

To find t_3 use

$$Q_f = \frac{2\lambda}{l}(t_2 - t_3)$$

Hence

$$t_3 = t_2 - \left(\frac{l}{2\lambda}Q_f\right)$$

> **EXAMPLE 10.7**
>
> Calculate the temperature gradient through a medium-weight concrete block wall 100 mm thick. The internal and external air temperatures are 20 °C d.b. and −1 °C d.b.

From Table 3.1 thermal conductivity $\lambda = 0.51$ W/m K, from Table 3.2 $R_{si} = 0.12\ m^2\ K/W$ and from Table 3.3 $R_{so} = 0.06\ m^2\ K/W$. Then the thermal transmittance is given by

$$U = \frac{1}{R_{si} + l/\lambda + R_{so}}$$

$$= \frac{1}{0.12 + 0.1/0.51 + 0.06}\ W/m^2\ K$$

$$= 2.66\ W/m^2\ K$$

For a wall area of 1 m^2

$$Q_f = 2.66 \times [22 - (-1)]\ W$$

$$= 61.16\ W$$

Using the numbered locations in Fig. 10.2

$$t_1 = 22\ °C \qquad t_5 = -1\ °C$$

$$t_2 = 22 - 61.16 \times 0.12 = 14.66\ °C$$

$$t_4 = 14.66 - \frac{0.1}{0.51} \times 61.16 = 2.67\ °C$$

and

$$t_5 = 2.67 - 61.16 \times 0.06 = -1\ °C$$

which agrees with the input data. Also,

$$t_3 = 14.66 - \left(\frac{0.1}{2 \times 0.51} \times 61.16 \right) = 8.66\ °\text{C}$$

which should be the temperature midway between t_2 and t_4 i.e. $(14.66 + 2.67)/2 = 8.67$ °C, which it is.

EXAMPLE 10.8

Calculate the temperature gradient through a cavity wall consisting of 13 mm lightweight plaster, 100 mm lightweight concrete block, 40 mm mineral fibre slab, 10 mm air space and 105 mm brickwork. Internal and external air temperatures are 22 °C and 0 °C.

From Table 3.1, the thermal conductivities are as follows: plaster, $\lambda_1 = 0.16$ W/m K; concrete, $\lambda_2 = 0.19$ W/m K; mineral fibre, $\lambda_3 = 0.035$ W/m K; brickwork, $\lambda_4 = 0.84$ W/m K. From Tables 3.2, 3.3 and 3.4, $R_{si} = 0.12$ m^2 K/W, $R_{so} = 0.06$ m^2 K/W and $R_a = 0.18$ m^2 K/W. Then

$$U = \left(R_{si} + \frac{l_1}{\lambda_1} + \frac{l_2}{\lambda_2} + \frac{l_3}{\lambda_3} + R_a + \frac{l_4}{\lambda_4} + R_{so} \right)^{-1}$$

$$= \left(0.12 + \frac{0.013}{0.16} + \frac{0.10}{0.19} + \frac{0.04}{0.035} + 0.18 + \frac{0.105}{0.84} + 0.06 \right)^{-1} \ \text{W/m}^2\,\text{K}$$

$$= 0.45\ \text{W/m}^2\,\text{K}$$

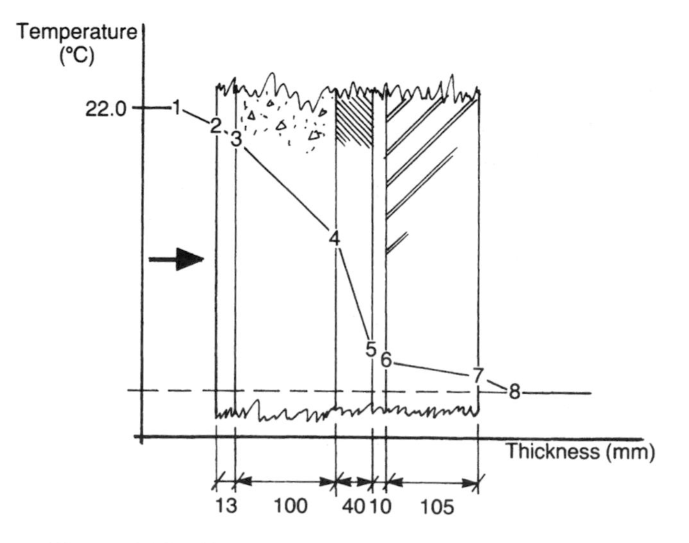

Figure 10.3 Temperature notation for Example 10.8

For a wall area of 1 m^2

$$Q_f = 0.45 \times (22 - 0) = 9.9 \text{ W}$$

Using the notation from Fig. 10.3, temperatures are calculated as follows:

$$t_2 = 22 - (0.12 \times 9.9) = 20.81\ ^\circ\text{C}$$

$$t_3 = 20.81 - \left(\frac{0.013}{0.16} \times 9.9\right) = 20.01\ ^\circ\text{C}$$

$$t_4 = 20.01 - \left(\frac{0.01}{0.19} \times 9.9\right) = 14.8\ ^\circ\text{C}$$

$$t_5 = 14.8 - \left(\frac{0.04}{0.035} \times 9.9\right) = 3.49\ ^\circ\text{C}$$

$$t_6 = 3.49 - (0.18 \times 9.9) = 1.71\ ^\circ\text{C}$$

$$t_7 = 1.71 - \left(\frac{0.105}{0.84} \times 9.9\right) = 0.47\ ^\circ\text{C}$$

$$t_8 = 0.47 - (0.06 \times 9.9) = 0.12\ ^\circ\text{C}$$

A slight error of 0.55% has occurred as a result of rounding all the results to two decimal places. Notice the larger temperature drops across the two main insulating materials.

Dew-point temperature gradient

Moist air passing through the structure from the high internal air vapour pressure to the lower external air vapour pressure will form a gradient of vapour pressures. The vapour pressure at any location is calculated from the mass flow rate of vapour G and vapour resistances in the same manner as for the thermal gradient.

The moist air dew-point temperature is calculated for each of these vapour pressures and another temperature gradient is drawn on the structural cross-section. If the thermally produced structure temperature equals or falls below the local air dew-point, then condensation will commence at that location. This information is used to decide whether a wall or roof will remain dry and whether a vapour barrier should be installed. The vapour barrier is fitted on the warm side of any zone of interstitial condensation.

Once the internal and external air vapour pressures, total vapour resistance and mass flow rate of vapour are known, the equation

$$G = \frac{p_{si} - p_{s2}}{R_v}$$

can be written as

$$p_{s2} = p_{s1} - R_v G$$

Note that this is of the same form as

$$t_2 = t_1 - RQ_f$$

Surface air films and cavities offer negligible resistance to the flow of moisture.

EXAMPLE 10.9

Calculate the dew-point gradient for the cavity wall in Example 10.8. Determine whether surface or interstitial condensation will take place. Internal and external percentage saturations are 50 and 100 respectively.

Referring to Fig. 10.2,

$$p_{s1} = p_{s2}$$

$$p_{s5} = p_{s6}$$

$$p_{s7} = p_{s8}$$

Thus

$$p_{s3} = p_{s1} - (\text{vapour resistance of plaster} \times G)$$

where p_{s3} and p_{s1} are the vapour pressures on each side of the plaster and

$$\text{vapour resistance of plaster} = r_v \times l$$

Similar equations are written for the other materials, with appropriate resistivities r_v and thicknesses l. The resistivities are plaster 50, concrete 30, mineral wool 6 and brickwork 40. From the CIBSE psychrometric chart, for internal air at 22 °C d.b., 50% saturation, $t_{dp} = 11.5$ °C, and for external air at 0 °C d.b., 100% saturation, $t_{dp} = 0$ °C. Then

$$p_{s1} = 600.245 \times \exp(0.0684 \times 11.5) \text{ Pa}$$
$$= 1318.09 \text{ Pa}$$
$$p_{s8} = 600.245 \times \exp(0.0684 \times 0) \text{ Pa}$$
$$= 600.245 \text{ Pa (because } e^0 = 1)$$

The vapour resistance R_v for this wall was calculated in Example 10.5:

$$R_v = 8.09 \text{ GN s/kg}$$

The mass flow of vapour per m^2 of wall area is

$$G = (1318.09 - 600.245)\ \frac{\text{N}}{\text{m}^2} \times \frac{\text{kg}}{8.09 \text{ GNs}} \times \frac{1 \text{ GN}}{10^9 \text{ N}}$$
$$= 88.7 \times 10^{-9} \text{ kg/m}^2 \text{ s}$$

The vapour pressure and corresponding dew-point temperature can now be calculated for each numbered point through the wall:

$$P_{s3} = 1318.09\,\frac{\text{N}}{\text{m}^2} - 50 \times 0.013\,\frac{\text{GN s}}{\text{kg}} \times \frac{10^9\ \text{N}}{1\ \text{GN}} \times \frac{88.7}{10^9}\,\frac{\text{kg}}{\text{m}^2\ \text{s}}$$

$$= (1318.09 - 57.7)\ \text{N/m}^2$$

$$= 1260.4\ \text{Pa}$$

$$t_{dp3} = \frac{1}{0.0684} \times \ln\left(\frac{p_{s3}}{600.245}\right)\ ^\circ\text{C}$$

$$= \frac{1}{0.0684} \times \ln\left(\frac{1260.4}{600.245}\right)\ ^\circ\text{C}$$

$$= 10.85\ ^\circ\text{C}$$

$$p_{s4} = 1260.4 - 30 \times 0.1 \times 88.7\ \text{Pa}$$

$$= 994.3\ \text{Pa}$$

$$t_{dp4} = \frac{1}{0.0684} \times \ln\left(\frac{994.3}{600.245}\right)\ ^\circ\text{C}$$

$$= 7.38\ ^\circ\text{C}$$

$$p_{s5} = 994.3 - 6 \times 0.04 \times 88.7\ \text{Pa}$$

$$= 973\ \text{Pa}$$

$$t_{dp5} = \frac{1}{0.0684} \times \ln\left(\frac{973}{600.245}\right)\ ^\circ\text{C}$$

$$= 7.06\ ^\circ\text{C}$$

$$p_{s7} = 973 - 40 \times 0.105 \times 88.7\ \text{Pa}$$

$$= 600.5\ \text{Pa}$$

$$t_{dp7} = \frac{1}{0.0684} \times \ln\left(\frac{600.5}{600.245}\right)\ ^\circ\text{C}$$

$= 0.006\ ^\circ\text{C}$ shows small calculation inaccuracy

$= 0\ ^\circ\text{C}$ the significant value

The dew-point temperature gradient ends with the input data and is superimposed upon a scale drawing of the thermally induced gradient shown in Fig. 10.4.

Owing to the high thermal resistance and very low vapour resistance of the mineral fibre slabs fixed in the cavity, under the design conditions as stated the material temperature drops below the moist air dew-point temperature midway through its thickness. Interstitial condensation will occur within the mineral fibre, air space and external brickwork. Ventilation of the remaining wall cavity allows evaporative removal of the droplets. Variation of internal and external air conditions will limit the duration of such temperature gradients. Periods of condensation will be very intermittent. Solar heat gains to the external brickwork will raise the temperature of the structure and help to reduce condensation periods. The walls most at risk are those always shaded from direct sunlight and having reduced wind exposure due to the proximity of nearby buildings.

When condensation takes place, a change of phase occurs as the water vapour turns into liquid. The calculations of water vapour transfer end at this discontinuity. The reason for undertaking the calculations was to establish if and where condensation is formed. The whole thickness of the layer where liquid forms is likely to be dampened owing to capillary attraction within porous solid material. The quantity of condensation that may be formed during a 60-day winter period can be assessed from the expected average air conditions. This aids the prediction of the physical damage that may be caused.

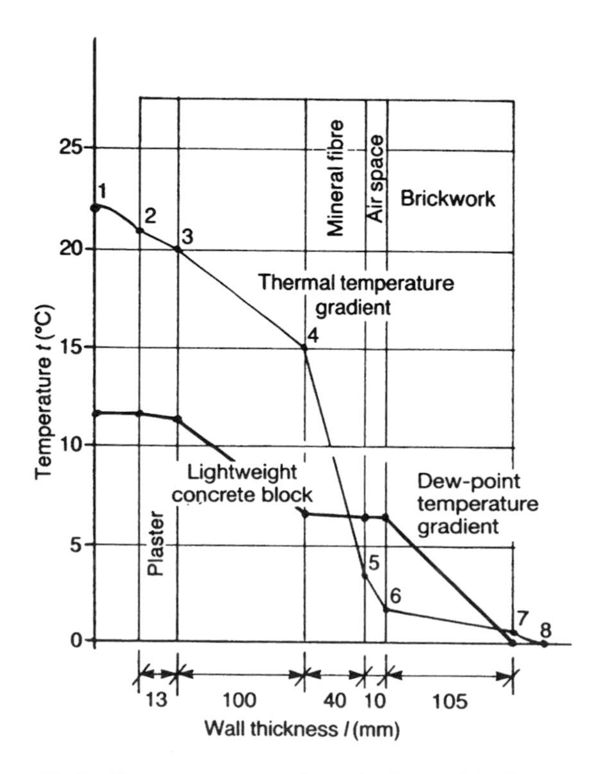

Figure 10.4 Temperature gradients in the wall in Example 10.9

Installation note

Added thermal insulation can cause conditions where condensation will take place owing to the lowered structural temperatures. The installation of a vapour barrier raises the overall vapour resistance and may be able to keep the dew-point gradient below the thermally produced temperatures.

1. Materials with a low thermal resistance but a high vapour resistance, such as aluminium foil, sheet plastic, roofing felt and gloss paint, are placed on the warm side of the structure.
2. Materials of high thermal and vapour resistance can be placed anywhere.
3. Materials of moderate vapour resistance but high thermal resistance should be placed on the cold side of the structure. Such materials will require the addition of a weather-resistant coating when applied to the outside of a building.

Thermal or acoustic insulation applied to industrial roofs can increase the possibility of the occurrence of condensation. Alternative schemes are shown in Figs 10.5 and 10.6 (Saunders, 1981).

In the cold-deck design, the roofing sheets remain at little more than the external air temperature. Ventilation through gaps at the eaves, junctions where the sheets overlap and holes around steel supports provides passageways for the ingress of moist air. Condensation on the underside of the cold deck will cause water to run down the steelwork and wet the ceiling. It is unlikely that a tight seal can be made between the vapour barrier and the steel supports to avoid this happening. Before

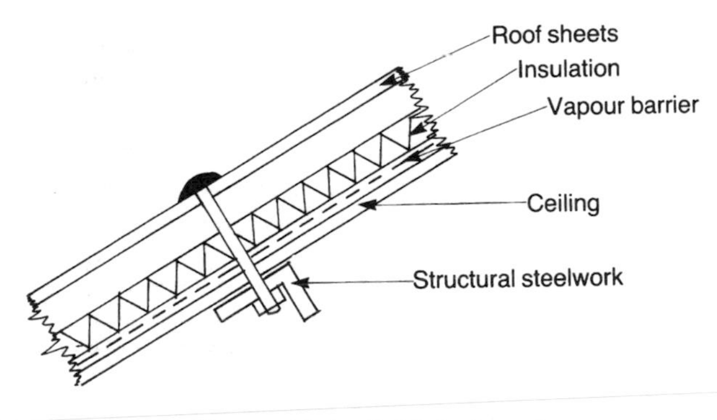

Figure 10.5 Cold-deck roof

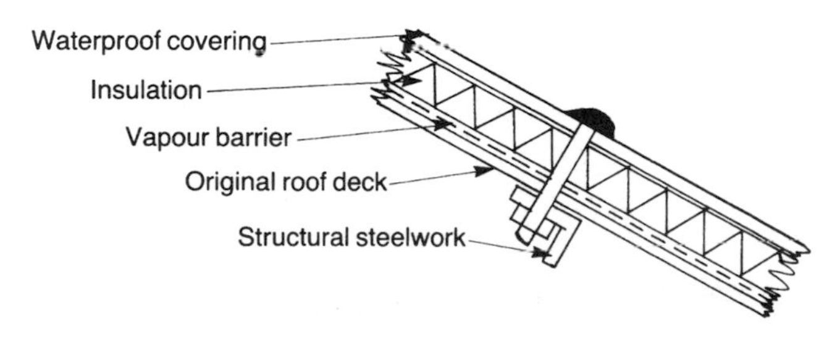

Figure 10.6 Warm-deck roof

insulation of an existing roof, its underside is maintained at above the local dew-point except during severe weather or when the heating plant is off.

The warm-deck design improves the weather resistance of the roof and raises the original underside surface temperature even further. Interstitial condensation is unlikely because of the vapour barriers.

EXAMPLE 10.10

Calculate the external air temperature that will cause condensation to form on the underside of a factory roof constructed from 10 mm corrugated asbestos cement sheet. Internal air conditions near the roof are 23 °C d.b., 30% saturation. The thermal conductivity of asbestos cement sheet is 0.4 W/m K. The roof has a severe exposure.

From the CIBSE psychrometric chart, the internal air dew-point is 5 °C. From Tables 3.2 and 3.3, $R_{si} = 0.1\ m^2\ K/W$ and $R_{so} = 0.02\ m^2\ K/W$. Let the unknown external air temperature be t_0. Then for a roof area of 1 m^2

$$Q_f = U\ (23 - t_0)\ W/m^2$$

For the external surface film

$$Q_f = \frac{1}{R_{si}}\ (23 - 5)\ W/m^2$$

$$= \frac{1}{0.1} \times 18\ W/m^2$$

$$= 180\ W/m^2$$

For the roof

$$U = \frac{1}{R_{si} + l/\lambda + R_{so}}$$

$$= \frac{1}{0.1 + 0.01/0.4 + 0.02}$$

$$= 6.9\ W/m^2\ K$$

Now

rate of heat flow through roof sheets = rate of heat flow through internal surface film

$$6.9\ \frac{W}{m^2\ K} \times (23 - t_0)\ K = 180\ \frac{W}{m^2}$$

$$26.0 = 23 - t_0$$

$$t_0 = -3.09\ °C$$

Questions

1. Describe how the following forms of condensation occur: temporary, permanent and interstitial.

2. List the sources of moisture in buildings.

3. What is the purpose of installing a vapour barrier and what effect does it have on the dew-point temperatures within a structure?

4. Discuss the use of thermal insulation in reducing the likelihood of condensation in walls and roofs.

5. State examples of thermal insulation increasing the risk that condensation occurs.

6. List the actions that could be taken to reduce the water vapour input to a dwelling.

7. Discuss the use of heating and ventilation in combating condensation problems.

8. Why might prefabricated concrete buildings suffer more from condensation than other constructions?

9. What sources of moisture would you look for when consulted about mould growth on a building?

10. Describe the constituent parts of the atmospheric pressure.

11. What drives water vapour from one area to another?

12. Describe the way in which moisture is alternatively stored and released by porous building materials.

13. What is the flow of vapour through a composite structure analogous to?

14. State the conditions under which water vapour will condense on or within a construction.

15. Calculate the temperature gradients through the following structures. Internal and external air temperatures are to be taken as 21 °C d.b. and −1 °C d.b. Assume that $t_a = t_e$. The answers should be expressed as the surface or interface temperatures in descending order from the warm side. Outside surfaces are taken as having a high emissivity and normal exposure. All air spaces are ventilated. The thermal conductivity of glass is 1.05 W/m K.

    (a) 6 mm single-glazed window.
    (b) 6 mm double-glazed window.
    (c) Cavity wall of 15 mm dense plaster, 100 mm lightweight concrete block, air space and 105 mm brick.
    (d) An industrial roof of 10 mm asbestos cement corrugated sheet which has been given an external coating of 50 mm phenolic foam. The thermal conductivity of asbestos sheet is 0.4 W/m K.

16. A shop window consists of 6 mm plate glass in an aluminium frame. The display area air temperature is expected to be 15 °C d.b. and to have a dew-point of 7 °C. Find the external air temperature that will start to produce condensation on the inside of the window. The window has normal exposure.

17. If a double-glazed window is to be fitted in the shop in Question 16, what could the external air temperature drop to before condensation starts?

18. A hospital ward is to be maintained at 24 °C d.b. and 80% saturation. The air dew-point is 20.5 °C. Thermal insulation is to be added to the inside of the existing wall to avoid surface condensation when the external air temperature falls to −5 °C d.b. The U value of the original wall is 1.9 W/m^2 K. Calculate the thickness of insulation material required if its thermal conductivity is 0.06 W/m K.

19. Calculate the thermal transmittance, temperature and dew-point gradients through a flat roof consisting of 25 mm stone chippings, 10 mm roofing felt, 150 mm aerated concrete, 75 mm wood wool/cement slabs, a ventilated 50 mm air space and 12 mm plasterboard. The roof has a sheltered exposure. Internal air conditions are 22 °C d.b., 50% saturation. External conditions are 2 °C d.b., 80% saturation. The stone chippings have a high emissivity when weathered and offer no resistance to the flow of water vapour. Plot a graph of the two temperature gradients and find if condensation is likely to occur.

20. Calculate the thermal transmittance, temperature and dew-point gradients through a wall consisting of 15 mm dense plaster, 100 mm medium-weight concrete blockwork, 40 mm glass fibre quilt, a ventilated 10 mm air space and 105 mm brickwork. The wall has a severe exposure. The average internal air conditions are 14 °C d.b., 60% saturation when the external conditions are 1 °C d.b., 80% saturation. Plot a graph of the two temperature gradients and find if condensation is likely to occur.

11 Lighting

Learning objectives

Study of this chapter will enable the reader to:

1. explain the use of day and artificial lighting;
2. use lux and lumen per square metre;
3. state normal lighting levels;
4. explain general and task illumination;
5. understand permanent supplementary artificial lighting of interiors (PSALI);
6. understand artificial lighting terminology;
7. assess the importance of the maintenance of lighting installations;
8. calculate the room index;
9. find utilization factors;
10. discuss the problem of glare;
11. calculate the number of lamps needed to achieve a design illumination level;
12. calculate lamp spacing for overall design;
13. calculate the electrical loading produced by the lighting system;
14. consider how luminaires should be oriented in relation to room layout and visual tasks to be performed;
15. understand the use of air-handling luminaires;
16. understand lamp colour-rendering and colour temperature;
17. know the range of available lamp types;
18. apply appropriate lamp types to designs;
19. understand the working principles of lamp types and their starting procedures.

Key terms and concepts

Introduction

Artificial illumination for both functional and decorative purposes is a major consumer of primary energy, and developed civilizations have become used to very high illumination standards with consequently high electricity consumption. The use of daylight is encouraged in order to reduce fuel consumption for lighting but this occurs at the expense of heating and cooling energy consumption at the building outer envelope, which is in contact with the external environment. A compromise solution is inevitable, and the building services engineer is at the centre of the calculations needed to minimize total energy consumption for all usages.

The factors involved in determining illumination requirements are discussed in relation to lighting levels for various tasks and the possible use of daylight. Lighting terms are introduced as are glare considerations. The lumen design method is demonstrated for office accommodation.

Lamp colour rendering is discussed, and the use of luminaires with air-conditioning systems. Lamp types, their uses and control arrangements are explained.

Natural and artificial illumination

Natural illumination by penetration of direct solar and diffuse sky visible radiation requires correctly designed passive architecture. Large glazed areas may provide sufficient daylighting at some distance into the building but can also cause glare, overheating and high heating and cooling energy costs.

The other extreme of vertical narrow slot windows limits energy flows while causing very unequal lighting levels near the room's perimeter. Reflected illumination from other buildings, particularly from those having reflective glazing or metallic architectural features, may cause annoyance. A careful consideration of all the, largely conflicting, variable elements is necessary if a comfortable internal environment is to be produced.

Artificial lighting is provided to supplement daylight on a temporary or permanent basis. Local control of lights by manual and/or automatic switches aids economy in electricity consumption. The colours rendered by objects on the working plane should match the colours under daylight. The working plane may be a desk, drawing board or display area.

Illumination intensity – illuminance, measured in lux – on the working plane is determined by the size of detail to be discerned, the contrast of the detail with its background, the accuracy and speed with which the task must be performed, the age of the worker, the type of space within which the task is to be performed and the length of time continuously spent on the task. The working plane is the surface being illuminated. Other areas are lit by overspill from it and by reflections from other room surfaces. Table 11.1 gives some typical values for illuminance commonly encountered and used for design.

Higher levels of illuminance may be provided for particularly fine detail tasks at the area of use by local, or task, illumination: for example, up to 3000 lx for inspection of small electronic components and 50 000 lx on a hospital operating table. Bright sunlight provides up to 100 000 lx. Local spot lighting for display purposes and exterior illumination are used to accentuate particular features of the working plane.

Permanent supplementary artificial lighting of interiors (PSALI) has become common in modern office accommodation, shops and public buildings. Figure 11.1 shows the constituents of the overall design illumination.

Table 11.1 Typical values of illuminance

Application	Illuminance (lx)
Emergency lighting	0.2
Suburban street lighting	5
Dwelling	50–150
Corridors	100
Rough tasks with large-detail storerooms	200
General offices, retail shops	400
Drawing office	600
Prolonged task with small detail	900

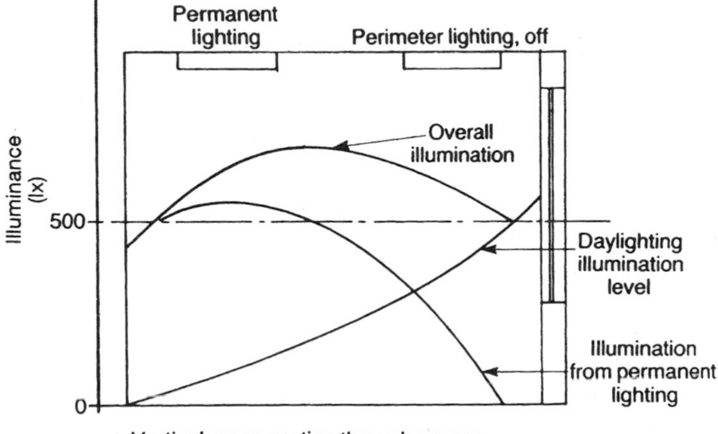

Figure 11.1 Permanent supplementary artificial lighting

The heat generated by permanent lighting can be extracted from the light fitting (luminaire) by passing the ventilation extract air through it, thus raising the air temperature to 30–35 °C, and then supplying this heated air to perimeter rooms in winter. Further air heating with finned-tube banks and automatic temperature control would be part of a normal ventilation or air-conditioning system. This can be termed a heat reclaim system incorporating regenerative heat transfer between the outgoing warm exhaust air and incoming cold fresh air.

The penetration of daylight into a building can be enhanced with north-facing roof-lights, skylights having motorized louvres which are adjusted to suit the sun's position and weather conditions, or mirrored reflectors which direct light rays horizontally into the building. An example is shown in Fig. 11.2.

Localized task illumination can be provided in addition to a background lighting scheme but may not necessarily produce a reduction in total power consumption. Tests (Ellis, 1981; Mckenna *et al.*, 1981) of combinations of overall lighting plus task illumination revealed a preference for two 40 W white fluorescent lamps 1200 mm long plus indirect background lighting.

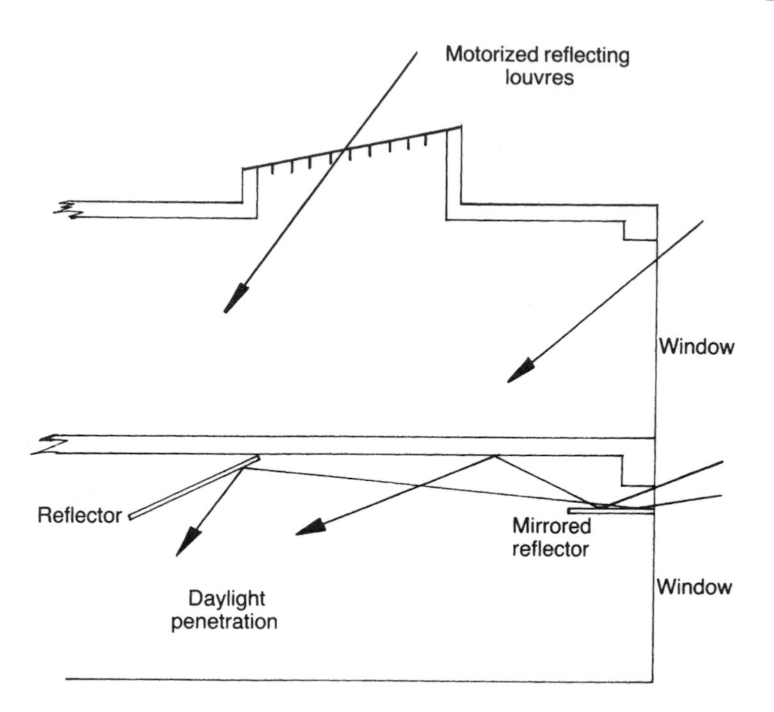

Figure 11.2 Use of daylight reflectors

Desk-mounted lamps produce a range of up to 9 : 1 in illuminance values across the working surface and form strong contrasts between surrounding and working surfaces which may result in discomfort glare. Direct dazzle from unshaded lamps, reflected glare, shadows around objects and hands, and heat radiation can cause discomfort.

Lighting is, obviously, a visual subject and it may be treated as such for design purposes. It is not necessarily helpful to study lighting only as applied mathematics. Designs are an artistic and engineering combination of architecture, interior design, decoration, illumination functionality, economical use of electrical energy, maintainability, safety, environmental health, controllability, prestige, and the overall and specific requirements of the user. Architects and engineers coordinate their skills to create an acceptable visual and technical solution.

To consider the design of lighting from a visual standpoint, acquire two or more cardboard boxes. These are to be used to represent rooms (W. Burt, Manchester University). The dimensions are unimportant as the principles of lighting apply to volumes of any size. They can be used to evolve answers to Questions 15–21 at the end of this chapter. This may be seen as a crude method of higher education; however, the use of scale models is a well-established artistic and scientific discipline. Computer-generated illumination plots are available for a similar design purpose but at much higher cost. The interior of the shoebox, or similar, can be covered or painted with dark or light colours. The source of light for the interior of the box is the indoor artificial illumination or daylight through a window of the user's location. Cut windows or roof-lights into the box that are in proportion to the room design that is to be modelled. Each window should be hinged so that different combinations of window opening, and night-time, can be

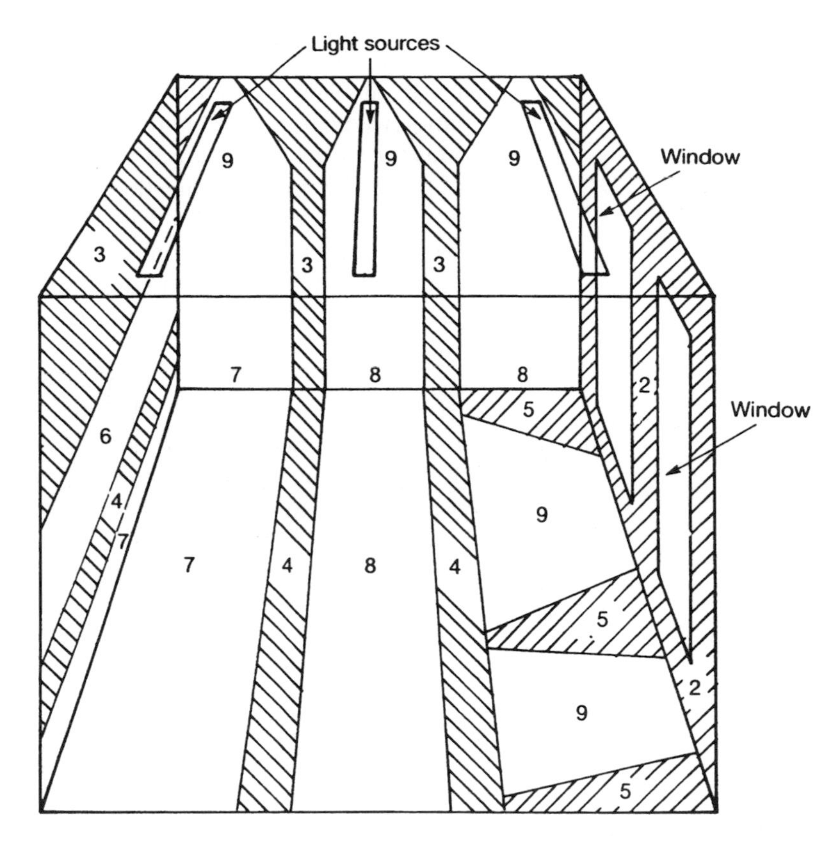

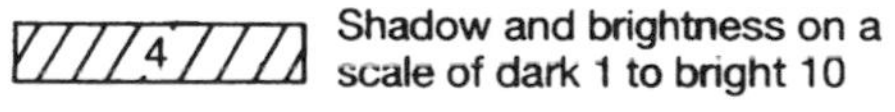

Figure 11.3 Observed illumination pattern

reproduced. Battery lamps could be utilized for the illumination arrangements. Rectangular slots in the flat ceiling of the box can represent linear fluorescent luminaires. Small circular holes in the ceiling would model spot lighting. Advanced users can place objects within the scaled room to establish the effects upon desks, furniture, partitions and artefacts. Cut a small viewing window in the end of the box. Use this aperture to observe and sketch the areas of light and shade within the room. Try various combinations of surface colour, day, artificial and permanent supplementary artificial lighting. Figure 11.3 is an example of an observed lighting pattern in a model room that has both side windows and a representation of recessed tubular fluorescent luminaires in the ceiling.

Definition of terms

Some of the terms that are used in lighting system design are as follows.

BZ classification: British Zonal Classification of 1–10 for the downward light emitted from a luminaire. The

	BZ class number relates to the flux that is directly incident upon the working plane in relation to the total flux emitted. BZI classification is for a downward directional luminaire. A BZ10 describes a luminaire that directs all its illumination upwards so that the room is illuminated by reflection from the ceiling
Contrast:	the difference in the light and dark appearance of two parts of the visual field seen simultaneously
Daylight factor:	the ratio of the natural illumination on a horizontal plane within the building to that present simultaneously from an unobstructed sky, discounting direct sunlight. A standard figure of 5000 lux is adopted for the external illuminance in the UK
Efficacy:	the luminous efficacy is the lamp light output in lumens per watt of electrical power consumption
Glare:	the discomfort or impairment of vision due to excessive brightness
Glare index:	a calculated numerical scale for discomfort glare
Illuminance:	the luminous flux density at a surface in lumens per square metre, $1/m^2$, lux. The surface is normally the working plane
Illumination:	the process of lighting an object or surface
Lightmeter:	a current-generating photocell which is calibrated in lumens per square meter, lux
Lumen, 1 m:	SI unit of luminous flux. It is the quantity of light emitted from a source or received by a surface. A 100 W tungsten filament lamp emits around 1200 lm
Lux, lx:	SI unit of illuminance; $1\ lx = 1\ lm/m^2$
Light loss factor, LLF:	the overall loss of light from the dirtiness of the lamp (0.8), luminaire (0.95) and the room surfaces (0.95). Clean conditions LLF may be 0.7 but 0.5 when equipment and room become soiled. Preferred to maintenance factor
Luminous intensity, *I* (candela):	the power of a source or illuminated surface to emit light in a given direction
Luminaire:	the complete apparatus that contains the lamp, the light emitter and the electrical controls
Maintenance factor, MF:	an allowance for reduced light emission due to the build-up of dust on a lamp or within a luminaire. Normally 0.8 but 0.9 if the lamps are cleaned regularly or if a ventilated luminaire is used. Light loss factor is preferred
Utilization factor, UF:	the ratio of the luminous flux received at the working plane to the installed flux

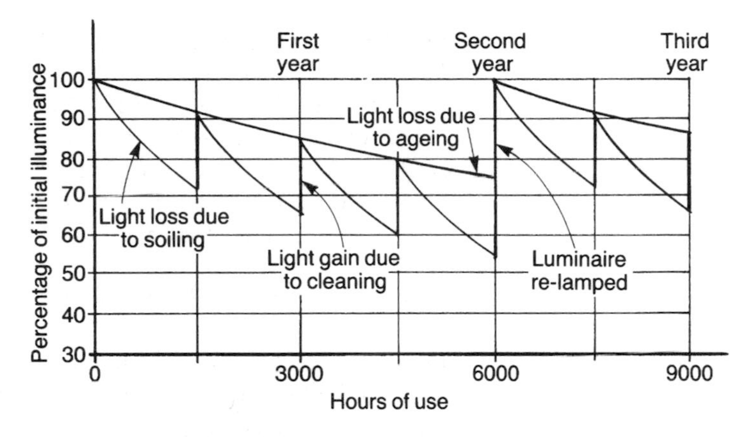

Figure 11.4 Overall fluorescent light fitting performance with maintenance

Maintenance

A planned maintenance schedule will include regular cleaning of light fittings and the lamp to ensure the most efficient use of electricity. Ventilated luminaires in air-conditioned buildings remain clean for quite long periods as the air flow through the building is mechanically controlled and filtered. The lamp also operates at a lower temperature, which prolongs its service and maximizes light output.

Because of gradual deterioration of the light output from all types of discharge lamps after their design service period, lamp efficacy could fall to half its original figure. Phased replacement of lamps after 2 or 3 years maintains design performance and avoids breakdowns. Figure 11.4 shows a typical illuminance profile for a tubular fluorescent light fitting with 6-monthly cleaning and a 2-year lamp-replacement cycle.

Utilization factor

The utilization factor is provided by the manufacturer and takes into account the pattern of light-distribution from the whole fitting, its light-distributing efficiency, the shape and size of the room for which it is being designed and the reflectivity of the ceiling and walls. Values vary from 0.03, where purely indirect distribution is employed, the room has poorly reflecting surfaces and all the light is upwards onto the ceiling or walls, to 0.75 for the most energy-efficient designs. Spot lighting can have a utilization factor of nearly unity.

The configuration of the room is found from the room index:

$$\text{room index} = \frac{lW}{H(l + W)}$$

where l is the room length (m), W is the room width (m) and H is the height of the light fitting above the working plane (m).

Table 11.2 Luminance factors for painted surfaces

Surface	Typical colour	Luminance factor range (%)
Ceiling	White, cream	70–80
Ceiling	Sky blue	50–60
Ceiling	Light brown	20–30
Walls	Light stone	50–60
Walls	Dark grey	20–30
Walls	Black	10
Floor	–	10

Table 11.3 Utilization factors for a bare fluorescent tube fitting with two 58 W 1500 mm lamps (%)

Luminance factors		Room index								
Ceiling	Walls	0.75	1	1.25	1.5	2	2.5	3	4	5
70	50	48	53	59	64	71	75	79	83	86
70	30	40	46	51	57	64	69	73	78	82
70	10	35	40	46	51	59	64	68	74	78
50	50	43	48	52	57	63	67	70	74	76
50	30	37	41	46	51	57	62	65	70	73
50	10	33	37	42	46	53	58	61	67	70
30	50	39	42	46	50	55	59	61	65	67
30	30	34	37	42	46	51	55	58	62	65
30	10	30	33	38	42	48	52	55	59	62

The ability of a surface to reflect incident light is given by its luminance factor from BS 4800: 1972. Sample values are given in Table 11.2.

Utilization factors for a light fitting comprising a white metal support batten and two 58 W white fluorescent lamps 1500 mm long (New Streamlite by Philips Electrical Limited) are given in Table 11.3. The data refer to bare fluorescent tubes suspended under the ceiling as used in commercial buildings. Enclosing the fitting with a plastic diffuser to improve its appearance usually lowers the utilization factor. This fitting has a BZ6 rating, operates at 240 V, consumes 140 W and has a power factor of 0.85 and a running current of 0.68 A.

Glare and reflections

Disability glare is when a bright light source prevents the subject from seeing the necessary detail of the task. Veiling reflections can be formed on windows and visual display unit (VDU) screens from nearby lamps. A limiting glare index is recommended for each application, for example general office 16, and this can be calculated (CIBSE, 1986).

To maximize contrast on the working plane, luminaires should be placed in rows parallel to the direction of view. The rows should be widely spaced to form work areas between them.

The zone of the ceiling that would cause glare or veiling reflections can be viewed with a mirror on the working plane from the normal angle of work. A luminaire or direct sunlight should not appear in the mirror.

Lumen design method

The number of light fittings is found from the total lumens needed at the working plane and the illumination provided by each fitting using the formula:

$$\text{number of fittings} = \frac{\text{lux} \times \text{working plane area m}^2}{\text{LDL} \times \text{UF} \times \text{MF}}$$

where LDL is the lighting design lumens produced by each lamp, UF is the utilization factor and MF is the maintenance factor. The high output New Streamlite luminaire with two Colour 84 fluorescent lamps, 1500 mm long, emits 5100 lumens measured at 2000 h of use. This is known as its lighting design lumens (LDL).

EXAMPLE 11.1

A drawing office 16 m × 11 m and 3 m high has a white ceiling and light-coloured walls. The working plane is 0.85 m above the floor. New Streamlite double-lamp luminaires are to be used and their normal spacing-to-height ratio SHR is 1.75. Calculate the number of luminaires needed and draw their layout arrangement. Find the electrical power consumption of the lighting system.

From Table 11.2 the luminance factors are 70 for the ceiling and 50 for the walls. A high standard of maintenance will be assumed, giving a maintenance factor of 0.9. The lighting design lumens is taken as 5100 lm for the whole light fitting.

From Table 11.1 the illuminance required is 600 lm/m^2. The height H of fittings above the working plane is

$$H = (3 - 0.85) \text{ m}$$

$$= 2.15 \text{ m}$$

$$\text{room index} = \frac{lW}{H(l + W)} = \frac{16 \times 11}{2.15 \times (16 + 11)} = 3.03$$

From Table 11.3, for a room index of 3,

$$\text{utilization factor} = 79\% = 0.79$$

$$\text{number of fittings} = 600\ \frac{\text{lm}}{\text{m}^2} \times \frac{16\ \text{m} \times 11\ \text{m}}{0.79 \times 0.9} \times \frac{\text{luminaire}}{500\ \text{lm}} = 29.12$$

The ratio of the spacing S between rows to the height H above the working plane is

$$\text{SHR} = \frac{S}{H} = 1.75$$

Therefore

$$S = 1.75\ H$$

$$= 1.75 \times 2.15\ \text{m}$$

$$= 3.76\ \text{m}$$

If it is assumed that windows are along one long side of the office and that rows of luminaires will be parallel to the windows, this will produce areas between rows where drawing boards, desks and VDU terminals can be sited to gain maximum benefit from side daylighting without glare and reflection. The perimeter rows of luminaires are spaced at about half of S, 1.74 m, from the side walls.

Three rows of 10 luminaires are required, as shown in Fig. 11.5, giving 30 luminaires and a slightly increased illuminance.

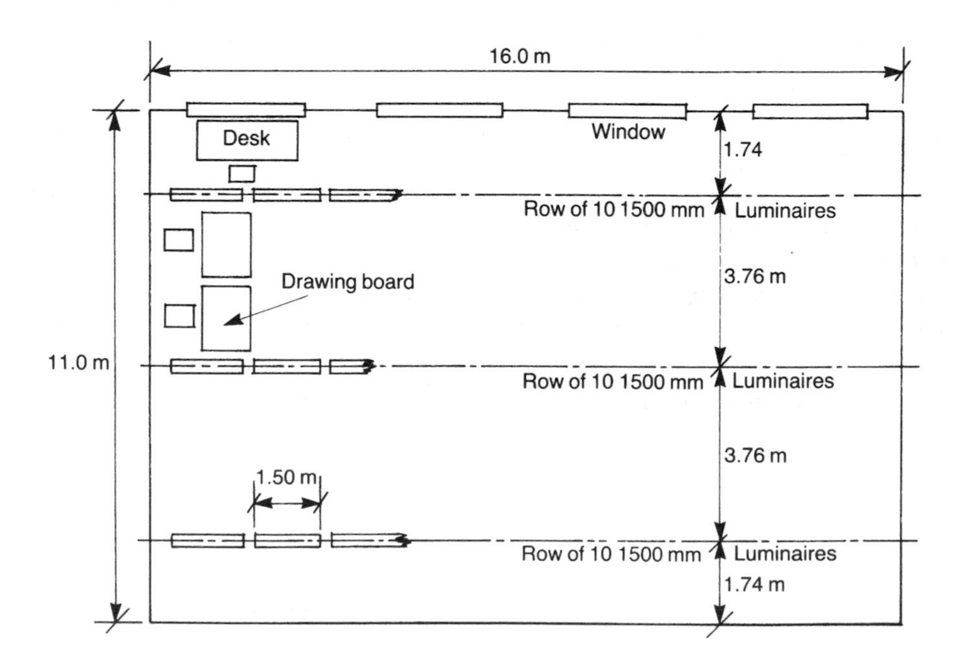

Figure 11.5 Arrangement of the luminaires in a drawing office in Example 11.1

The electrical power consumption of each luminaire is 140 W. For the room the power consumption will be 30 × 140 W, i.e. 4200 W, which is

$$\frac{4200\ \text{W}}{16\ \text{m} \times 11\ \text{m}} = 23.86\ \text{W/m}^2 \text{ floor area}$$

Air-handling luminaires

Luminaires that are recessed into suspended ceilings are ideally placed to be extract air grilles for the ventilation system. The heat generated is removed at its source and the lamp can be maintained at its optimum operating temperature to maximize light output and colour-rendering properties. Dust build-up should also be less in an air-conditioned building where all the circulating air is filtered in the plant room. Figure 11.6 shows the air flow through a luminaire that has ventilation openings and mirrored reflectors.

Up to 80% of the electrical energy used by the light fitting can be absorbed by the ventilation air as it passes through. Air flow rates are around 20 l/s through a 1500 mm twin tube fluorescent luminaire and a temperature increase of about 8 °C is achieved. Extract air at about 30 C can be produced and the heat it contains can be reclaimed for use in other parts of the building.

Colour temperature

Colour temperature is a term used in the description of the colour-rendering property of a lamp. Colours of surfaces under artificial illumination are compared with the colours produced by a black body heated to a certain temperature and radiating in the visible part of the spectrum between the ultraviolet and infrared bands. Any colour that matches that shown by the heated black body is said to have a colour temperature equal to the temperature of the black body. A candle has a colour temperature of 2000 K and blue sky has a colour temperature of 10 000 K. Correlated colour temperature is that temperature of the heated black body at which its colour most closely resembles that of the artificial source.

The colour-rendering index Ra8 is used to compare the colour-rendering characteristics of various types of lamp. Eight test colours are illuminated by a

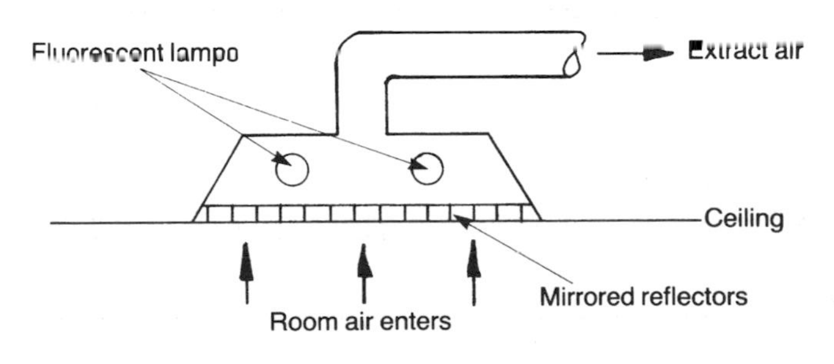

Figure 11.6 Ventilated luminaire

reference source, which is a black body radiator of 5000 K correlated colour temperature or 'reconstituted' daylight if more than 5000 K is needed. These eight colours are then illuminated by the test lamp. The average of the colour differences produced between the source and test lamps provides a measure of the colour-rendering properties of the test lamp. An Ra8 of 50 corresponds to a warm white fluorescent lamp. An Ra8 of 100 would be produced by a lamp that radiated identically with the reference source.

Lamp types

A summary of lamp types, their performances and applications is given in Table 11.4.

General lighting service (GLS) tungsten filament lamps are inexpensive, give good colour matching with daylight and last up to 2000 h in service. They can be controlled by variable-resistance dimmers and are used in a supplementary role to higher-efficacy illumination equipment. Tungsten halogen spot or linear lamps have a wide variety of display and floodlighting applications.

Miniature fluorescent lamps (SL) can be used as energy-saving replacements for GLS lamps. Folded-tube and single-ended types are available. A typical folded

Table 11.4 Lamp data

Lamp	Lamp designation BS colour	Efficacy (lm/W)	Colour temperature	Colour-rendering index Ra8	Applications
Incandescent	GLS, tungsten	18	2900	100	Interiors
Incandescent	GLS tungsten-halogen	22	3000	100	Interior displays, outdoors
Fluorescent	White	80	3500	50	Industrial
Fluorescent	Natural	85	4000	85	Commercial
Fluorescent	Warm white	85	3000	85	Social
Discharge lamps					
Low-pressure sodium	SOX	183	2000	—	Roads, car parks, floodlighting
High-pressure sodium	SON	112	2250	29	Floodlighting, exteriors, large hall interiors
Mercury fluorescent	MBF	50	4300	47	Roads, floodlights, factory interiors
Metal halide	MBI	85	4400	70	Industrial interiors, floodlighting
Mercury-blended tungsten fluorescent	MBTF	25	3700	—	Road, floodlights, factory interiors

lamp, SL18 18 W, produces 900 lumens at 100 h, has a correlated colour temperature of 2700 K, an Ra8 of 80 and a service period of 5000 h. Its lumen output is equivalent to a 75 W GLS filament lamp.

Low-pressure mercury-vapour-filled fluorescent tubular lamps (MCF) are the most common. The tube diameter is 38 mm. Electrical excitation of the mercury vapour produces radiation, which causes the tube's internal coating to fluoresce. The colour produced depends on the chemical composition of the internal coating. High-efficacy 26 mm diameter lamps (TLD) are filled with argon or krypton vapour and have a phosphor internal coating. A circuit diagram for a switch-start fluorescent lamp is shown in Fig. 11.7.

The glow-switch starter S has bimetallic electrodes, which pass the lamp electrode preheating current when starting cold. Upon becoming warmed, the bimetallic electrodes move into contact and establish a circuit through the lamp electrodes. Making this contact breaks the starter circuit, whose electrodes cool and spring apart after about a second, subjecting the lamp to mains voltage. Twin lamp and starterless controls are used in new installations to minimize running costs.

Luminaires are designed to run in air of 5–25 °C d.b., and their service period will be reduced at higher temperatures. Circuit breakers or high rupturing capacity (HRC) fuses, rather than rewirable fuses, should be used for circuit protection. All fluorescent luminaires make an operating noise, which may be noticeable against very low background noise levels. Some radio interference is inevitable but this diminishes with distance from the radio set.

Low-pressure sodium discharge lamps emit light from electrically excited sodium vapour in a glass discharge tube. High-pressure sodium discharge lamps comprise a discharge tube of sintered aluminium oxide containing a mixture of mercury and sodium vapour at high pressure.

Mercury fluorescent lamps consist of electrically excited mercury vapour in a quartz discharge tube, which emits ultraviolet radiation, an infrared component from the mercury arc, and visible light. A phosphor internal coating fluoresces to produce the desired colour. Metal halide lamps contain metal halides in a quartz discharge tube, which gives a crisp white light. MBTF lamps are the mercury fluorescent type with additional tungsten filaments, which need no control gear and

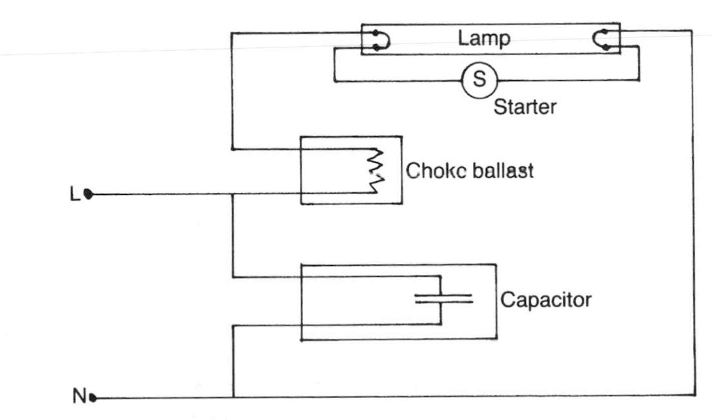

Figure 11.7 Switch-start circuit diagram for fluorescent and discharge lamps

give light immediately. Other discharge lamps incorporate current-limiting ballasts and power-factor-correcting capacitors in a similar arrangement to that in Fig. 11.7.

EXAMPLE 11.2

A windowless office is to be illuminated for 15 h per day, for 6 days per week for 50 weeks per year. The floor is 20 m long and 12 m wide. An overall illumination of 450 lx is to be maintained over the whole floor. The total light loss factor for the installation is 70%. The designers have the choice of using 100 W tungsten filament lamps, which have an efficacy of 12 lm/W and need replacing every 3000 h, or 65 W tubular fluorescent warm white lamps, which have an initial output of 5400 lm and are expected to provide 12 000 h of service. The room layout requires an even number of lamps. Electricity costs 8 p/kWh. The tungsten lamps cost £1 each while the flourescent tubes cost £10 each. Compare the total costs of each lighting system and make a recommendation as to which is preferable, stating your reasons.

For the lighting system

$$\text{lighting hours} = 15 \times 6 \times 50 = 4500 \text{ h/yr}$$

$$\text{floor area} = 20 \text{ m} \times 12 \text{ m} = 240 \text{ m}^2$$

$$\text{installed lumens} = \frac{450 \times 240 \times 100}{70}$$

$$= 154\ 286 \text{ lm}$$

For the tungsten lamps

$$\text{input power} = \frac{154\ 286}{12 \times 1000}$$

$$= 12.857 \text{ kW}$$

$$\text{number of lamps} = \frac{12.857 \times 1000}{100}$$

$$= 129 \text{ lamps}$$

next even number is 130 lamps

$$\text{installed power} = \frac{130 \times 100}{1000}$$

$$= 13 \text{ kW}$$

$$\text{electricity cost} = \frac{13 \times 4500 \times 8}{100}$$

$$= £4680/\text{yr}$$

The average annual cost for replacing the lamps can be found by multiplying the number of installed lamps by the anticipated hours of use, dividing by the lamp manufacturer's rated average life hours and then multiplying by the replacement cost for each lamp. In this case

$$\text{lamp cost} = \frac{130 \times £1 \times 4500}{3000}$$

$$= £195/\text{yr}$$

These tungsten lamps expire within a year, so there will be annual expenditure. A new lighting system that has lamps providing reliable service for more than a year will not produce replacement expenditure in the first year or two. The owner needs to budget for the eventual replacement costs by assessing the average annual cost. A planned maintenance programme will have lamps replaced, and luminaires cleaned, prior to expiry. This work may be performed out of normal working hours for the building to avoid disturbance to its normal functions.

$$\text{total annual cost} = £(4680 + 195) = £4875/\text{yr}$$

For the flourescent lamps

$$\text{number of lamps} = \frac{154\,286}{5400}$$

$$= 29 \text{ lamps}$$

next even number is 30 lamps

$$\text{input power} = \frac{30 \times 65}{1000}$$

$$= 1.95 \text{ kW}$$

$$\text{electricity cost} = \frac{1.95 \times 4500 \times 8}{100}$$

$$= £702$$

$$\text{lamp cost} = \frac{30 \times £10 \times 4500}{12\,000}$$

$$= £112.5/\text{yr}$$

$$\text{total annual cost} = £(702 + 112.5) = £814.5/\text{yr}$$

Both methods of lighting require at least the annual cleaning of lamps and luminaires.

The reasons for using fluorescent lamps are as follows.

1. They produce an annual cash saving of £(4875 − 814.5) = £4060.5/yr.
2. They only need changing (12 000/4500) after 2 years 8 months of use.
3. Tungsten lamps need changing (12 × 3000/4500) every 8 months.
4. There is less heat emission from fluorescent lighting, particularly radiant heat, so the air-conditioning cooling load is lower.
5. They give better colour rendering.
6. They give better diffused light distribution.

Control of lighting services

The energy that is consumed by artificial lighting systems is both an expensive use of resources and a high monetary cost. A minimum level of illumination may be desirable for the security of personnel or monitoring of the building and its contents for the detection of intruders. The changes from daylighting to full artificial lighting and then to security illumination can be achieved with manual and automatically timed operation of switches. This usually leaves unoccupied areas illuminated. A light-sensitive photocell can be used to detect illumination level and an automatic controller may be programmed to reduce the use of the electrical lighting system. The presence of the occupants, or an intruder, can be detected by passive infrared, acoustic, ultrasonic or microwave-radar-based

Figure 11.8 Presence detectors (reproduced by courtesy of Ex-Or Ltd)

systems. Figure 11.8 shows a selection of detectors that are used for the control of lighting and water services (Ex-Or Limited).

The detector and control system needs to be sufficiently fast-acting and sensitive so that the occupant is not stranded within a darkened room and suffers injury or fear. It is equally important that only those lights that are actually needed are switched on, and not for the entire room or space to be illuminated when only one person enters to use a small area. Local control of the light switching may be preferable to a system that is operated from a remote energy-management system computer. The local system should be faster in operation and will be less subject to distribution system or computer breakdown.

The design of a control scheme for an occupied space may include a minimum number of luminaires, which are switched on from a time switch or by the occupant to provide safe access. Groups of luminaires that are near windows may be controlled from local photocell detectors to ensure that the perimeter lighting remains off as long as possible. The internal parts of the space may be operated from automatic presence detectors. Data on the length of operation of each lighting unit can be transmitted to the computer-based building-management system so that real-time usage and electrical power consumption can be recorded.

Questions

1. Explain, with the aid of sketches, how interiors can be illuminated by daylight. State how natural illumination is quantified.

2. State the relationship between the visual task and the illuminance required, giving examples.

3. Sketch and describe how supplementary artificial lighting is used to achieve the desired illuminance.

4. Discuss the use of localized task-illumination systems in relation to the illumination level provided, reflection, energy conservation, shadows and user satisfaction.

5. Define the terms used in lighting system design.

6. Draw a graph of illumination provided versus service period for an artificial illumination installation to show the effect of correct maintenance procedure.

7. Calculate the room index for an office 20 m × 12 m in plan, 3 m high, where the working plane is 0.85 m above floor level.

8. State the luminance factors for a room having a cream ceiling and dark grey walls.

9. Find the utilization factor for a bare fluorescent tube light fitting having two 58 W, 1500 mm lamps in a room 5 m × 3.5 m in plan and 2.5 m high. The working plane is 0.85 m above floor level. Walls and ceiling are light stone and white respectively.

10. Sketch satisfactory arrangements for natural and artificial illumination in modern general and drawing offices, a library and a lounge. Comment particularly on how glare and reflections are controlled.

11. A supermarket of dimensions 20 m × 15 m and 4 m high has a white ceiling and mainly dark walls. The working plane is 1 m above floor level. Bare fluorescent tube light fittings with two 58 W, 1500 mm lamps are to be used, of 5100 lighting design lumens, to provide 400 lx. Their normal spacing-to-height ratio is 1.75 and total power consumption is 140 W. Calculate the number of luminaires needed, the electrical loading per square metre of floor area and the circuit current. Draw the layout of the luminaires.

12. Discuss how the use of air-handling luminaires improves the performance of the lighting installation and makes better use of energy.

13. Explain how the rendering of colours by illumination systems is measured.

14. Compare the energy efficiency and colour-rendering of different lamp types, stating suitable applications for each.

15. Use the cardboard box small-scale models of rooms to investigate the visual design of lighting systems.

    (a) Cut different shapes and locations of windows and roof lights such that they all have the same open area.
    (b) Colour the internal surfaces differently by means of dark, light, removable and reflective sheets of materials.
    (c) Cut slots and holes into the ceiling to model different designs of strip fluorescent and filament lighting layouts; replaceable ceilings with different designs are helpful.
    (d) View the interior of the room under various daylighting and artificial lighting arrangements.
    (e) Make three-dimensional sketches of what you see of the lighting layouts produced, showing the shading. Write the lighting level found on each area on a scale of 1 (dark) to 10 (bright).

16. Using the models created for Question 15, answer the following.

    (a) What is the effect of *quantity* of daylight on the *quality* of the daylighting system created?
    (b) What effect do the colours of the room interior surfaces have on the quality of the lighting produced?
    (c) What colours should the room surfaces be?

    Justify your views in relation to the use of the room, its maintenance costs and design of the decoration.

17. Using the models created for Question 15, answer the following.

    (a) What patterns of illumination are produced on the end walls by differently spaced rows of strip lighting?
    (b) What are the best spacing arrangements between rows of strip or circular lamps? These depend upon what is being illuminated, so state the objectives of the lighting design first.

18. Create an approximate scale model of the interior of a room that is known to you. Experiment with three combinations of daylighting and artificial lighting to find the best overall lighting scheme for the tasks to be performed in the room.

19. Write a technical report to explain how the reflectance of room surfaces, the location, dimensions and shape of glazing, the spacing of rows of luminaires and their height above the working plane are related to the efficient use of electrical energy in the overall lighting design.

20. Put small boxes and partitions into a scale model of a room to represent furniture, desks, horizontal and vertical working planes. Carry out an experimental investigation of the problems that arise for the lighting designer.

21. State how the lighting design can be made to feature particular parts of the interior of the building and the parts that should be featured for safety and appearance reasons.

22. Analyse the costs of these competitive lighting systems and recommend which is preferable, stating your reasons.

    A heavy engineering factory is to be illuminated for 15 h per day, for 5 days per week for 50 weeks per year. The floor size is 120 m long and 80 m wide. An overall illumination of 250 lx is to be maintained over the whole floor. The overall light loss factor for the installation is 63%. The designers have the choice of using 150 W tungsten–halogen lamps, which produce 2100 lm and need replacing every 2000 h, 80 W tubular fluorescent lamps, which produce 6700 lm and are expected to provide 12 000 h of service, and 250 W high-pressure sodium lamps, which produce 27 500 lm and are expected to last for 24 000 hours. The lighting layout needs an even number of lamps. Electricity costs 7.2 p/kWh. The tungsten lamps cost 90p each, the fluorescent tubes cost £10.50 each and the sodium lamps cost £61 each. Replacing any lamp takes two people 2 min, and their combined labour rate is £17 per hour. The hire cost of scaffolding is £120 per 8 hour day.

23. A lecture theatre is to be illuminated for 8 h per day, for 5 days per week for 30 weeks per year. The floor is 32 m long and 16 m wide. An overall illumination of 350 lx is to be maintained over the whole floor. An even number of lamps is to be used.

The utilization factor for the installation is 0.73 and the maintenance factor is 0.7.

The designers have the choice of using 100 W tungsten filament lamps, which have luminous efficacy of 10 lm/W and need replacing every 2000 h, 100 W quartz halogen low-voltage lamps in reflectors, which have an efficacy of 95 lm/W and provide 23 000 hours' use, and 65 W tubular fluorescent lamps, which have an efficacy of 57 l/MW and are expected to provide 7500 h of service. Electricity costs 8 p/kWh. The tungsten lamps cost 85p each, the halogen cost 29 each and the fluorescent tubes cost £11.25 each. Compare the total costs of each lighting system and make a recommendation as to which is preferable, stating your reasons.

12 Gas

Key terms and concepts

Learning objectives

Study of this chapter will enable the reader to:

1. calculate the flow of gas required by an appliance;
2. identify gas pressures;
3. know how to measure gas pressure;
4. calculate gas pressures;
5. calculate gas pressure drops in pipelines;
6. calculate the equivalent length of a gas distribution pipe system;
7. use gas-pipe-sizing tables;
8. choose suitable gas pipe sizes;
9. describe gas service entry;
10. understand the working of a gas meter;
11. identify the space requirement for gas meters;
12. explain the flue requirements for gas appliances;
13. describe gas flue systems;
14. understand fan-assisted flue systems;
15. apply appropriate gas flue systems to designs,
16. understand how gas combustion is controlled and regulated;
17. know how gas systems are operated safely.

Introduction

Gas services are provided to most buildings, and safety is of paramount importance. Economy is also important, and the versatility and controllability of gas are appreciated. It is used for heating, hot-water production, refrigeration for small and large cooling loads, electrical power generation, cooking and decorative heating.

Gas is converted from its primary fuel state into useful energy at the point of use. Its distribution energy loss is accounted for in the standing charge to the final consumer and, although it is charged for, does not appear to be related to the load as for water pipes or electricity cables. Gas is conveyed in a one-way pipe system and is not returned to the supplier as are electricity and water.

The use of natural gas and, ultimately, an artificially produced substitute from coal and oil is a highly efficient use of primary resources, and all efforts are directed at continuing this trend.

This chapter introduces the calculation of gas flow rate into a load and the gas pressure measurement that is used to monitor the flow rate and condition. The sizing of pipework depends on the gas pressure of the incoming service and that required by the final gas-burning appliance in order to maintain the correct combustion rate. Incoming gas service provisions, metering and particular flue systems are explained. Gas is more suitable than any other fuel for low-level flue gas discharge provided that sufficient dilution with fresh air is available. Typical methods of gas burner control and pressure reduction are explained.

Gas pipe sizing

The gas flow rate required for an appliance can be found from the manufacturer's literature or calculated from its heat output and efficiency:

$$\text{appliance efficiency } \eta = \frac{\text{heat output into water or air}}{\text{heat input from combustion of gas}} \times 100\%$$

Most gas appliances have an efficiency of 75%.

The gross calorific value (GCV) of natural gas (methane) is 39 MJ/m^3. If the appliance heat output is SH kW, then the gas flow rate Q required is

$$Q = \frac{\text{SH kW}}{\eta} \times \frac{1 \text{ kJ}}{1 \text{ kWs}} \times \frac{1 \text{ MJ}}{10^3 \text{ kJ}} \times \frac{1}{\text{GCV}} \frac{\text{m}^3}{\text{MJ}} \times \frac{10^3 \text{ l}}{1 \text{ m}^3}$$

$$= \frac{\text{SH}}{\eta \times \text{GCV}} \text{ l/s}$$

The maximum allowable gas pressure drop due to pipe friction between the gas meter outlet and the appliance will normally be 75 Pa. Gas pressure in a pipe is measured with a U-tube water manometer as shown in Fig. 12.1. The water in the U-tube manometer is displaced by the gas pressure until the gas and water pressures in the two limbs are balanced. The atmosphere acts equally on both limbs under normal circumstances and all pressures are measured above atmospheric. These are called gauge pressures.

Methane has a specific gravity (SG) of 0.55 and so its density is

$$\rho_{gas} = \text{SG} \times \rho_{air}$$

The standard density of air at 20 °C d.b. and 1013.25 mb barometric pressure is 1.2 kg/m^3. Thus

$$\rho_{gas} = 0.55 \times 1.2 \text{ kg/m}^3$$

$$= 0.66 \text{ kg/m}^3$$

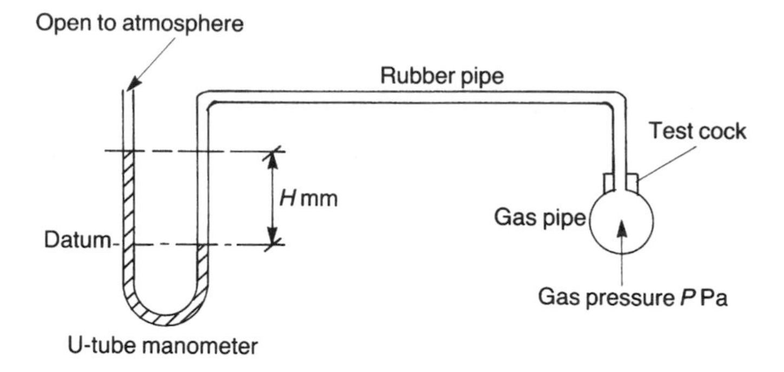

Figure 12.1 Measurement of gas pressure

The water column H in the manometer shows the net effect of the gas pressure P and the different density of the column H of gas in the right-hand limb. At the datum, the pressure exerted by the left-hand column equals the pressure exerted by the right-hand column.

$$P_{\text{water}} = P - \text{pressure of column } H \text{ of gas}$$

The pressure at the base of a column of fluid is ρgH where $g = 9.807\ \text{m/s}^2$. The density of water ρ_w is $10^3\ \text{kg/m}^3$. Therefore

$$\rho_w gH = P - \rho_{\text{gas}}\ gH$$

$$P = \rho_w gH - \rho_{\text{gas}}\ gH$$

$$= gH(\rho_w - \rho_{\text{gas}})$$

Thus

$$P = 9.807\ \frac{\text{m}}{\text{s}^2} \times H\ \text{mm} \times \frac{1\ \text{m}}{10^3\ \text{mm}} \times (1000 - 0.66)\ \frac{\text{kg}}{\text{m}^3}$$

$$= \frac{9.807 \times 999.34}{1000} - H\ \frac{\text{kg}}{\text{m s}^2} \times \frac{1\ \text{N s}^2}{1\ \text{kg m}} \times \frac{1\ \text{Pa m}^2}{1\ \text{N}}$$

$$= 9.801H\ \text{Pa}$$

A sufficiently good approximation is $P = gH$ Pa, where H is in millimetres of water gauge.

EXAMPLE 12.1

A gas-fired boiler has a heat output of 40 kW and an efficiency of 75%. Calculate the flow rate of gas required.

$$Q = \frac{\text{SH}}{\eta \times \text{GCV}} = \frac{40}{0.75 \times 39} = 1.37\ \text{l/s}$$

EXAMPLE 12.2

Calculate the gas pressure if a water manometer shows a level difference of 50 mm.

$$P = 9.807 \times 50 \text{ Pa} = 490.35 \text{ Pa}$$

EXAMPLE 12.3

A gas-fired warm-air unit requires a gas pressure of 350 Pa at the burner test cock. What reading will be found on a water manometer?

$$P = gH \text{ Pa}$$

and so

$$H = \frac{P}{g} = \frac{350}{9.807} = 36 \text{ mm}$$

Gas pressure may be expressed in millibars (mb) as

$$1 \text{ bar} = 100\ 000 \text{ Pa}$$

$$1 \text{ bar} = 100 \text{ kPa}$$

$$1 \text{ mb} = \frac{100 \text{ kPa}}{1000} = 0.1 \text{ kPa}$$

$$1 \text{ mb} = 100 \text{ Pa}$$

The gas pressure loss rate $\Delta p/EL$ along a pipeline is needed to find the pipe diameter. EL is the equivalent length of the measured straight pipe, bends and fittings. An initial estimate can be made for the flow resistance of pipe fittings by adding 25% to the straight lengths L m of pipe.

EXAMPLE 12.4

Calculate the pressure loss rate in a gas pipeline from the meter to a water heater. The pipe has a measured length of 22 m.

$$EL = (1.25 \times 22) \text{ m} = 27.5 \text{ m}$$

The allowable pressure loss Δp is 75 Pa. Thus

$$\frac{\Delta p}{EL} = \frac{75 \text{ Pa}}{27.5 \text{ m}} = 2.727 \text{ Pa/m}$$

EXAMPLE 12.5

A pressure loss rate of 3.5 Pa/m is to be used for sizing a gas pipeline in a house. Calculate the maximum length of pipe that can be used if the resistance of the pipe fittings amounts to 20% of the installed pipe run.

$$\frac{\Delta P}{EL} = \frac{75 \text{ Pa}}{EL \text{ m}} = 3.5 \text{ Pa/m}$$

Thus

$$EL = \frac{75}{3.5} = 21.43 \text{ m}$$

and

$$EL = 1.2\ l$$

Hence

$$l = \frac{21.43}{1.2} = 17.86 \text{ m}$$

This is the maximum length of run.

The gas pressure of the incoming service will be up to 5 kPa and this is reduced by a governor at the meter inlet to give 2 kPa in the installation within the building. For large gas-burning equipment the gas pressure may have to be increased with a booster. A boosting system comprises an electrically driven reciprocating compressor, a high-pressure storage tank and automatic pressure and safety controls. Pipe diameters can be found from Table 12.1.

EXAMPLE 12.6

Find the pipe size required for a gas service carrying 1.4 l/s and having a pressure loss rate of 2.7 Pa/m.

Table 12.1 Flow of methane (natural gas) in copper pipes

	Gas flow rate Q (l/s) for pipe diameters of			
$\Delta p/EL$ (Pa/m)	15 mm	22 mm	28 mm	32 mm
1	0.08	0.31	0.69	1.22
2	0.16	0.47	1.05	1.84
3	0.21	0.59	1.33	2.34
5	0.29	0.81	1.8	3.15
7	0.35	0.98	2.2	3.83
10	0.44	1.21	2.7	4.71

Source: Reproduced from *IHVE Guide* (CIBSE, 1986 [IHVE, 1970]) by permission of the Chartered Institution of Building services Engineers.

From Table 12.1, using the nearest pressure drop below 2.7 Pa/m, in this case 2 Pa/m, a 32 mm copper pipe can carry 1.84 l/s and would be suitable.

Gas service entry into a building

The gas service pipe from the road main should slope at 1 in 20 up to the entry point to the building, at right angles to the road main and entering the building at the nearest convenient place. Ground cover of 375 mm is maintained and new pipework is made of plastic. When old steel services are renewed, the plastic pipe can be run inside the steel.

A meter compartment can be built into the external wall in housing installations and the service clipped to the wall under a cover. This facilitates meter reading without entry to the property. Computer monitoring of energy meters using a telephone link to the supply authority will eventually replace manual reading.

Where the meter compartment is inside the building, the service should pass through the foundations in a pipe sleeve, plugged to stop the ingress of moisture and insects but allowing for some movement. A 300 mm square pit is provided in a concrete floor to allow the service to rise vertically to the meter. The pit can subsequently be filled with concrete.

The meter compartment must not be under the only means of escape in the event of a fire in a building where there are two or more storeys above the ground floor unless it is located in an enclosure having a minimum fire resistance of half an hour.

Gas service pipes, meters and appliances should always be in naturally ventilated spaces, as dilution with outside air is the best safety precaution against the accumulation of an explosive mixture with air. Early detection of leaks is essential, but ventilation assists the dilution of leaks. Gas detectors can be provided as an additional precaution.

Domestic credit meters pass up to 10 l/min, 0.17 l/s and are 212 mm wide, 270 mm high and 155 mm deep. Their overall space requirement is approximately double the width and height measurements for pipework, valve and filter. Industrial meters have flanged steel pipework up to 100 mm in diameter and a bypass to allow uninterrupted gas flow in the event of meter breakdown. A 500 l/s meter is 2 m wide, 2.25 m high and 1.6 m deep. Due allowance must be made for doorways and access for replacing the meter during the building's use. A separate meter room is recommended, which should be secure, accessible, illuminated and weatherproof with no hot pipes or surfaces.

Manufactured town gas came from the conversion of coal or oil. It had a high hydrogen content and flame speed but its cross-calorific value was half that of methane. In future, substitute natural gas (SNG) may be manufactured from hydrocarbons as indigenous reserves become exhausted. SNG will come from the chemical conversion of coal, tar sand or crude oil and will have characteristics similar to those of methane.

Gas pipes or meters should usually be spaced 50 mm from electrical cables, conduits, telecommunications cables or other conductors. Electric and gas meters may be accommodated in a single compartment if a fire-resistant partition separates them.

Figure 12.2 shows a cross-section through a positive displacement meter under operating conditions. The meter is divided into three compartments by the horizontal valve plate near the top and the vertical division plate. Bellows formed by a metal disc surrounded by a leather diaphragm are located on each side of the division plate.

The gas enters the upper chamber X through the inlet port, from which it is led to the inside or outside of the bellows depending on the position of the slide valves. In the diagram on the left-hand side, gas passes into the bellows Y, which it expands, forcing the gas in chamber Z to flow through the valve to the outlet port T_2. The outlet ports are connected to the meter outlet pipe.

When Y is fully to the right-hand side, the sliding valve closes the compartment, as shown in part A, but the left-hand valve moves so that gas then flows from X into the bellows S, forcing the gas in R to flow through T_1. Part B shows the cycle continuing, with the left-hand sliding valve closing the compartment and the right-hand sliding valve moving across so that the gas in X passes into Z, forcing gas in Y to flow through T_2. Part C completes the cycle, with the right-hand sliding valve closing the compartment and the left-hand sliding valve opening to allow gas in X to flow into R, forcing the gas in S to flow through T_1.

Gas flow to the appliance causes the bellows to move sideways and this movement is connected via linked rods, levers and gear wheels to the sliding valves and the meter dials.

The Gas Act 1986 compels British Gas to supply, and to continue to supply, gas to any premises within 25 yards (23 m) of a gas-distributing main. The cost of making a connection, laying the service pipe and maintaining it up to a maximum length of 9 m on public land is met by British Gas, and any remainder on public land, and all on private land, by the applicant.

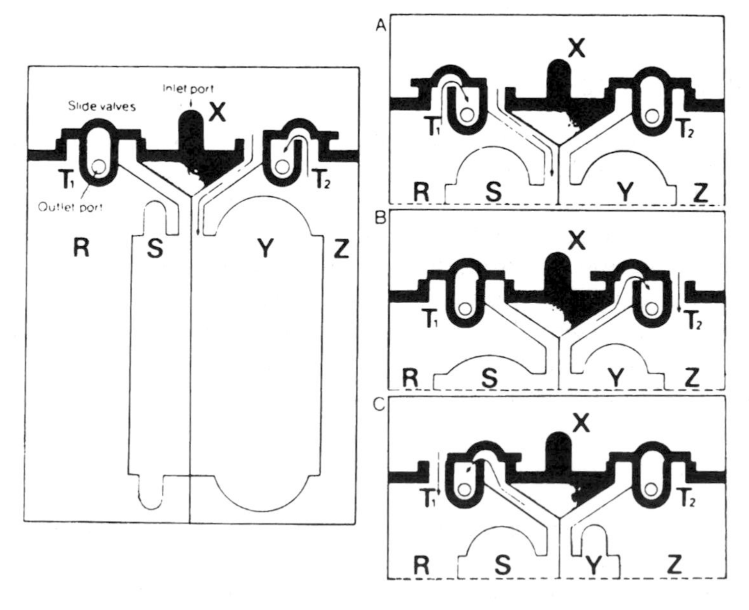

Figure 12.2 Operation of a gas meter

Flue systems for gas appliances

Gas appliances can be flued by a wide variety of methods, as the products of combustion are mainly water vapour, carbon dioxide, nitrogen and oxygen, at a temperature of about 95 °C after the draught diverter. The function of the draught diverter is to discharge flue products into the boiler room during a down-draught through the chimney. Such reverse flows occur infrequently for a few seconds during adverse wind conditions. Diversion ensures that the correct combustion process is not interrupted. It stops the pilot flame from being blown out, with consequential shut-down of the appliance until manual ignition is arranged. The draught diverter also dilutes the products of combustion by entrainment of room air into the flue. A carbon dioxide concentration of 4% by volume is found in the secondary flue after the diverter. The primary flue pipes are those before the diverter. Flue systems are described below.

Brick chimney

New masonry chimneys must be lined with vitrified clay or stainless steel pipe. Existing chimneys may incorporate a stainless steel flexible flue liner, which can be pulled through an existing chimney with a rope and rounded plug. The liner has the same diameter as the appliance flue outlet, often 125 mm for domestic appliances, and is built into the top of the chimney with a plate to form a sealed air space between the liner and the brickwork. This acts as thermal insulation to maintain flue gas temperature. If the flue gases were allowed to cool to below about 25 °C condensation of the water vapour would occur and deterioration of the metal and brickwork would reduce serviceability. Asbestos cement or glazed earthenware pipes can be built into new chimneys for protection of the brickwork. A cowl is fitted to the flue to reduce the ingress of rain and the possibility of down-draughts.

Free-standing pipe

Figure 4.22 shows a typical free-standing pipe flue, which can be used for a gas appliance. The pipe will be either asbestos cement or double-walled stainless steel with thermal insulation between the inner and outer pipes. A flue pipe taken through a roof is fitted with a lead slate to weatherproof the junction. The terminal should be 600 mm from the roof surface and clear of windows or roof-lights. An internal flue from a small domestic appliance can be connected to a ridge terminal. An externally run asbestos cement flue pipe has a branch tee junction at its emergence through the wall. A 25 mm copper drain pipe takes condensation to a drain gulley.

Balanced flue

Figure 12.3 shows the balanced-flue system used for boilers, warm air units, convectors and water heaters. It is used for appliance ratings up to around 30 kW. External wind pressure is applied equally to the combustion air inlet and the flue gas outlet parts of the combined terminal. The only pressure difference

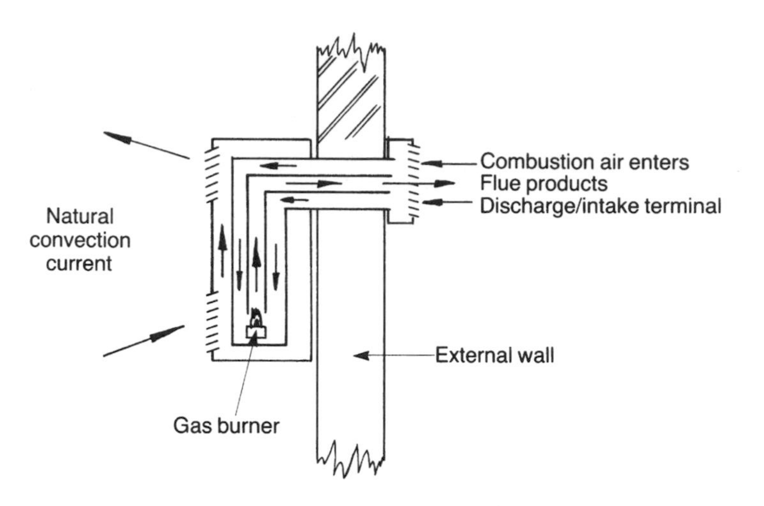

Figure 12.3 Balanced-flue gas appliance

causing air flow through the appliance is that caused by combustion. The flue terminal should not be underneath a window or within 0.5 m of a doorway or openable window. It should not be located in a confined corner where external air flow might be restricted. Fan-assisted balanced flues have been used and these allow more flexibility in siting the appliance further away from the terminal. Balanced-flue appliances are also called room-sealed appliances.

Se-ducts and U-ducts

Room-sealed appliances in multi-storey flats are connected to a vertical precast concrete shaft extending from the fresh air inlet grille at ground level to a terminal on the roof. Combustion air is taken from the duct by each heater and its flue products are passed into the shaft. The duct is sized so that sufficient ventilation is provided for the whole installation. With a U-duct a separate combustion air inlet duct takes air from the roof downwards to the lowest appliance, and then the upward duct acts in the same manner as the Se-duct.

Shunt duct

Precast concrete wall blocks, 100 mm wide, with a rectangular flue passage are built into partition walls or the inner leaf of a cavity wall. A continuous flueway is formed for each heater, often a gas fire, to ceiling level. An asbestos pipe then connects to a ridge terminal. Several flues built into a wall side by side are called a shunt duct system.

Fan-diluted flue

Fan-diluted flues are mainly used in commercial buildings where a conventional flue pipe and terminal could not be used or would be unsightly: for example, in a shopping precinct. Fresh air enters a galvanized sheet metal duct, which passes through the boiler plant room and discharges back into the atmosphere. A centrifugal or axial flow fan in the duct is started before the boilers are ignited and an air flow switch cuts the burners off in the event of fan failure. Secondary flue pipes from the boilers are connected into the duct on the suction side of the fan. Dilution of the combustion products takes place and the discharge air from the system may contain as little as 0.5% carbon dioxide and be down to 30 °C. Any condensing moisture is carried by the high-velocity airstream and is dispersed as steam into the atmosphere.

The air inlet and discharge louvres should be positioned on the same external wall to balance wind pressures on each. The discharge can be made into a shopping arcade or covered walkway to make use of the available heat. Careful

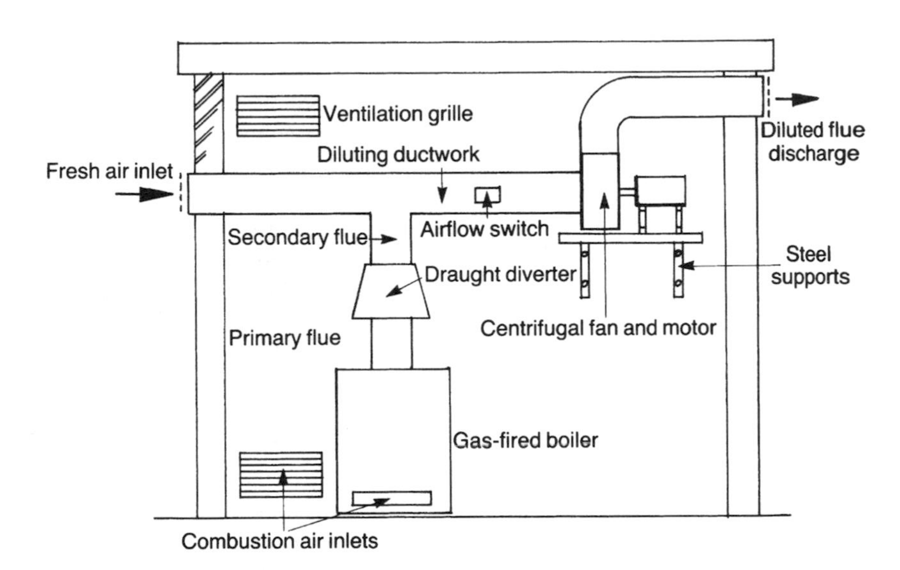

Figure 12.4 Fan-diluted flue (reproduced courtesy of Aidelle Products, High Wycombe)

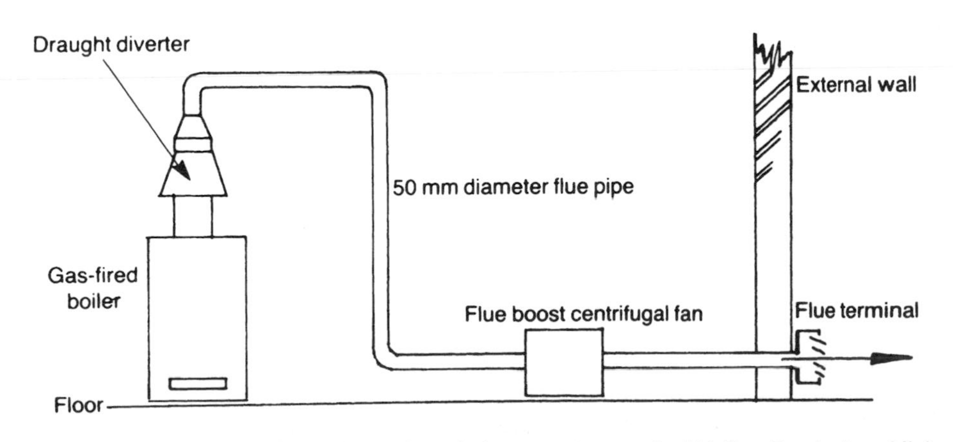

Figure 12.5 Boosted flue (reproduced by courtesy of Aidelle Products, High Wycombe)

fan selection is essential to avoid creating a noise nuisance. Figure 12.4 shows a fan-diluted flue installation for one boiler.

Boosted flue

A domestic boiler may have a booster centrifugal fan fitted into its flue pipe to allow a long horizontal run or even a downwards run. The pipe diameter can be smaller than the boiler flue outlet diameter and the fan pressure rise is used to overcome the frictional resistance. A typical installation is shown in Fig. 12.5.

Ignition and safety controls

Natural gas is burnt in an aerated burner in which half the air needed for combustion is entrained into the gas pipeline by a nozzle and venturi throat. This premixed gas and air goes to the burner, which is often a perforated plate through which the mixture passes. Further mixing occurs above the plate and the flame is ignited by a permanently lit pilot jet. A sheet of clear blue flame is established over the top of the burner plate or matrix. Large gas boilers, over 45 kW, use forced-draught burners in which gas and air are blown under pressure into a swirl chamber where the flame is established, with a fair amount of noise.

Gas burner control of appliances under 45 kW is achieved as follows. The pilot flame, which is ignited manually or with a piezoelectric spark, heats a thermocouple circuit whose electrical voltage and current holds open the flame-failure solenoid valve. In the event that the pilot flame is extinguished, the flame-failure solenoid becomes de-energized because the thermocouple is no longer heated, and the main gas supply is stopped. As long as the pilot is alight, the control thermostat in the boiler waterway or warm-air unit outlet duct is able to ignite the burner by opening its own solenoid valve. Figure 12.6 shows the

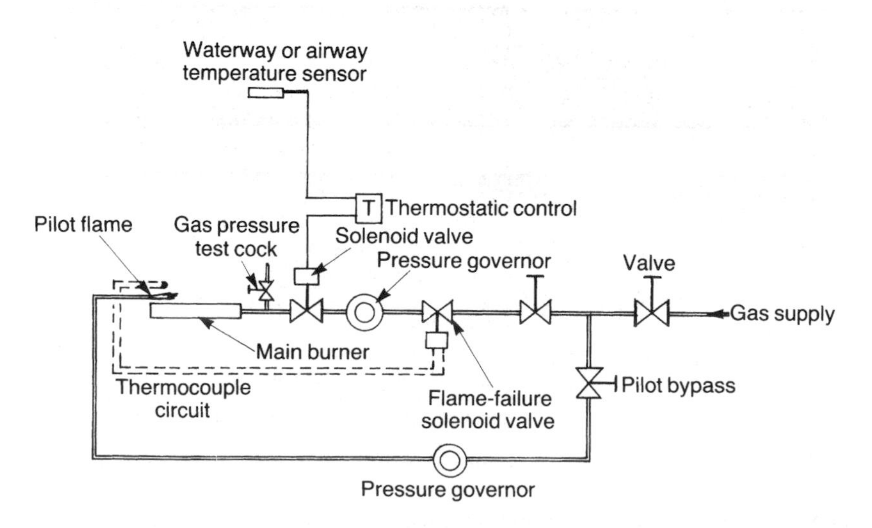

Figure 12.6 Gas burner controls

diagrammatic arrangement of the control system for a gas burner of less than 45 kW. A combination gas valve is used to incorporate some of these functions into one unit.

The governor, shown in Fig. 12.7, maintains a constant gas pressure to the burner by means of a synthetic nitrile rubber diaphragm which rises when the inlet pressure is increased. The diaphragm is connected to a valve, which closes when the diaphragm rises. This action increases the resistance to gas flow and maintains the outlet gas pressure at the set value. An adjustable spring is used to set the downstream pressure appropriate to the gas flow rate required by the burner.

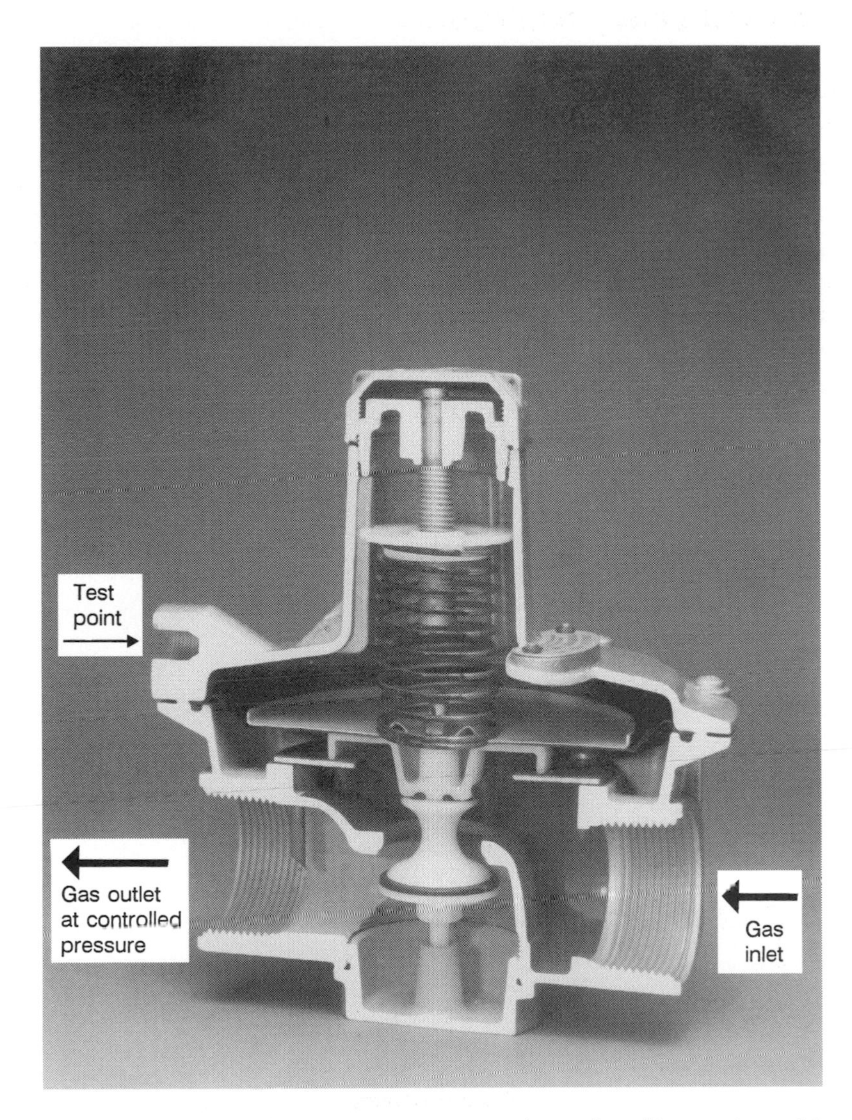

Figure 12.7 Spring-loaded gas pressure governor (reproduced by courtesy of Jeavons Engineering Company)

Questions

1. A gas-fired water heater has a heat output of 30 kW at an efficiency of 75% and a gas pressure of 1225 Pa. Calculate the gas flow rate required at the burner and the reading on a U-tube manometer in millimetres water gauge at the outlet from the pressure governor.

2. Express gas pressures of 55 mm H_2O, 350 N/m^2, 75 Pa, 1.5 kPa and 1.05 bar in millibars.

3. The pipe from a gas meter to a boiler is 18 m long and has elbows that cause a resistance equivalent to 25% of the measured length. Calculate the maximum allowable pressure loss rate for the pipeline.

4. If the maximum allowable pressure loss rate in a pipeline can be 2.3 Pa/m and the resistance of the pipe fittings amounts to 20% of straight pipe, what is the maximum length of pipe that can be used?

5. A gas boiler of 43 kW heat output and 75% efficiency is supplied from a meter by a pipe 23 m long. The resistance of the fittings amounts to 25% of the pipe length. Find the gas supply pipe size needed.

5. Calculate the actual gas pressure drop through a 22 mm pipe carrying 0.81 l/s when the pipe length is 12 m and the fittings resistance amounts to 20% of its length.

7. Sketch and describe the gas service entry and meter compartment arrangements for housing.

8. Explain how a gas meter measures gas flow rate and total quantity passed during a year.

9. List the methods of flueing gas appliances and compare them in relation to their application, complexity and expected cost.

10. Explain, with the aid of sketches, the sequence of operation of safety and efficiency controls on gas fired appliances.

13 Electrical installations

Learning objectives

Study of this chapter will enable the reader to:

1. understand how electricity is generated and distributed;
2. know the difference between single- and three-phase electricity;
3. distinguish line, neutral and earth conductors;
4. calculate the resistance of conductors;
5. understand the temperature effect of a current;
6. calculate current and power in electrical circuits;
7. know how to measure current and voltage;
8. use power factor;
9. calculate series and parallel circuit resistances;
10. find the current capacity of cables;
11. choose cable sizes;
12. calculate permissible cable lengths;
13. understand temporary electrical installations for construction sites;
14. calculate the total electrical loading in kilovolt-amperes and amperes for an installation;
15. estimate the total cost of electricity likely to be consumed in an installation during normal use;
16. choose the correct fuse rating;
17. understand the operation of fuses and circuit-breakers;
18. know the distribution of electricity within buildings;
19. identify the use of isolating switches, distribution boards and meters;
20. understand earth bonding of services;
21. know the types of cable, conduits and their applications;
22. understand the principles of ring circuits;
23. understand how electrical systems are tested;
24. be aware of telecommunications cables;
25. design lightning conductor systems;
26. identify the graphic symbols used on drawings.

Key terms and concepts

Introduction

The safe and economical use of electricity is of paramount importance to the building user and the world as it is the most highly refined form of energy available. Electricity production consumes up to three times its own energy value in fossil fuel, and electricity in its distributed form is potentially lethal.

In this chapter the handling methods and safety precautions for utilizing electricity are explained and a range of calculations, which can easily be performed by the services designer or constructor prior to employing specialist help, is introduced.

Electricity distribution

The electrical power generating companies supply electrical power into the national 400 kV grid system of overhead bare wire conductors. This very high voltage is used to minimize the current carried by the cables over long distances. Step-down transformers reduce the voltage in steps down to 33 kV, when it can be supplied to industrial consumers and to other transformer stations on commercial and housing estates.

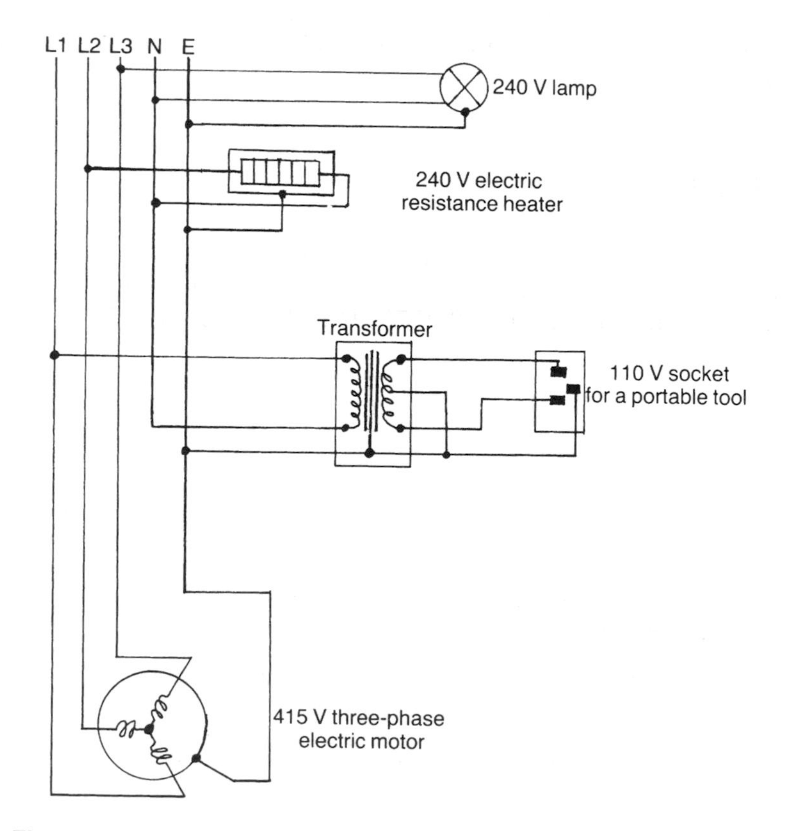

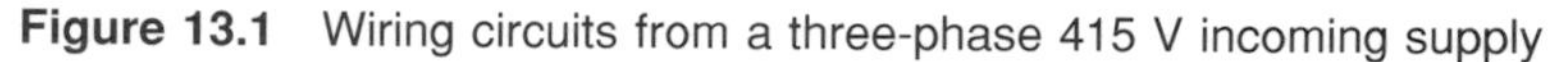

Figure 13.1 Wiring circuits from a three-phase 415 V incoming supply

The electricity-generating alternator rotates at 50 Hz (3000 rev/min) and has three coils in its stator. The output voltages and currents from each coil are identical but are spaced in time by one-third of a revolution, 120 °. Each coil generates a sine wave or phase voltage that has the same heating effect as a 240 V continuous direct current supply. This is its root mean square (RMS) value. The RMS value of the three phases operating together is 415 V.

Figure 13.1 shows the connections to a three-wire three-phase 415 V, 50 Hz alternating current supply entering a non-domestic building. Various circuits of different voltages are supplied from the incoming mains. Equal amounts of power are fed into each phase, and so it is important that power consumption within a building is equally shared by each line. The neutral wire is a live conductor in that it is the return path to the alternator for the current which has been distributed.

A balanced load, such as a three-phase electrical motor driving an air-conditioning fan, water pump or lift motor, does not produce a current in the neutral wire. This is because an alternating current flows alternately in the forward and backward directions along the line wire. The overall effect of three driving coils in the motor is a balance in the quantity and direction of the current taken from the line conductors. There is no net return current in the neutral wire from such a balanced load. Single-phase electrical loads, which are not in balance, produce a net current in the neutral conductor.

The casings of all electrical appliances are connected to earth by a protective conductor, the earth wire, connected to the earthed incoming service cable of the electricity supply authorities or an earth electrode in the ground outside the building. Gas and water service pipes are bonded to the earth by a protective conductor.

Circuit design

The resistance R ohms (Ω) of an electrical conductor depends on its specific resistance $\rho \Omega$ m, its length lm and its cross-sectional area A m^2. The specific resistance of annealed copper is 0.0172 $\mu\Omega$ m (μ, micro stands for 10^{-6}) at 20 °C.

$$R = \rho \frac{l}{A}\ \Omega$$

EXAMPLE 13.1

Calculate the electrical resistance per metre length at 20 °C of a copper conductor of 2.5 mm^2 cross-sectional area.

$$R = \frac{0.0172}{10^6}\ \Omega\ \text{m} \times \frac{1\ \text{m}}{2.5\ \text{mm}^2} \times \frac{10^6\ \text{mm}^2}{1\ \text{m}^2}$$

$$= 0.0069\ \Omega$$

The resistance of a cable increases with increase in temperature and the temperature coefficient of resistance α of copper is 0.004 28 Ω/Ω °C at 0 °C. If the resistance of the conductor is R_0 at 0 °C, then its resistance at another temperature R_t can be found from

$$R_t = R_0\ (1 + \alpha\ t)\ \Omega$$

where t is the conductor temperature (°C).

EXAMPLE 13.2

Find the resistance of a 2.5 mm^2 copper conductor at 40 °C.

R_0 is not known but the resistance of this conductor at 20 °C was found in Example 13.1 and t can represent the increase in temperature above this value. A graph of resistance versus temperature would reveal a straight line of slope α.

$$\begin{aligned} R_{40} &= R_{20}\ (1 + \alpha \times 20)\ \Omega \\ &= 0.0069 \times (1 + 0.004\,28 \times 20)\ \Omega \\ &= 0.0075\ \Omega \end{aligned}$$

The relation between applied voltage, electric current and resistance is given by Ohm's law:

$$I \text{ amps} = \frac{V \text{ volts}}{R \text{ ohms}}$$

Figure 13.2 shows how an ammeter and a voltmeter are connected into a circuit to measure power consumption. The load may be an electrical resistance heater or tungsten filament lamp, in which case the power consumption in watts is found from

$$\text{power in watts} = V \text{ volts} \times A \text{ amps} \times \text{cox}\ \phi$$

for single phase and

$$\text{power in watts} = V \text{ volts} \times A \text{ amps} \times \sqrt{3} \times \cos \phi$$

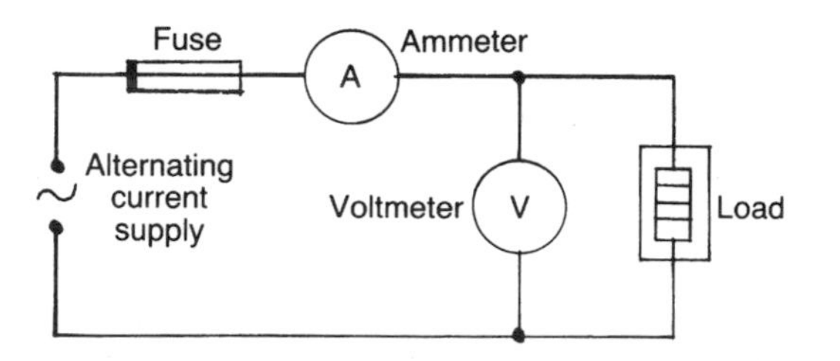

Figure 13.2 Measurements of power consumption in an electrical circuit

for three phase, where $\cos\phi$ is the power factor (zero to unity). An electrical resistance heater and tungsten lamps are purely resistive loads whose power factor, $\cos\phi$, is unity, 1.

Loads such as electric motors and fluorescent lamps have a property known as inductance, which causes the current to lag behind the voltage that is producing that current. This is due to an electromotive force (e.m.f.), i.e. a voltage, which opposes the incoming e.m.f. The densely packed electromagnetic windings of a motor have a high inductance and thus the 50 Hz cyclic variation of voltage and current is opposed by the 'inertia' of the equipment. The opposing e.m.f. comes from the expanding and collapsing magnetic fields of the input power around the conductors.

The lag angle $\phi°$ between peak voltage and peak current means that the instantaneous available power is less than the product of the two peaks. Power factor is the term used to differentiate between useful output power in watts and the input instantaneous product of voltage and current to the load:

$$\text{power factor} = \frac{\text{useful power in watts}}{\text{input volts} \times \text{amps}} = \frac{\text{watts}}{\text{volt} \times \text{amps}}$$

$$= \frac{\text{real power (to do work)}}{\text{apparent power}}$$

$$\text{power factor} = \frac{\text{kW}}{\text{kVA}}$$

$$= \frac{\text{kilowatts}}{\text{kilovolt-amps}} = \frac{\text{kW real}}{\text{kVA (apparent)}}$$

Capacitors have an electrical storage capability, which is used to overcome the effects of inductance. Power factors of electrical equipment are commonly 0.85 and these can be improved to 0.95 by the addition of power-factor-correcting capacitors.

Several loads to a circuit may be connected either in series or in parallel with each other. For series-connected resistances

$$R = R_1 + R_2 + R_3 + \cdots$$

For parallel-connected resistances

$$\frac{1}{R} = \frac{1}{R_1} + \frac{1}{R_2} + \frac{1}{R_3} + \cdots$$

The resistance of electrical cables must be sufficiently low that the cables do not become significant sources of heat and run at temperatures that could be a fire hazard or damage their electrical insulation. Such heat generation would generally be wasted energy. The maximum voltage drop in a cable permitted in the Institution of Electrical Engineers (IEE) Wiring Regulations is 4% of the nominal supply voltage from the consumer's intake terminal to any point in the installation at full load current.

Cables that are grouped together, run in conduits or are covered with thermal insulation, say in a roof space, can operate at a temperature above the 30 °C ambient condition assumed in the selection of their size and their electrical insulation. Where their temperature is likely to rise above this value, their current-carrying capacity is reduced by appropriate rating factors during design of the system. Care should be taken to allow natural cooling of all cable routes. The current-carrying capacity has to be 1.33 times the design current for cables partly surrounded by thermal insulation and twice the design current if they are wholly surrounded. This will generally mean an increase by one or two cable sizes.

EXAMPLE 13.3

Calculate the power consumption and resistance of a 240 V filament lamp if it has 1.5 A passing through it.

$$\text{power in watts} = \text{volts} \times \text{amps}$$

$$\text{power} = 240 \text{ V} \times 1.5 \text{ A}$$

$$= 360 \text{ W}$$

From Ohm's law

$$I = \frac{V}{R}$$

and so

$$R = \frac{V}{I}$$

$$= \frac{240}{1.5}$$

$$= 160 \ \Omega$$

EXAMPLE 13.4

PVC insulation on a conductor carrying 415 V has an electrical resistance to earth of 500 MΩ. What leakage current could flow through the PVC when the cable is laid on an earthed metal support? ($1 \text{ M}\Omega = 10^6 \ \Omega$.)

The difference between line and earth is 240 V. From Ohm's law

$$I = \frac{V}{R}$$

$$= \frac{240 \text{ V}}{500 \times 10^6 \ \Omega}$$

$$= 0.48 \times 10^{-6} \text{ A}$$

$$= 0.48 \ \mu\text{A}$$

EXAMPLE 13.5

Compare the currents carried by an overhead line at 400 kV and 33 kV for the transmission of 10 MW of power for unity power factor, cos $\phi = 1$.

For a three-phase system, using

$$\text{watts} = \text{volts} \times \text{amps} \times \sqrt{3} \times \cos\phi$$

$$\text{current in amps} = \frac{\text{watts}}{\text{volts}} \times \frac{1}{\sqrt{3}} \times \frac{1}{\cos\phi}$$

For 400 kV

$$\text{current} = \frac{10 \times 10^6 \text{ W}}{400 \times 10^3 \text{ V}} \times \frac{1}{\sqrt{3}} = 14.4 \text{ A}$$

For 33 kV

$$\text{current} = \frac{10 \times 10^6 \text{ W}}{33 \times 10^3 \text{ V}} \times \frac{1}{\sqrt{3}} = 175 \text{ A}$$

This demonstrates the advantages of high-voltage transmission of electrical power as smaller cable sizes can be used for the long distances involved.

Cable capacity and voltage drop

The maximum current-carrying capacities and actual voltage drops according to the IEE Regulations for Electrical Installations (16th Edn, 1991) for unenclosed copper cables which are twin-sheathed in PVC, clipped to the surface of the building, are given in Table 13.1. Flexible connections to appliances may use 0.5 mm^2 conductors for 3 A and 0.75 mm^2 conductors for 6 A loads. The

Table 13.1 Electrical cable capacities

Nominal cross-sectional area of conductor (mm^2)	Maximum current rating (A)	Voltage drop in cable (mV/Am)
1	15	44
1.5	19.5	29
2.5	27	18
4	36	11
6	46	7.3
10	63	4.4
16	85	2.8

maximum voltage drop allowed is 4% of the 240 V nominal supply (Jenkins, 1991).

EXAMPLE 13.6

Find the maximum lengths of 1 mm^2, 1.5 mm^2 and 2.5 mm^2 copper cable which can be used on a 240 V circuit to a 3 kW immersion heater.

$$\text{current} = \frac{3000 \text{ W}}{240 \text{ V}} = 12.5 \text{ A}$$

$$\text{allowed voltage drop} = \frac{4}{100} \times 240 = 9.6 \text{ V}$$

$$\text{maximum length or run} = \frac{\text{maximum voltage drop allowed mV}}{\text{load current A} \times \text{voltage drop mV/A m}}$$

For 1 mm^2 cable

$$l = \frac{9.6 \times 10^3}{12.5 \times 44} \text{ m} = 17.5 \text{ m}$$

For 1.5 mm^2 cable

$$l = \frac{9.6 \times 10^3}{12.5 \times 29} \text{ m} = 26.5 \text{ m}$$

For 2.5 mm^2 cable

$$l = \frac{9.6 \times 10^3}{12.5 \times 18} \text{ m} = 42.7 \text{ m}$$

Construction site distribution

A list is made of all electrical equipment to be used on site in order to assess the maximum demand kilovolt-amperes and cable current rating required. An estimate of the cost of electricity for running the site may also be made.

EXAMPLE 13.7

A building site is to have the following electrical equipment available for use:

(a) tower crane, electric motors totalling 13 kW at 415 V;
(b) sump pump, 2 kW at 240 V;
(c) 50 tungsten lamps of 100 W each at 240 V;
(d) 8 floodlamps of 500 W each at 240 V;
(e) 10 hand tools of 750 W at 110 V.

1. Find the total kilovolt-amperes to be supplied to the site if the power factor of all rotary equipment is 0.85.
2. Find the electrical current rating for the incoming supply cable to the site.
3. Estimate the cost of electricity consumed on the site during a 12-month contract.

For rotating machinery

$$\text{power, } VA = \frac{\text{W}}{0.85}$$

For single-phase current

$$\text{line current} = \frac{VA}{V}\ \text{A}$$

For three-phase current

$$\text{line current} = \frac{VA}{V \times \sqrt{3}}\ \text{A}$$

The results of the calculations are given in Table 13.2. Hence the answers required are as follows.

1. The total kilovolt-amperes required for site is 35.471 kVA.
2. The incoming supply cable capacity should be $(35\,470/415 \times \sqrt{3}) = 49.4$ A.
3. Assume that the crane, pump and tools are running for 25% of an 8 h working day, 5 days per week for 52 weeks, 10 of the tungsten lamps are for security lighting 16 h every night, and the remaining 40 tungsten lamps and the floodlamps are used for 3 h per day, 5 days per week for the winter period of 20 weeks. The crane, pump and tools are working for

$$0.25 \times 8\ \frac{\text{h}}{\text{day}} \times 5\ \frac{\text{days}}{\text{week}} \times 52\ \text{weeks} = 520\ \text{h}$$

Table 13.2

Equipment	Power (W)	Number	VA	V (V)	A (A)
Tower crane	13 000	1	15 294.12	415	21.28
Sump pump	2 000	1	2 352.94	240	9.80
Lamps	100	50	5 000	240	20.83
Flood lamps	500	8	4 000	240	16.67
Hand tools	750	10	8 823.53	110	80.21
Total			35 470.59		

The security lamps are working for

$$16\,\frac{\text{h}}{\text{day}} \times 7\,\frac{\text{days}}{\text{week}} \times 52\text{ weeks} = 5842\text{ h}$$

The other lamps are working for

$$3\,\frac{\text{h}}{\text{day}} \times 5\,\frac{\text{days}}{\text{week}} \times 20\text{ weeks} = 300\text{ h}$$

The total number of kilovolt-ampere hours is found from

$$\text{kVAh} = \text{number of appliances} \times \text{kVA per appliance} \times \text{operation hours}$$

as shown in Table 13.3.

If electricity costs 5.65 p per unit (kVA h) then the estimated cost for the 1-year contract will be

$$\text{cost} = 5.65\,\frac{\text{P}}{\text{kVA h}} \times 21\,972.8\text{ kVA h} \times \frac{£1}{100\text{ p}}$$

$$= £1241.46$$

Adequate safety in the use of electricity on site is essential and a legal obligation upon employers. The area electricity supply authority must be contacted before any site work, to establish the locations of overhead and

Table 13.3

Equipment	Power (kW)	Number	kVA each	Hours	kVA h
Tower crane	13	1	15.29	520	7950
Sump pump	2	1	2.35	520	1222
Tungsten lamps, security	0.1	10	0.10	5824	5824
Lamps	0.1	40	0.10	300	1200
Flood lamps	0.5	8	0.50	300	1200
Tools	0.75	10	0.88	520	4576
				Total power used =	21 972

underground power cables. Assume that all lines are live. Overhead lines are not insulated except at their suspension points. The Electricity At Work Regulations 1989 need to be consulted.

Roadways for site vehicles should be made underneath overhead cables by the erection of clearly marked goalposts, on each side of the cable route, through which traffic must pass. These goalposts form the entrance and exit from the danger area and are spaced at 1.25 jib lengths of the mobile crane to be used on site, or at a minimum of 6 m either side of the cables. Entry to the roadway other than through the goal posts is barred with wooden fencing or tensioned ropes with red and white bunting at high and low levels.

Underground cables that become exposed during excavations must remain untouched until the electricity authority has given advice. Safe working clearances will be ascertained at this time.

Hand lamps and tools are operated from a transformer at 110 V to reduce the damage caused by an electric shock. For work within tunnels, chimneys, tanks or drains, 25 V lamps, or battery lamps, are advised. Each portable appliance should be checked by the operator before use and also inspected and tested by a competent person at intervals not exceeding 7 days. Records of maintenance and safety checks should be kept.

A weatherproof cubicle is provided at the edge of the site by the main contractor for the electricity authority's temporary fuses and main switch. Site distribution cables are supported from hangers on an independent wire suspended between poles around the edge of the site, with spur branches to site accommodation and work areas. The minimum clear heights under cables should be 4.6 m in positions inaccessible to vehicles, 5.2 m in positions accessible to vehicles and 5.8 m across roads.

The site programme for the main contractor is as follows.

1. Arrange a precontract meeting between the executives responsible for the work.
2. List electrical requirements for all temporary plant.
3. Prepare layout drawings showing equipment siting and electrical loads, including site offices, stores, canteen, sanitary accommodation and illumination. Carefully site equipment to minimize interference with the construction work.
4. Apply to the electricity supply authority for a temporary supply to the site, stating maximum kilovolt-ampere demand and voltage and current requirements.
5. Provide the electrical distribution equipment: 415 V, three-phase, 50 Hz for fixed plant and movable plant fed by trailing cables; 240 V, single phase, 50 Hz for site accommodation and site illumination; 110 V, single-phase, 50 Hz for portable lighting and tools; 50 V or 25 V, single-phase, 50 Hz for portable lamps to be used in confined spaces and damp areas.
6. Once site accommodation is in place, ensure that a satisfactory semi-permanent electrical system is provided.
7. A competent electrician is to carry out all site work. His name, designation and location are to be prominently displayed on site, so that faults, accidents and alterations can be expedited. All plant and cables are regularly inspected and tested.

8. Display the electricity regulations placard and the first aid instruction card.
9. Appraise the use of electrical equipment and distribution arrangements weekly to ensure that the most efficient use is made of the system. Idle equipment is returned to the supervised store.

Distribution equipment for site use is housed in weatherproof rugged steel boxes on skids. Built-in lifting lugs facilitate crane or manual transportation. The main items are as follows.

Supply incoming unit (SIU): a unit to house the electricity supply authority's incoming cable, service fuses, neutral link, current transformers and meters. Outgoing circuits of 100 A or more are controlled by triple pole and neutral (TPN) switches for three phase and either cartridge fuses or residual current devices to break the circuits in the event of a current flowing to earth from a fault in a wire or item of plant

Main distribution unit (MDU): a cubicle, which may be bolted to the SIU, providing a number of single- and three-phase outlets through weatherproof plugs and sockets. Each outlet is protected by a residual current device

Transformer unit (TU): a unit providing 110 V single- and three-phase supplies with 65 V between any line and earth for safety in the use of hand tools. It may be coupled to an SIU. Each outlet circuit is protected by a residual current device

Outlet unit (OU): easily portable distribution box fed from an MDU or TU for final connection of subcircuits to tools, lighting or motors. Each circuit is protected by a residual current device and clearly labelled with its voltage, phases and maximum current capacity

Extension outlet unit (EOU): similar to an OU but fed from a 16 A supply and circuit protection is by a cartridge fuse. It may have up to four 16 A outlets on a cubicle metal box

Earth monitor unit (EMU): flexible cables supplying electricity to movable plant; may incorporate a separate pilot conductor in addition to the protective conductor to earth. A small current passes through the portable plant and the EMU via the pilot and protective conductors. If the earth conductor is broken, the EMU current is interrupted and the circuit is automatically isolated at its circuit-breaker

Semi-permanent installations are run in metal-sheathed or armoured flexible cable. The metal sheath is permanently earthed as is the earth wire. An

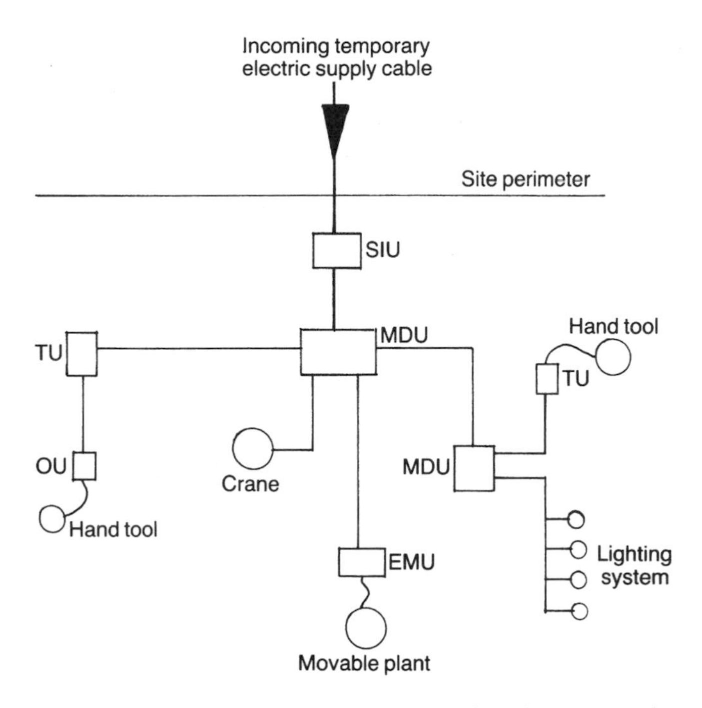

Figure 13.3 Distribution of electricity during site construction

oversheath of PVC or an oil-resisting and flame-retardant compound is provided.

Connections from outlet units to hand-operated tools and lighting systems are made in tough rubber-sheathed (TRS) flexible cable. Walkways and ladders must be kept clear of cables and the cables must be kept 150 mm from piped services. Cables under site roadways are installed in a temporary service duct, such as drain pipework, at a depth of 600 mm and with markers at each end of the crossing. Figure 13.3 shows an arrangement of a site's electrical distribution.

Safety cut-outs

An electric shock is sustained when part of the human body establishes contact between a current-carrying conductor and earth. It is also likely from contact with two conductors of different phase. Voltages of less than 100 V have proved fatal under certain circumstances. The size of the current depends on the applied voltage and the body's resistance to earth.

Rubber shoes or flooring greatly increase the resistance of the shock circuit. Body resistance with damp skin is around 1100 Ω at 240 V; thus a current of 218 mA could flow. At 55 V, body resistance is 1600 Ω and this could produce 34 mA. A current of 1–3 mA is generally not dangerous and can just be perceived. At 10–15 mA acute discomfort and muscle spasm occurs, making release from the conductor difficult. A current of 25–50 mA causes severe muscle spasm and heart fibrillation, and will probably be fatal.

Prolonged exposure to a shock current causes burns from the heating-effect of the supply. If electric shock occurs, switch off the supply without contacting any metal component. If necessary, begin resuscitation and summon qualified medical assistance immediately.

During normal operation, current flows from the 240 V (or other nominal value) line, through the appliance and along the neutral conductor back to the power station alternator. 240 V is the nominal drop of voltage across the appliance. Should either the line or the neutral conductors come into contact with a conducting material that is earthed, owing to a wiring fault, the current will choose the lower-resistance path to earth on its return journey to the earthed alternator. Immediately, a higher current will flow and the appliance has become a shock hazard.

The increased heating effect of the fault current can be used to melt a rewirable or cartridge fuse at the appliance, its fault current being 60% above the stated continuous rating. High rupturing capacity (HRC) cartridge fuses have silver elements in a ceramic tube, which is packed with granulated silica. They allow for the high starting currents required for electric motors. The correct fuse rating must be used for each appliance to avoid damage to cables and buildings from overheating through the use of too high a fuse capacity. Fuse ratings are quoted in Table 13.4 and are found from

$$\text{fuse current rating} = 1.6 \times \frac{\text{appliance input VA}}{\text{circuit voltage}}$$

A faster-acting protection, with greater reliability, whose operation can be tested is provided by a miniature circuit-breaker (MCB), which opens switch contacts upon detection of an excess current. Circuit faults that cause a leakage to earth are detected by a residual current circuit-breaker (Fig 13.4).

During normal operation, the line and neutral coils around the electromagnetic core generate equal and opposite magnetic fluxes, which cancel out. A current leakage to earth at the appliance reduces the current in the neutral conductor and a residual current is generated in the core by the line coil. This residual current generates an e.m.f. and current in the detector circuit, which in turn energizes the trip solenoid, which opens the double-pole switch, or TPN switch in a three-phase circuit, and isolates the appliance.

Residual currents of 30 mA are set for sensitive applications and outdoor equipment. They are frequently used in addition to fuses. Pressing the test switch

Table 13.4 Fuse ratings for 240 V single phase and unity power factor

Power consumption (W)	Fuse required (A)
120	0.8
240	1.6
720	4.8
1200	8
3120	20.8
3600	24
7200	48

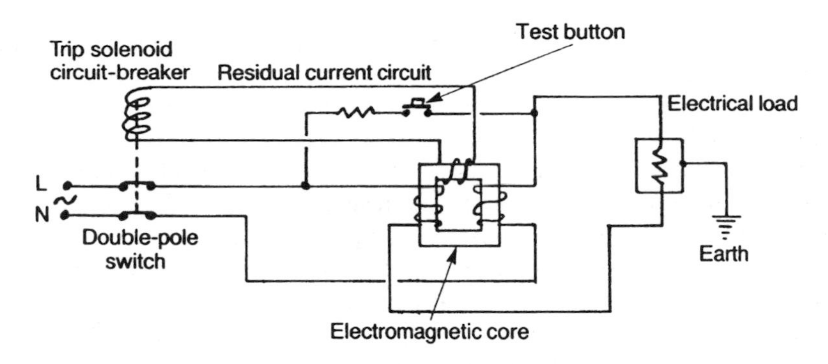

Figure 13.4 Residual current circuit breaker

short-circuits the line coil and a residual current is generated in the core, tripping the residual current device for test purposes.

Fuses and circuit-breakers are selected for their time–current characteristics in relation to the risk that is being protected against. Figure 13.5 shows the performance curve for a cartridge fuse to British Standard 1361 : 1971 type 1 having a 15 A continuous rating for domestic installations and a comparative

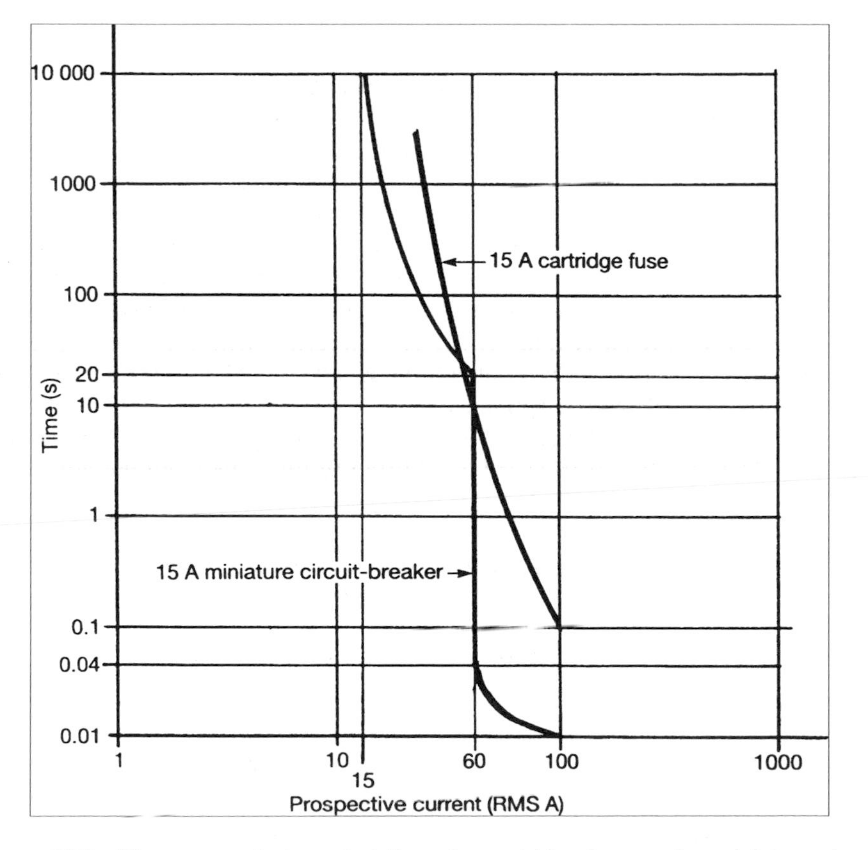

Figure 13.5 Time–current characteristics of a cartridge fuse and a miniature circuit-breaker

miniature circuit-breaker (IEE Wiring Regulations for Electrical Installations, 16th Edn, 1991). The horizontal and vertical scales of the graph are logarithmic. It can be seen that both devices will pass the design maximum 15 A current without opening the circuit. The cartridge fuse is designed to melt if a current of 46 A were to flow for a period of 5 s, or 97 A for 0.1 s. Other combinations of heating effect are in proportion. These points lie to the left of the fuse curve and do not cause it to break.

The MCB is designed to open the circuit when 60 A flows for between 0.1 and 5 s. A lower current value will take in excess of 20 s to open the contacts. A fault in the protected circuit that causes the current to rise above 60 A will open the MCB in less than 0.1 s. A miniature circuit-breaker and a residual current device (RCD) may be combined in one moulded casing to protect against excess current, short circuit and a leakage to earth. The MCB has a bimetallic strip thermal and magnetic trip mechanism. The speed of operation of an RCD is typically 20 ms (Midland Electrical Manufacturing Company Limited).

Electrical distribution within a building

The incoming cable, residual current device and meter are the property of the electricity supply authority. Underground cables are at a depth of 760 mm under roads, and enter the building through a large radius service duct of 100 mm internal diameter. A drainpipe can be used for this purpose, laid through the foundations and rising directly to the meter compartment. External meter compartments can be used. The meter should not be exposed to damp or hot conditions and the electricity supply authority's advice should be sought. Figure 13.6 shows a distribution system for a dwelling.

Each circuit has a fuse or circuit-breaker and the fused distribution board connects the neutral and earth protective conductors to the supply cable. Appliances have a cartridge fuse at their connecting plug. Three-phase distribution in a large building is shown in Fig. 13.7. Switches are used to enable separation of individual circuits as well as appliances. A fuse or circuit-breaker is always fitted on the live line so that the incoming current is disconnected.

Power socket outlets are fitted into a ring circuit as shown in Fig. 13.8. Care must be taken not to overload the circuit by connecting appliances whose total current consumption would exceed the 30 A limit, particularly in kitchens.

Types of cable used in distribution systems are as follows.

PVC-insulated and -sheathed

PVC-insulated and -sheathed cables consist of copper conductors of multi-stranded or solid wire having sizes from 1 mm^2 to 16 mm^2 cross-sectional area. Single-, twin- or three-core cables, with or without earth wires, are used. They are among the cheapest cables available and can be pulled through conduits, trunking or holes bored in floor joists. Such holes are drilled 50 mm below the floorboards. Ambient temperature limits for the cable are 0–65 °C and the cable must not exceed 70 °C in use. Colour coding of the insulation ensures

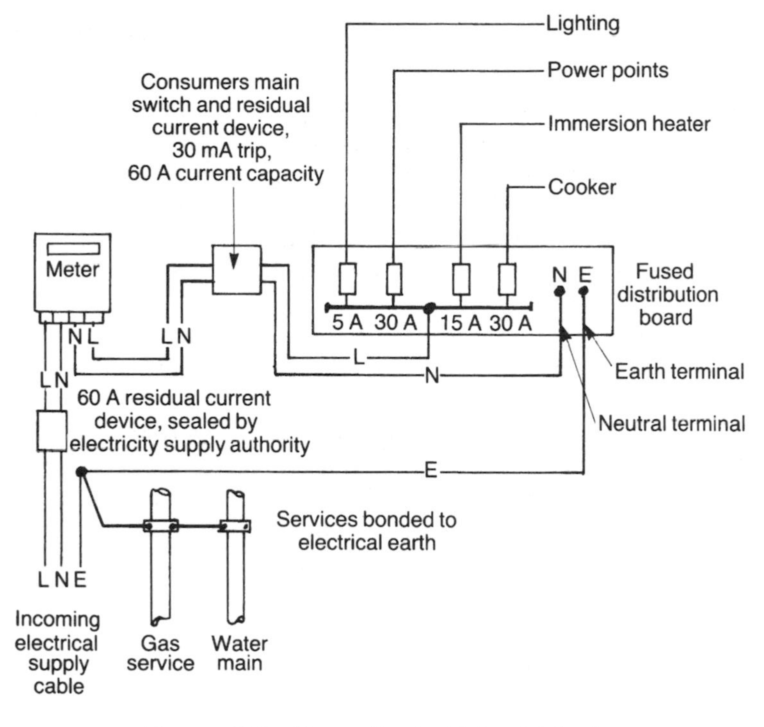

Figure 13.6 Domestic electricity distribution

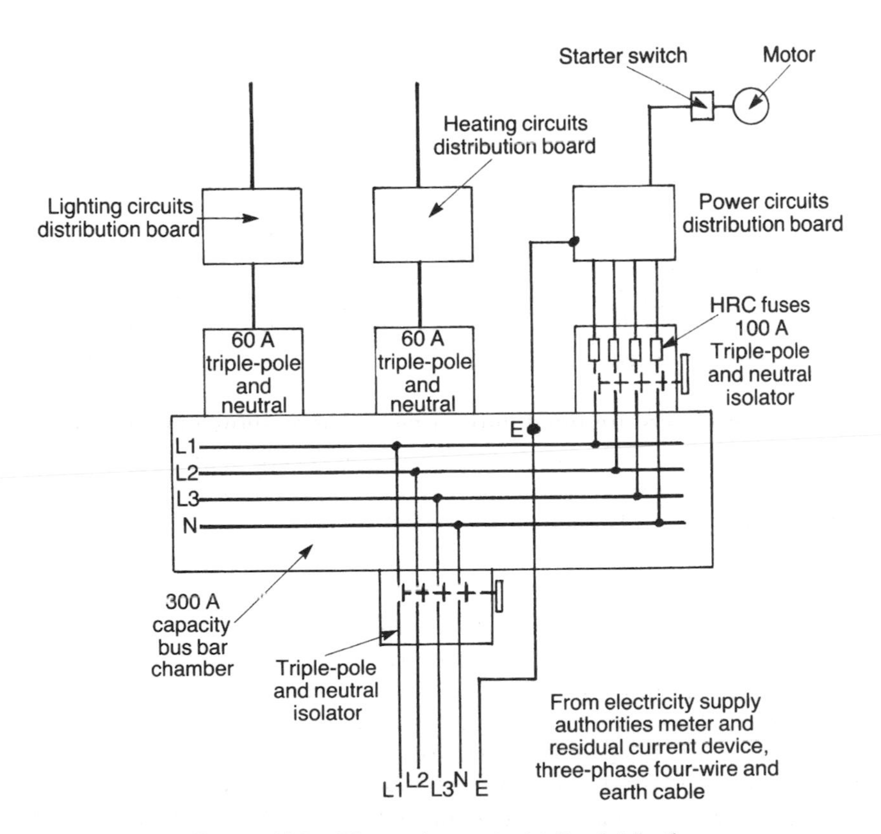

Figure 13.7 Three-phase electricity distribution

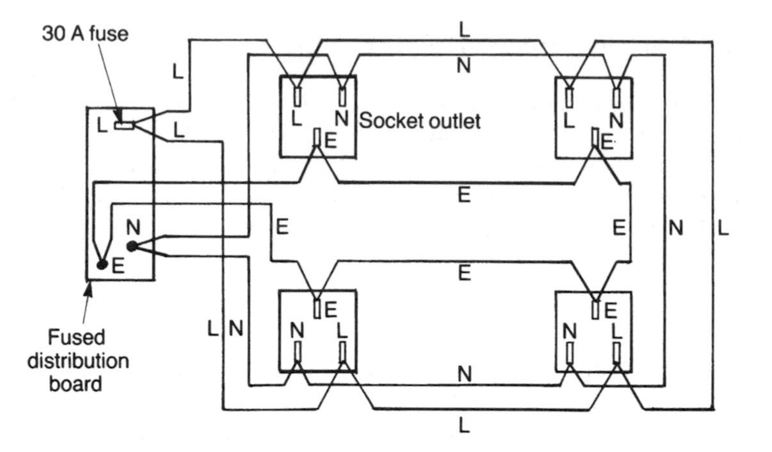

Figure 13.8 Ring main to socket outlets

correct polarity at terminals. The earth, or protective conductor, is always green, neutral is always black and the line conductor is red on single phase. Three-phase line conductors are one red, one yellow and one blue.

PVC is a flexible, non-hygroscopic, tough, durable, corrosion-resistant and chemically inert material, which is used for both electrical insulation and conduits. Installations are tested for earth leakage through their insulation at regular intervals and cables are replaced after about 20 years because of ageing of the PVC.

Mineral-insulated copper-sheathed

Solid copper conductors surrounded by compressed magnesium powder are factory-fitted inside a copper tube or sheath. A PVC oversheath gives extra protection for cables that are to be buried in building materials. These cables are used for both internal and external wiring and withstand severe conditions, even continuing to operate during a fire. They can be operated continuously at temperatures of up to 250°C compared with only 70 °C for PVC-covered cables.

The soft copper cable is supplied in rolls and is run continuously from the distribution board to the switch or power point. Screwed gland joints are designed to exclude dampness from the hygroscopic insulant. One sheath may encapsulate up to 19 1.5 mm^2 conductors. They are non-ageing and unlikely to require replacement during the building's period of service.

Particular applications are fire alarm systems, in petrol filling stations and within boiler plant rooms. The copper sheath is used as the earth protective conductor and also withstands severe abuse, flattening or twisting without short-circuiting the conductors. The cables can be bent by hand or machine, and conduit fittings are not needed. The overall diameters of mineral-insulated cables are much less than those of other types of comparably rated cable system. Only a thin plaster covering is needed if the cables are not to be surface mounted.

Armoured PVC-insulated, PVC-sheathed

Copper or aluminium conductors in PVC insulation, a PVC bedding, galvanized steel wire armour and a PVC sheath are used for heavy-current cables to large machinery and mobile plant on site. The cable can be run on the building surface, laid on the ground or put in a trench. Screwed gland nuts are used to bond the armour to the appliance casing.

Busbar

Bare copper or aluminium rectangular bar conductors are supported on insulators within a sheet metal duct or trunking. Vertical service shafts within buildings may have a rising busbar system with tap-off points at each floor level for the horizontal distribution of power with insulated cables. Small busbar distribution systems can be used in retail premises and within raised floors in office buildings. These provide flexibility in the siting of outlet boxes.

Overhead distribution in factories allows connections at any point along the busbar with plug-in boxes, allowing machinery to be moved at a later date. The trunking acts as the protective conductor. Three-phase 415 V supplies are distributed at current ratings of 100–600 A, branches being at 30, 60 or 100 A. An armoured PVC cable can supply the incoming end of the busbar. Where the system passes through a fire-resistant partition in the building, a fire barrier is fitted across the inside of the trunking.

Other cable insulants

Flexible external and special application cables are used as follows.

Butyl rubber: for up to 85 °C continuous plus higher overload rating. Additional heat-resisting glass fibre wrapping increases the continuous temperature to 100 °C. Flexible cord use

Ethylene propylene rubber (EP rubber): an elastomer with similar properties to butyl rubber. Has improved resistance against the effects of water and long-term ageing. Retains its flexibility down to −70 °C

Silicone rubber: withstands −75 °C and a wide variety of chemical and oxidizing agents, weak acids, salts and vegetable oils. It retains its insulation and elasticity at working temperatures of up to 150 °C

Polychloroprene (PCP): for general purpose heat-resisting, oil-resisting and flame-retardant (HOFR) use in the presence of oil and petrol

Chlorosulphonated polyethylene (CSP): heavy duty use in aggressive atmospheres in laundries or in the presence of oil and petrol

Conduit and trunking

Circular conduit systems are used to carry insulated cables and should last the service period of the building. The space occupied by the cable must not exceed 40% of the cross-sectional area of the inside of the conduit to allow for ventilation to remove the heat generated by cable resistance.

Materials used are light- or heavy-gauge steel, depending upon exposure to damp or explosive fumes. The external conduit will be galvanized. Lug grip connections are used for light-gauge pipework and screwed joints for heavy-gauge pipework. Pipe sizes are 16, 20, 25 and 32 mm. The conduits are used as the protective conductor.

PVC conduit, using solvent weld joints, is lighter and easier to handle and does not corrode but requires the cable to incorporate the protective conductor. Its upper temperature limit is 60 °C.

Rectangular galvanized sheet steel or PVC trunking is used where large-cable carrying capacity is needed. These must not be filled to greater than 45% of their cross-sectional area with cables. Surface-mounted trunking can be incorporated into the interior decoration and up to three separate cable compartments are used for different services, including telecommunications, computer, power and lighting cables.

Trunking may be installed under raised timber flooring, within the concrete floor slab or screed, in a grid, branch duct or perimeter distribution arrangement. Outlets that are raised or flush with the floor are provided to suit either fixed or movable office layouts.

Testing

Inspection and testing of an electrical installation is carried out before it is put into service and at regular intervals during use. The main reasons for this are to ensure that its operation will be entirely safe, in accordance with the demands put upon it, and energy efficient. The work entails the following tests.

Verification of correct polarity

A visual inspection is carried out of all fuses and switches to check that they are fitted into a line conductor. The centre contact of each screw lampholder is connected to the line conductor. Plugs and sockets must be correctly connected and wire rigidly held.

Tests of effective earthing

There are four separate tests.

Test of the protective conductor

A 40 V 50 Hz supply of up to 25 A is injected into the earth conductor. Its resistance is not to exceed 1 Ω. An impedance test meter is used.

Earth loop impedance test

A line–earth loop impedance test meter is attached to a 13 A three-pin plug. This is plugged into each power socket and the meter injects a current into the earth protective conductor. The current flows along the supply authority's cable sheath to the local transformer and back to the power socket along the line conductor.

Test of residual current devices

A test transformer providing 45 V is connected to a socket outlet. A short-circuit current is passed from the neutral to the protective conductor, causing the residual current device to trip instantaneously.

Measurement of consumer earth electrode resistance

Where this is used, a test electrode is put into the ground and a steady 50 Hz current is passed between the electrode and the consumer's earth electrode to determine its circuit resistance.

Insulation resistance tests

An insulation test meter is connected between the line and protective conductors. A 500 V direct current is applied to this circuit by the meter and an electrical resistance of 0.5 M Ω or more must be shown.

Test of ring circuit continuity

Each ring circuit is tested for resistance at the distribution board with an ohmmeter. Probes are connected to each side of the line conductor ring and a zero resistance proves a continuous circuit. The test is repeated on the neutral and protective conductor circuits and spur branches.

Tests on installations must be carried out in accordance with the IEE Wiring Regulations by a competent person, who should preferably be a professionally qualified electrical engineer having installation experience. IEE Completion and Inspection Certificates are issued by the engineer.

Telecommunications

Cables between the switchboard and socket for each telephone are accommodated within vertical and horizontal service ducts spaced 50 mm from alternating current cables to avoid speech interference. Alternatively, a partitioned chamber can be reserved throughout the cable trunking.

Lightning conductors

Rules are provided (BS Code of Practice 326: 1965) to determine whether a protection system is required. This depends on building construction, degree of isolation, height of the structure, topography, consequential effects and lightning prevalence. Recommendations on system types, including those for temporary structures, are given.

Copper and aluminium 10 mm rod, 25 mm × 3 mm strip, PVC-covered strip, copper strand and copper braid are used for conductors. The air terminal is sited above the highest point of the structure and a down conductor is bolted to the outside of the building so that side flashing between the lightning conductor and other metalwork will not occur.

Ground termination is with a series of earth rods driven to depths of up to 5 m, cast iron or copper plates 1 m square horizontally or vertically oriented 600 mm below ground, or a copper lattice of flat strips 3 m × 3 m at a depth of 600 mm. Where large floor areas containing earth rods are to be concreted, a precast concrete inspection pit is built over the rod location.

The electrical resistance to earth of the whole system is not to exceed 10 Ω (Butler, 1979a). Calculation of the ground earthing resistance R requires a knowledge of the earth type (BS Code of Practice 1013: 1965) and resistivity. Typical values of earth resistivity are 10 Ωm for clay, 50 Ωm for chalk, 100 Ωm for clay shale and 1000 Ω for slatey shales. The resistance of a rod electrode in earth is

$$R = \frac{0.37\,\rho}{l}\log\left(\frac{4000l}{d}\right)\ \Omega$$

where l is the earth rod length (m), d is the earth rod diameter (mm) and ρ is the resistivity of the soil (Ωm). A number of rods are connected in parallel and spaced 3.5 m apart to provide the required resistance.

EXAMPLE 13.8

Design a lightning conductor system for a building 30 m high in an area where thunderstorms are expected. The ground has a high chalk content and rod electrodes 4 m long are to be used. The conductors are to be 25 mm × 3 mm copper strip. The specific resistance ρ of copper is 0.0172 $\mu\Omega$ m.

length of conductor = air terminal + down conductor + ground lead

Take the length of the conductor as 40 m.

$$\text{resistance } R \text{ of conductor } R = \rho\frac{l}{A}\ \Omega$$

where A is the conductor cross-sectional area (m^2); hence

$$R = 0.0172 \times 10^{-6}\ \Omega\text{m} \times \frac{40\ \text{m}}{(0.025 \times 0.003)\ \text{m}^2}$$

$$= 0.0092\ \Omega$$

$$\text{resistance } R \text{ of earth electrode} = \frac{0.37\,\rho}{l}\log\left(\frac{4000\,l}{d}\right)\ \Omega$$

where the earth resistivity ρ is 50 Ω m, the electrode length l is 4 m and the electrode diameter d is 10 mm; hence

$$R = \frac{0.37 \times 50\ \Omega\text{m}}{4\ \text{m}} \times \log\left(\frac{4000 \times 4}{10}\right)\ \Omega$$

$$= 4.625 \times \log 1600$$

$$= 14.819\ \Omega$$

The resistance of one electrode in the ground plus the down conductor is greater than the 10 Ω allowed, and so we find the combined resistance of two

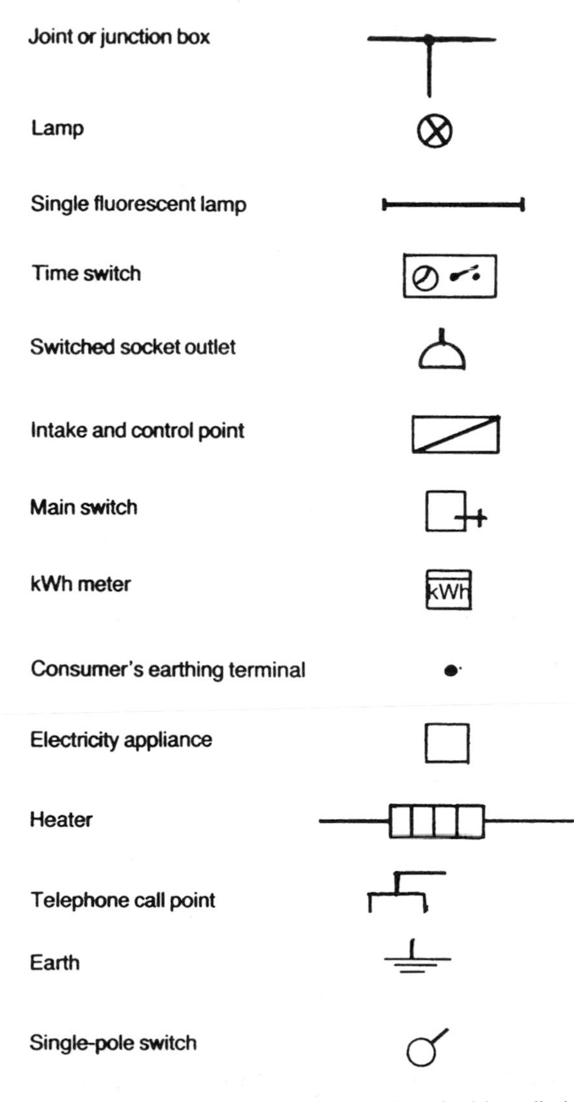

Figure 13.9 Drawing symbols for electrical installations

electrodes connected in parallel:

$$\frac{1}{R} = \frac{1}{R_1} + \frac{1}{R_2} = \frac{1}{14.8} + \frac{1}{14.8} = 0.135$$

$$R = \frac{1}{0.135} = 7.4\ \Omega$$

Two electrodes and the down conductor connected in series have a total resistance of

$$R = (7.4 + 0.0092)\ \Omega = 7.4\ \Omega$$

This is less than the 10 Ω allowed and is satisfactory. The resistance of the lightning conductor is negligible in relation to that of the earth electrodes. The calculations have been made on the assumption that the lightning discharges in a direct current. Lightning energy can produce a current to earth of 20 000 A for a few milliseconds. It causes physical damage to building structures, starts fires in combustible materials and injury to people, sometimes fatal.

Graphical symbols for installation diagrams

Some of the symbols used on drawings of electrical installations are listed in Fig. 13.9 as in accordance with BS 3939: 1985, *Guide for Graphical Symbols.*

Questions

1. Explain how electricity is generated and transmitted to the final user.

2. List the sources of energy used for the generation of electricity and state their immediate and long-term benefits.

3. Explain, with the aid of sketches, the meaning of the terms 'single-phase' and 'three-phase' electricity supplies and show how they are used within buildings.

4. What does 'balancing the phases' mean?

5. Calculate the electrical resistance per metre length at 20 °C of a copper conductor of a 10 mm^2 cross-sectional area.

6. Find the electrical resistance of a copper conductor 1.5 mm^2 in cross-sectional area if its total length is 25 m and its temperature is 20 °C.

7. 28 m of copper conductor 4 mm^2 in cross-sectional area is covered with thermal insulation, which causes the cable temperature to rise to 45 °C. Calculate the percentage increase in electrical resistance compared with its value at a cable temperature of 20 °C.

8. Sketch the methods of connection used for measurements of current, voltage drop and power consumption in an electrical resistance heater on an alternating current circuit.

9. State the function of power factor correction in alternating current circuits.

10. Calculate the apparent power, in kilovolt-amperes, of an electric motor which is connected to a 415 V a.c. three-phase supply and has a current flow of 17.5 A.

11. Calculate the resistance of a 3 kW immersion heater on a 240 V a.c. circuit.

12. What current, in milliamperes, would flow to earth during an insulation resistance test when a 500 V d.c. e.m.f. is applied between the line and protective conductors and the resistance is found to be 1.75 MΩ?

13. Show by sample calculation why smaller cables can be used for long-distance power transmission when very high voltages are used.

14. A 33 kV supply to a factory carries 250 A per phase or line. Calculate the usable electrical power in the factory if the average power factor is 0.68.

15. Find the maximum length of 6 mm^2 cable that can be used if the maximum current-carrying capacity is to be utilized on a 240 V circuit.

16. A building site is to have the following electrical equipment in use each day:

    (a) concrete mixer, 5 kW, 4 h, 415 V;
    (b) sump pump, 1.5 kW, 6 h, 240 V;
    (c) 20 lamps, 150 W each, 4 h, 240 V;
    (d) 5 floodlamps, 300 W each, 4 h, 240 V;
    (e) 6 hand tools, 750 W each, 5 h, 110 V.

    The power factor of the machinery is 0.8. Site work takes place 5 days per week for 28 weeks. Electricity costs 6p per unit. Find

    (i) the total kilovolt-amperes and the line current of the required temporary incoming supply system, and
    (ii) the cost of the electricity used on site during the contract.

17. Sketch and describe the safety precautions taken to avoid contact with both overhead and underground electricity cables during site construction work.

18. Sketch a suitable arrangement of temporary wiring, control and safety equipment on a site where the following items are employed: tower crane, sump pump, five floodlamps, security lighting and circuits on each of three floors for hand lamps and tools.

19. List the site programme for the main contractor in the installation and operation of temporary site electrical services.

20. Sketch and describe the characteristics of rewirable and cartridge fuses and residual current devices.

21. Show how an underground electrical service cable enters a building. Sketch the arrangement of electricity distribution within a typical residence.

22. List the cable and conduit systems used for electricity distribution and state their applications.

23. Briefly describe the methods of testing electrical installations.

24. State the requirements of telecommunications installations.

25. Sketch and describe a lightning conductor installation for a city centre office block.

26. Design a lightning conductor system for a 60 m high building on clay shale using 4.5 m earth rods. The down conductor is to be a copper rod 10 mm in diameter.

14 Room acoustics

Key terms and concepts

Learning objectives

Study of this chapter will enable the reader to:

1. know the potential sources of sound and vibration within buildings;
2. know what is meant by noise;
3. understand how sound travels through a building;
4. understand what is meant by sound power, sound pressure wave, sound power level and sound pressure level;
5. know how to calculate sound pressure levels for normal building services design examples;
6. use sound levels at the range of frequencies commonly used in building services engineering;
7. understand how sound and vibration are transmitted through buildings;
8. be able to identify the need for sound attenuation and vibration isolation;
9. understand and use the decibel unit of measurement of sound energy;
10. know the meaning and use of direct and reverberant sound fields;
11. calculate the sound pressure level in a plant room, a space adjacent to the plant room, in the target occupied room and in the external environment outside the plant room;
12. use logarithms to base 10 in acoustic calculations;
13. understand the principle of sound absorption;
14. calculate the sound absorption constant for a room at different frequencies;
15. know the sound absorption coefficients for some common building materials and constructions;
16. understand and use reverberation time and attenuation;
17. calculate sound pressure levels at different frequencies within a plant room;
18. know what a reverberant room and an anechoic chamber are;
19. use directivity index, sound absorption coefficients, mean absorption coefficient and room absorption constant;

20. understand the behaviour of equipment at resonant conditions and how to minimize or avoid its occurrence;
21. calculate and use the sound pressure level in a plant room;
22. calculate the sound pressure level experienced at an external location from a plant room;
23. calculate the sound pressure level generated in a room or space that is adjacent to a plant room;
24. calculate the sound pressure levels at different frequencies that are produced in the target occupied room;
25. understand, calculate and use noise rating data;
26. know how the acoustic design engineer relates the noise output from plant and systems to the human response;
27. be able to calculate noise rating curves;
28. know the noise rating criteria used for building services design;
29. plot noise rating curves, plant and system sound pressure levels and find a suitable design solution;
30. know the formulae used in practical acoustic design work;
31. be able to carry out sound pressure level and noise rating design calculations, try different solutions to attenuate plant noise and be able to produce a practical design to meet a design brief.

Introduction

This chapter uses the worksheet file DBPLANT.WKS to find the noise rating that will be produced within an occupied room by direct transmission through the building from the noise-producing plant. The plant noise source creates sound pressure levels within the plant room. The plant room noise can pass through an intermediate space, such as a corridor, and then into the target occupied space. Sound can be transmitted from the plant room to a recipient outdoors for an environmental impact noise rating.

Sufficient reference data is provided on the worksheet for examples within this chapter and for some real applications. Reference data from any source can easily be added. This chapter allows for most practical examples of mechanical plant to be assessed quickly and without having to deal with the equations themselves. Data is provided for frequencies from 125 Hz to 4000 Hz as this range is likely to cover the important noise levels for comfort. The range of frequencies can be added to should the need arise. The reader may wish to study the principles of acoustics in the appropriate text books and the references made as it is not the intention of this chapter to teach the subject in its entirety. However, it is the purpose of this chapter to provide an easily understandable method of analysing practical noise applications. Consequently, the reader should not find it difficult to enter correct data and acquire suitable results for educational reasons and in practical design office cases. The worksheet DBDUCT.WKS is used to calculate the noise rating in the target room that is produced by noise being transmitted from the air-conditioning fan through the ductwork system. Further examples of spreadsheet applications and explanation of spreadsheet use are provided in *Building Services Engineering Spreadsheets* (Chadderton, 1997b).

Acoustic principles

The building services design engineer is primarily concerned with controlling the sound produced by items of plant such as boilers, supply air and exhaust air fans, refrigeration compressors, water pumps, diesel or gas engine-driven electrical generating sets and air compressors. An excess of sound that is produced by the plant, above that which is acceptable to the recipient, is termed noise. All the mechanical service equipment and distribution systems to be installed within an occupied building are capable of generating noise.

Sound travels through an elastic, compressible medium, such as air, in the form of waves of sound energy. These waves of energy are in the form of variations in the pressure of the air above and below the atmospheric air pressure. The human ear receives these air pressure fluctuations and converts the vibration generated at the eardrum into electric impulses to the brain. What we understand to be recognizable language, music and noise is the result of human brain activity. Animals and the mythical person from another planet do, or may, process what we determine as normal sounds and come to a different conclusion from those of us who are conditioned to life on Earth. These variations in the pressure of the atmospheric air are very small when measured in the Pascal or millibar values that engineers use. A scale of measurement that relates to the subjective response of the human ear is used. Although absolute units of measurement are taken and normal calculation procedures are adopted, it is important to remember that the smallest unit of sound is that which can be detected by the human ear. The waves of air pressure which pass through the atmosphere are measured in relative pressure units. The acoustic energy of the source which caused the air pressure waves has an acoustic power, or rate of producing energy, in the same way that all thermodynamic devices have a power output. There are two ways of assessing the output and transmission of acoustic energy:

1. source sound power: Watts;
2. sound wave atmospheric pressure variation: Pascals.

Sound waves are generated at different frequencies measured in cycles per second, Hertz (Hz). The plant which produces the noise has components that rotate, move and vibrate at a range of different speeds, or frequencies of rotation. The flowing fluid is vibrated by the passage of fan or pump blades and it transfers the plant vibration through to downstream parts of the connected services systems. The fluid is either water, oil, air, gas, refrigerant or steam, and can either simply transmit the plant vibrations and noise or add to them by means of its own pulsations due to its turbulent flow. Turbulence means that a fluid flow contains recirculatory parcels of fluid in the form of eddy currents. These parcels of swirling eddy currents move in all directions, i.e. along with the general direction of the main flow, but also in the reverse direction and transversely across the main flow. Viewing wave action on a beach or a fast river flow from a bridge or at a bend, reveals the nature of turbulent flow. The turbulent eddy currents occur at a range of frequencies, parcels of recirculating fluid per second, depending upon the overall diameter of the eddy current. The physical movement of the swirling fluid can vibrate the containing water pipe or air duct, causing vibration and noise. Obstructions in the air or waterway occasioned by sharp edges, dampers, grilles, temperature sensors, and changes in duct cross-section, can cause the turbulent fluid to shear into additional swirling eddy currents and produce more vibration and noise. Turbulent fluid can vibrate air ducts, pipes and terminal heat exchange units. The structure of the building transmits noise by the vibration of its solid material particles and continuous steel frame and reinforcing bars within concrete framework. Acoustic energy is transferred between pressure waves in the air and vibration through solid materials in either direction. The vibration of fans, compressors, engines and pumps is controlled by mounting them on coiled steel springs, rubber feet and rubber matt. Fluid pipes and air ducts are separated from fans, air-handling units and pumps with flexible connections. These minimize, or stop, plant vibration being transferred to the reticulation system. Fluid-borne noise is reduced by selective absorption with a porous lining to the air duct. Sound waves are absorbed into the thickness of the lining material through a perforated surface material which protects the absorber from fluid damage and erosion. Sound energy is dissipated within the absorber by multiple reflections among the fibrous material.

Sound power and pressure levels

Sound power and pressure levels are measured over a range of frequencies that are representative of the response of the human ear to sounds. The unit of measurement of sound is the Bel (B). The smallest increment of sound that the human ear can detect is one-tenth of a Bel, one decibel (dB). This means that the smallest change in sound level that is perceptible by the human ear is 1 dB, so any decimal places that are produced from calculations using sound power or pressure level are not relevant. A calculated sound level of 84.86 can only be 84 dB as the 0.86 decimal portion is not detectable by the ear. The 'A' scale of measurement gives a weighting to each frequency in the range 20 Hz to 20 kHz in the same ratio as can be heard. For example, the human ear is more sensitive to sounds at 1000 Hz than at higher frequencies.

The acoustic output power of a machine is termed its sound power level, SWL dB. Think of SWL as the sound watts level of the acoustic output power of the machine. The value of acoustic power in watts from building services plant is very small, much less than one watt of power. The word level is used because it is not the actual value of the number of watts that is normally used; it is the sound level produced in acoustic units of measurement, dB, that are taken for practical use. The manufacturer of the plant provides the sound power levels produced by a particular machine from test results and predictions for known ranges of similar equipment. The sound power level of a machine at the range of frequencies from 125 Hz to 8000 Hz is required by the building services design engineer in order to assess the acoustic affects upon the occupied spaces of the building. The overall sound power level for a range of frequencies is also quoted by the manufacturer of a machine.

Sound pressure level

A sound field is created by the sound power output from a machine within a plant room. It is made up of a direct sound field, i.e. directly radiated sound, and a reverberant sound field, i.e. general sound that reflects uniformly from the hard surfaces around the room. The direct sound field reduces with the inverse square of the distance from the sound source and is not normally of importance as it only applies to very short distances from the sound source. The reverberant sound field results from the average value of the sound pressure waves passing around the room. These waves try to escape from the plant room and find their way into the occupied spaces where the air-conditioning engineer is attempting to create a quiet and comfortable environment. The sound pressure level, SPL dB, of the total sound field, direct plus reverberant, that is generated within a room from a sound source of sound power level SWL dB, is found from

$$\text{SPL} = \text{SWL} + 10 \times \log\left(\frac{Q}{4 \times \pi \times r^2} + \frac{4}{R}\right) \text{ dB}$$

(CIBSE [1985] and Sound Research Laboratories Limited), where,

SPL = sound pressure level produced in room	dB
SWL = sound power level of acoustic source	dB
log = logarithm to base 10	dimensionless
Q = geometric directivity factor	dimensionless
r = distance from sound source to the receiver	m
R = room sound absorption constant	m^2

Logarithms to base 10, $\log_{10}$, are used throughout the calculation of acoustic values. A sound source that radiates sound waves uniformly in all directions through unobstructed space, will create an expanding spherical sound field and have a dimensionless geometric directionality factor Q of 1. A sound source that is on a plane surface, radiates all its sound energy into a hemispherical sound field moving away from the surface. This has a directionality factor Q of 2, that is, twice the sound energy passes through a hemisphere. Similarly, if the sound source occurs at the junction of two adjacent surfaces that are at right angles to each other, such as the junction of a wall and ceiling, Q is 4. When there are three adjacent surfaces at the sound source, such as two walls and a ceiling, Q is 8. Distance r is that from the sound source to the receiving person, surface or measurement location, such as an air outlet duct from the plant room or outdoor air grille.

Absorption of sound

The room sound absorption constant, R m^2, is found from the total surface area of the enclosing room, S m^2, and the mean sound absorption coefficient of the room surfaces, $\overline{\alpha}$, at each of the relevant frequencies:

$$R = \frac{S \times \overline{\alpha}}{1 - \overline{\alpha}}$$

where

$\overline{\alpha}$ = mean absorption coefficient of room surfaces
S = total room surface area m^2

Mean absorption coefficient, $\overline{\alpha}$, is found from the area and absorption coefficient for each surface of the enclosing space. All the absorbing surfaces within the space, such as seats and people in a theatre, are included in the overall sound absorbing ability of the room:

$$\overline{\alpha} = \frac{A_1 \times \alpha_1 + A_2 \times \alpha_2 + A_3 \times \alpha_3}{A_1 + A_2 + A_3}$$

where

A_1 = surface area of surface number 1 m^2
α_1 = absorption coefficient of surface number 1

Table 14.1 Absorption coefficients of common materials

Material	Absorption coefficient at					
	125 Hz	250 Hz	500 Hz	1000 Hz	2000 Hz	4000 Hz
25 mm plaster, 18 mm plasterboard, 75 mm cavity	0.3	0.3	0.6	0.8	0.75	0.75
18 mm board floor on timber joists	0.15	0.2	0.1	0.1	0.1	0.1
Brickwork	0.05	0.04	0.04	0.03	0.03	0.02
Concrete	0.02	0.02	0.02	0.04	0.05	0.05
12 mm fibreboard, 25 mm cavity	0.35	0.35	0.2	0.2	0.25	0.3
Plastered wall	0.01	0.01	0.02	0.03	0.04	0.05
Pile carpet on thick underfelt	0.07	0.25	0.5	0.5	0.6	0.65
Fabric curtain hung in folds	0.05	0.15	0.35	0.55	0.65	0.65
15 mm acoustic ceiling tile, suspended 50 mm mineral fibre wool or glass fibre matt	0.5	0.6	0.65	0.75	0.8	0.75
50 mm polyester acoustic blanket, metallized film	0.25	0.55	0.75	1.05	0.8	0.7
50 mm glass fibre blanket, perforated surface finish	0.15	0.4	0.75	0.85	0.8	0.85

Materials absorb different amounts of sound energy at each frequency due to the frequency of natural vibration of their fibres and the method of their construction. Stiff, dense materials, such as brickwork walls, absorb sound by molecular vibration. Highly porous materials, such as glass wool, have large air passageways that allow the sound waves to penetrate the whole of the material thickness quickly. The strands of glass wool are vibrated by the sound waves and the sound energy is dissipated as heat. Dense materials are very efficient at absorbing acoustic energy. The reduction in sound level between the surfaces of a sound barrier is proportional to the mass of the barrier. The absorption coefficients of some common surface materials are given in Table 14.1. This data is repeated on the worksheet from line 201.

Reverberation time

Reverberation time is the time in seconds taken for a sound to decrease in value by 60 dB. This effectively means the time taken for the sound source to decay to an imperceptible level, as a sound pressure of 30 dB is very quiet to the human ear. An echo is produced by sound waves bouncing, or reverberating, from one or more hard surfaces and this may last for several seconds. A room that has a long reverberation time sounds noisy, lively and it allows echoes. A room having a short reverberation time, less than 1 s, sounds dull and there is no echo. The ultimate in short reverberation time is found in the anechoic chamber that is used for the acoustic testing of equipment. The walls and ceiling of the chamber are lined with thick acoustic absorbent wedges. The floor is a suspended wire mesh, and beneath the floor more absorbent wedges complete the coverage of all the room surfaces. The sound source radiates outward and upon reaching the surfaces is instantly absorbed, allowing no reverberation or echo. This is as close to a free field test method as can be achieved because there is no reverberant field caused by reflected sound waves.

An interesting example of a large semi-anechoic chamber is the car testing facility at Gaydon, England. The four walls and the ceiling are covered with acoustic wedges, while the floor is a plain concrete surface. This simulates an open road, hemispherical acoustic field under laboratory repeatable conditions (CIBSE, 1995).

Reverberation time of a room is found from

$$\text{reverberation time } T = \frac{0.161 \times V}{S \times \overline{\alpha}}$$

(CIBSE [1986] and Sound Research Laboratories Limited).

EXAMPLE 14.1

A plant room for an air-conditioning fan is 4 m × 3 m in plan and 2.5 m high. It has four brickwork walls, a concrete floor and a pitched sheet steel deck roof having 50 mm thickness of glass fibre and an aluminium foil finish to the underside. Ignore the effects of the metal plant, air ductwork and the door into the plant room. Calculate the room constant and the reverberation time for the plant room.

The surface absorption coefficients are selected from Table 14.1. It can be seen that there will be a different room constant and reverberation time for each frequency. The solution is presented in Table 14.2.

$$\text{Room volume } V = 4 \times 3 \times 2.5 \text{ m}^3$$

$$= 30 \text{ m}^3$$

floor area 12 m^2

ceiling area 12 m^2

wall area 35 m^2

$$\text{Room surface area } A = (2 \times 4 \times 3) + (4 + 4 + 3 + 3) \times 2.5 \text{ m}^2$$

$$= 59 \text{ m}^2$$

For 125 Hz, the mean absorption coefficient is,

$$\overline{\alpha} = \frac{12 \times 0.02 + 12 \times 0.15 + 35 \times 0.05}{12 + 12 + 35}$$

$$= 0.064$$

$$\text{room constant } R = \frac{S \times \overline{\alpha}}{1 - \overline{\alpha}} \text{ m}^2$$

$$= \frac{59 \times 0.064}{1 - 0.064} \text{ m}^2$$

$$= 4.03 \text{ m}^2$$

Table 14.2 Solution to Example 14.1

Surface	Absorption data at frequency					
	125 Hz	250 Hz	500 Hz	1 kHz	2 kHz	4 kHz
Floor α	0.02	0.02	0.02	0.04	0.05	0.05
Ceiling α	0.15	0.4	0.75	0.85	0.8	0.85
Walls α	0.05	0.04	0.04	0.03	0.03	0.02
Floor (S $\times$ α)	0.24	0.24	0.24	0.48	0.6	0.6
Ceiling (S $\times$ α)	1.8	4.8	9.0	10.2	9.6	10.2
Walls (S $\times$ α)	1.75	1.4	1.4	1.05	1.05	0.7
$\overline{\alpha}$	0.064	0.109	0.18	0.199	0.191	0.195
Room constant R m^2	4.03	7.21	12.95	14.66	13.93	14.29
Reverberation T s	1.28	0.75	0.45	0.41	0.43	0.42

$$\text{reverberation time } T = \frac{0.161 \times V}{S \times \overline{\alpha}} \text{ s}$$

$$= \frac{0.161 \times 30}{59 \times 0.064} \text{ s}$$

$$= 1.28 \text{ s}$$

EXAMPLE 14.2

An air-conditioning centrifugal fan has an overall acoustic output power level SWL of 87 dB on the 'A' scale. The fan is to be installed centrally within the air-handling plant room described in Example 14.1. Calculate the sound pressure level that will be produced in the plant room at 1000 Hz when the fan is operating, close to the fan and also generally within the room.

Room absorption constant from Example 14.1 at 1000 Hz,

$$R = 14.66 \text{ m}^2$$

The fan is on the centre of a concrete floor in the plant room. Sound pressure waves leaving the fan will radiate into a hemispherical field above floor level. The sound waves are concentrated into half of a completely free field. The directivity, Q, of the sound field is 2. A person within the plant room can stand in the range of 100 mm to 2 m away from the fan. A typical distance between the fan and the recipient is 1 m. The room sound pressure level is calculated for 100 mm and 1 m distances from the sound source. When

$$r = 100 \text{ mm}$$

$$\text{SPL} = \text{SWL} + 10 \times \log_{10} \left(\frac{Q}{4 \times \pi - r^2} + \frac{4}{R} \right) \text{ dB}$$

$$= 87 + 10 \times \log_{10} \left(\frac{2}{4 \times \pi \times 0.1^2} + \frac{4}{14.66} \right) \text{ dB}$$

$$= 87 + 10 \times \log_{10} (16.188) \text{ dB}$$

$$= 87 + 10 \times 1.2092 \text{ dB}$$

$$= 99 \text{ dB}$$

The smallest change in sound level that is perceptible by the human ear is 1 dB, so the decimal places are not relevant. The plant room sound pressure level at 100 mm radius from the fan is 99 dB. At 1 m from the fan, the recipient experiences a sound pressure level of

$$r = 1 \text{ m}$$

$$\text{SPL} = 87 + 10 \times \log_{10}\left(\frac{2}{4 \times \pi \times 1^2} + \frac{4}{14.66}\right) \text{dB}$$

$$= 87 + 10 \times \log_{10}(0.432) \text{ dB}$$

$$= 87 + 10 \times -0.3645 \text{ dB}$$

$$= 83 \text{ dB}$$

The direct sound field diminishes with distance from the source. The reverberant sound field establishes the general room sound pressure level when the recipient is sufficiently far away from the source.

EXAMPLE 14.3

The spectrum of sound power levels produced by the centrifugal fan being installed in the 4 m × 3 m × 2.5 m high plant room in Example 14.1 is 78 dB at 125 Hz, 82 dB at 250 Hz, 86 dB at 500 Hz, 87 dB at 1 kHz, 70 dB at 2 kHz, and 60 dB at 4 kHz. Use the surface absorption data from Example 14.1 and calculate the room sound pressure level at a radius of 1.5 m from the fan for each frequency from 125 Hz to 4 kHz.

At 125 Hz,

SWL is 78 dB.

$$r = 1.5 \text{ m}$$

$$Q = 2$$

$$R = 4.03 \text{ m}^2$$

$$\text{SPL} = \text{SWL} + 10 \times \log_{10}\left(\frac{Q}{4 \times \pi \times r^2} + \frac{4}{R}\right) \text{dB}$$

$$= 78 + 10 \times \log_{10}\left(\frac{2}{4 \times \pi \times 1.5^2} + \frac{4}{4.03}\right) \text{dB}$$

$$= 78 \text{ dB}$$

The results are shown in Table 14.3.

Table 14.3 Fan sound spectrum in Example 14.3

	Data at frequency					
Item	125 Hz	250 Hz	500 Hz	1 kHz	2 kHz	4 kHz
Room constant *R* m^2	4.03	7.21	12.95	14.66	13.93	14.29
Reverberation *T* s	1.28	0.75	0.45	0.41	0.43	0.42
Fan SWL dB	78	82	86	87	70	60
Room SPL dB	78	79	81	82	65	55

Plant sound power level

The design engineer requires to know the sound power level of the, potentially, noise-producing items of plant. These plant items will be the supply air fan, extract air fan, exhaust fans from toilets, kitchens and some store rooms, fan coil units in ceiling spaces above occupied rooms, packaged air-handling units incorporating fans, direct refrigerant expansion outdoor condensing units, direct refrigerant expansion packaged air-conditioning roof-mounted units, gas- and oil-fired boilers, packaged air conditioners and heat pumps within rooms, external cooling towers and dry-air-cooled heat exchangers, refrigeration compressors and water chilling refrigeration plant. In addition to these major items of plant, supply air grilles, extract air grilles, room terminal air-handling units, dampers, air volume control boxes and fan-powered variable air volume control boxes, can also generate noise. The manufacturer of these items will provide the results of acoustic test data for the building services design engineer. Current acoustic data, rather than catalogue information, is acquired and the manufacturer then becomes responsible for the numbers used. The designer needs the sound power level at each frequency that is to be analysed. These are normally 125 Hz to 4000 Hz. Often the critical frequency for design will be 1000 Hz and this corresponds to a sensitive band in the human ear response.

For the worked examples and questions within this book, sound power levels are provided, either in the form of a discrete value for each frequency, or a single value for all frequencies for the plant item. Figure 13.1 gives an indication of the variations in sound power level from a single value for centrifugal fans, axial fans, refrigeration compressors, cooling towers, fan coil units and boilers. The reader will find the spectral sound power level by subtracting the variances from the single value quoted in the example or question. This data is not to be used in real design work as is provided for illustration purposes only. The numbers that were used to produce Fig. 14.1 are listed in Table 14.4. Figure 14.1 is also provided as a chart on the worksheet file.

Transmission of sound

The sound pressure within a space will cause the flow of acoustic energy to an area that has a lower acoustic pressure. Sound energy converts into structural vibration and passes through solid barriers. A reduced level of sound pressure is

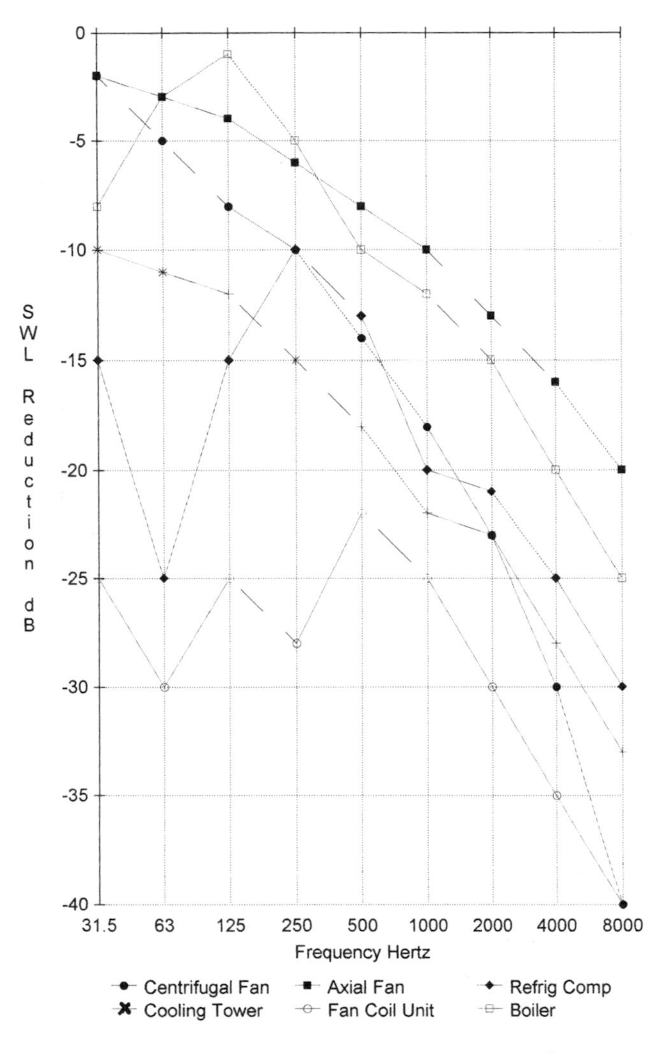

Figure 14.1 Plant SWL dB, spectral variation

Table 14.4 Illustrative sound power level variances from Figure 14.1

Plant item	Sound power level dB variance at frequency								
	31.5 Hz	63 Hz	125 Hz	250 Hz	500 Hz	1 kHz	2 kHz	4 kHz	8 kHz
Centrifugal fan	−2	−5	−8	−10	−14	−18	−23	−30	−40
Axial fan	−2	−3	−4	−6	−8	−10	−13	−16	−20
Refrigeration compressor	−15	−25	−15	−10	−13	−20	−21	−25	−30
Cooling tower	−10	−11	−12	−15	−18	−22	−23	−28	−33
Fan coil unit	−25	−30	−25	−28	−22	−25	−30	−35	−40
Boiler	−8	−3	−1	−5	−10	−12	−15	−20	−25

established in the adjacent space due to the attenuation of the separating partition, wall, floor or ceiling. Air passageways through the separating partition act as sound channels that have little, or no, sound-reducing property, or attenuation. The reader can validate this effect by partially opening a window when the outdoor sound level is substantial. Compare the open and closed window performance when a train, lorry or high traffic volumes are present. A well-air-sealed single-glazed window imposes a sound reduction of 30 dB on external noise but a poorly sealed or open window has little attenuation.

Sound reduction by a surface is from the reflection of sound waves striking the surface and by the absorption of sound energy into a porous material. Absorbed acoustic energy is dissipated as heat within the solid components of the absorber. Dense materials are often efficient sound attenuators. The exception is metal. Sound travels easily through metals for great distances due to their molecular vibration. When the imposed sound frequency coincides with a natural frequency of vibration of the metal, resonance occurs and an increased sound level may be generated. This happens in particular when the shape of the metal creates an air space for the sound waves to resonate within, such as in a bell, an empty tank or a pipe. The structural steel within a building, service pipework, air ducts and railways lines can all transfer noise and vibration over long distances.

The sound pressure level generated within a room by mechanical plant, or sound systems for entertainment, will be passed through sound barrier materials and constructions such as walls and the ceiling, to adjacent spaces, occupied rooms and to the external environment around the building. The sound pressure levels received at each frequency depend upon the barrier attenuation, distance between the sound source and the recipient and the acoustic properties of the receiving space. The low-frequency sound waves, below 1000 Hz, are more difficult to attenuate than those above 1000 Hz. This is because the commonly used building and sound absorbing materials and vibration isolating rubber all have a low natural frequency of vibration. They will resonate at a frequency often as low as 100 Hz. A material loses its attenuation property at the resonant frequency. Worse still, of course, is that when a rotary machine passes through or runs at its natural frequency of vibration, during start-up procedures, additional noise can be generated and the amplitude of its vibration may escalate to the point of physical destruction. It is vital that variable speed controllers run the rotary machine speed through its resonant frequency band as quickly as possible to minimize noise and vibration. Attenuation materials such as brick, concrete, timber and acoustic fabric are good at absorbing sounds at the higher frequencies. The human ear is most sensitive to sounds around 1000 Hz, making this the critical frequency for the acoustic design engineer.

Sound pressure level in a plant room

The sound source space is normally the mechanical services plant room. The reverberant sound pressure level in a plant room can be taken as

$$\text{SPL}_1 = \text{SWL} + 10 \times \log(T_1) - 10 \times \log(V_1) + 14 \text{ dB}$$

(Sound Research Laboratories Limited; see also, Smith, Peters and Owen [1985]), where

SPL_1 = sound pressure level in plant room	dB
SWL = sound power level of source mechanical plant	dB
T_1 = rerberation time of plant room	s
V_1 = volume of plant room	m^3

The reverberant sound pressure level is independent of the measurement location within the room. When a sound pressure level is required at a known location, the earlier equation is used with the radius from the source, r m,

$$\mathrm{SPL} = \mathrm{SWL} + 10 \times \log_{10}\left(\frac{Q}{4 \times \pi \times r^2} + \frac{4}{R}\right) \text{ dB}$$

EXAMPLE 14.4

A refrigeration compressor has an overall sound power level of 86 dB on the 'A' scale. The plant room has a reverberation time of 2 s and a volume of 70 m^3. Calculate the plant room reverberant sound pressure level.

$$\mathrm{SWL} = 86 \text{ dBA}$$

$$T_1 = 2 \text{ s}$$

$$V_1 = 70 \text{ m}^3$$

$$\begin{aligned}\mathrm{SPL}_1 &= \mathrm{SWL} + 10 \times \log(T_1) - 10 \times \log(V_1) + 14 \text{ dB} \\ &= 86 + 10 \times \log(2) - 10 \times \log(70) + 14 \text{ dB} \\ &= 86 + 3 - 18 + 14 \text{ dBA ignoring decimal places} \\ &= 85 \text{ dBA}\end{aligned}$$

Outdoor sound pressure level

The sound pressure level in the outdoor environment immediately external to the plant room, can be taken as

$$\mathrm{SPL}_2 = \mathrm{SPL}_1 - B + 10 \times \log(S_2) - 20 \times \log(d) + \mathrm{DI} - 17 \text{ dB}$$

(Sound Research Laboratories Limited), where

SPL_2 = outdoor air sound pressure level	dB
SPL_1 = sound pressure level in source room	dB
B = sound reduction index of exterior wall or roof	dB
S_2 = surface area of external wall or roof	m^2
d = distance between plant room surface and recipient	m
DI = directivity index	dB

EXAMPLE 14.5

A refrigeration compressor generates an overall sound pressure level of 85 dBA within a plant room. The plant room has an external wall of 12 m^2 that has an acoustic attenuation of 30 dB. Sound radiates from the plant room wall into a hemispherical field that has a directivity index of 2 dB. Bedroom windows of an hotel are at a distance of 4 m from the plant room wall. Calculate the external sound pressure level at the hotel windows.

$$SPL_1 = 85 \text{ dBA}$$

$$B = 30 \text{ dBA}$$

$$S_2 = 12 \text{ m}^2$$

$$d = 4 \text{ m}$$

$$DI = 2 \text{ dB}$$

$$\begin{aligned} SPL_2 &= SPL_1 - B + 10 \times \log(S_2) - 20 \times \log(d) + DI - 17 \text{ dB} \\ &= 85 - 30 + 10 \times \log(12) - 20 \times \log(4) + 2 - 17 \text{ dB} \\ &= 85 - 30 + 10 - 12 + 2 - 17 \text{ dB} \\ &= 38 \text{ dBA} \end{aligned}$$

Sound pressure level in an intermediate space

The sound which is generated within a plant room may be transferred into an intermediate space within a building before being received in the target occupied room. Such intermediate spaces are corridors, store rooms, service ducts or roof voids. While it may not be important what the sound pressure level is within the intermediate space, the acoustic performance of this space affects the overall transfer of sound to the target occupied area. When the intermediate space is very large and has thermally insulated surfaces, for example, in a roof space, a considerable attenuation is possible. The sound pressure level in such an intermediate room or space can be taken as

$$SPL_3 = SPL_1 - SRI + 10 \times \log(S_4) + 10 \times \log(T_2) - 10 \times \log(0.16 \times V_2) \text{ dB}$$

(Sound Research Laboratories Limited), where,

SPL_3 = sound pressure level in intermediate space	dB
SPL_1 = sound pressure level in plant room	dB
SRI = sound reduction index of common surface	dB
S_4 = area of surface common to both rooms	m^2
T_2 = reverberation time of intermediate space	s
V_2 = volume of intermediate space	m^3

EXAMPLE 14.6

A showroom has floor dimensions of 25 m × 10 m and a height of 3.6 m to a suspended tile ceiling. The average height of the ceiling void is 1.8 m. An air-conditioning system has distribution ductwork in the roof void above the suspended acoustic ceiling tiles. The air-handling plant room is adjacent to the roof void and there is a common plant room wall of 5 m × 2.5 m high in the roof void. The sound pressure level in the plant room is expected to be 50 dB. The reverberation time of the roof void is 0.8 s. The plant room wall adjoining the roof void has a sound reduction index of 10 dB. Calculate the sound pressure level that is produced within the roof void as the result of the air-handling plant room noise.

$$\mathrm{SPL}_3 = 50 \text{ dB}$$

$$\mathrm{SRI} = 10 \text{ dB}$$

$$S_4 = 12.5 \text{ m}^2$$

$$T_2 = 0.8 \text{ s}$$

$$V_2 = 25 \times 10 \times 1.8 \text{ m}^3$$

$$= 450 \text{ m}^3$$

$$\mathrm{SPL}_3 = \mathrm{SPL}_1 - \mathrm{SRI} + 10 \times \log(S_4) + 10 \times \log(T_2) - 10 \times \log(0.16 \times V_2) \text{ dB}$$

$$= 50 - 10 + 10 \times \log(12.5) + 10 \times \log(0.8) - 10 \times \log(0.16 \times 450) \text{ dB}$$

$$= 50 - 10 + 10 + 0 - 11 \text{ dB}$$

$$= 39 \text{ dB}$$

Sound pressure level in the target room

The sound pressure level in the target occupied room or space can be taken as

$$\mathrm{SPL}_4 = \mathrm{SPL}_3 - \mathrm{SRI} + 10 \times \log(S_5) + 10 \times \log(T_3) - 10 \times \log(0.16 \times V_3) \text{ dB}$$

(Sound Research Laboratories Limited), where

SPL_4 = sound pressure level in target room	dB
SPL_3 = sound pressure level in adjacent room	dB
SRI = sound reduction index of common surface	dB
S_5 = area of surface common to both rooms	m^2
T_3 = reverberation time of target room	s
V_3 = volume of target room	m^3

The target room may be adjacent to, or close to, the plant room, or it may not be influenced by the plant room other than by the transfer of noise through the interconnected air-ductwork system. Analysis of the ductwork route for noise transfer is calculated separately and is not covered in this book.

EXAMPLE 14.7

The showroom in Example 14.6 has floor dimensions of 25 m × 10 m and a height of 3.6 m to a suspended tile ceiling. The reverberation time of the showroom is 0.5 s. The air-conditioning plant room generates a sound pressure level of 39 dB in the roof space. The acoustic tile ceiling has a sound reduction index of 12 dB. Calculate the sound pressure level that is produced within the showroom by the air-conditioning plant.

$$SPL_3 = 39 \text{ dB}$$

$$SRI = 12 \text{ dB}$$

$$S_5 = 250 \text{ m}^2$$

$$T_3 = 0.5 \text{ s}$$

$$V_3 = 25 \times 10 \times 3.6 \text{ m}^3$$

$$= 900 \text{ m}^3$$

$$SPL_4 = SPL_3 - SRI + 10 \times \log(S_5) + 10 \times \log(T_3) - 10 \times \log(0.16 \times V_3) \text{ dB}$$

$$= 50 - 12 + 10 \times \log(250) + 10 \times \log(0.5) - 10 \times \log(0.16 \times 900) \text{ dB}$$

$$= 50 - 12 + 23 - 3 - 21 \text{ dB}$$

$$= 37 \text{ dB}$$

Noise rating

The human ear has a different response to each frequency within the audible range of 20 Hz to 20 000 Hz. It has been found that a low-frequency noise can be tolerated at a greater sound pressure level than a high-frequency noise. Noise rating (NR) curves are used to specify the loudness of sounds. Each curve is a representation of the response of the human ear in the range of audible frequencies.

The design engineer makes a comparison between the sound pressure level produced in the room at each frequency and the noise rating curve data at the same frequency. When all the noise levels within the room fall on or below a noise rating curve, that noise rating is attributed to the room. Noise rating curves for NR 25 to NR 50 are shown on Fig. 14.2. The values are plotted from,

$$SPL = NR_f \times B_f + A_f \text{ dB}$$

SPL = sound pressure level at frequency f and noise rating NR	dB
NR_f = noise rating at frequency f Hz	dimensionless
B_f and A_f = physical constants	dB
f = frequency	Hz

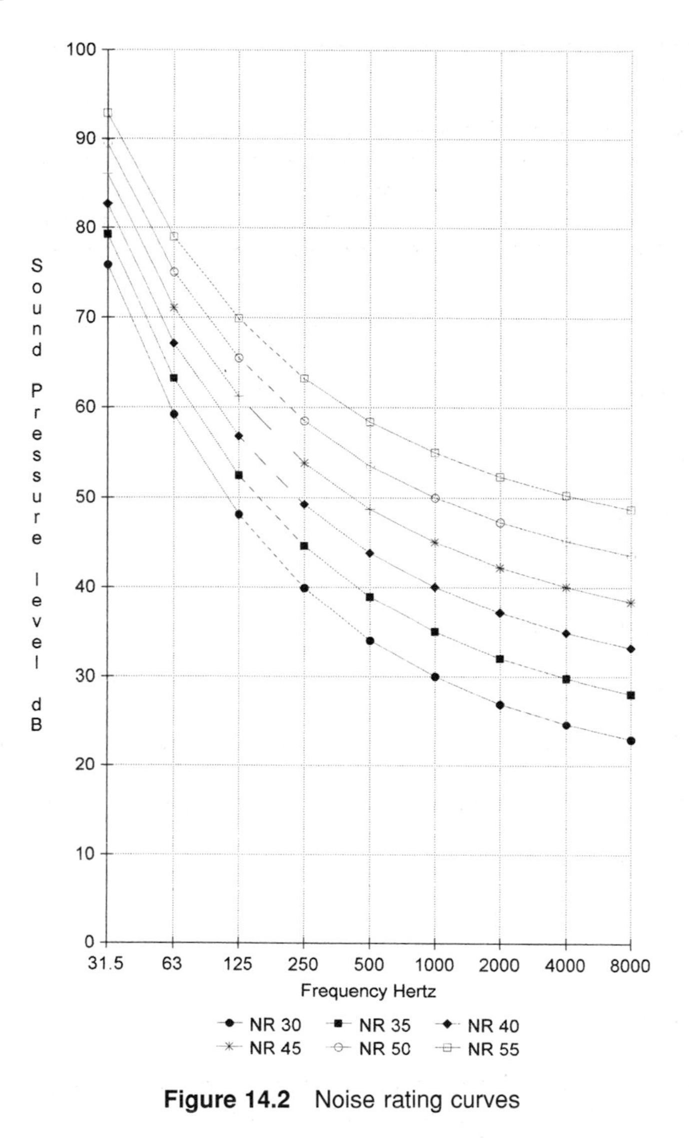

Figure 14.2 Noise rating curves

The values of the physical constants to calculate noise rating are shown in Table 14.5 (Australian Standard AS 1469–1983).

The normal applications of noise rating are shown in Table 14.6.

EXAMPLE 14.8

An air-conditioning fan produces the sound spectrum shown in Table 14.7 within an occupied room. Calculate the sound pressure levels for noise ratings NR 35, NR 40, NR 45, NR 50 and NR 55, and plot the noise rating curves for the frequency range from 31.5 Hz to 8000 Hz. Plot the room sound pressure levels on the same graph and find which noise rating is not exceeded.

Table 14.5 Physical constants for noise rating calculation

Frequency f Hz	A_f dB	B_f dB
31.5	55.4	0.681
63	35.5	0.79
125	22.0	0.87
250	12.0	0.93
500	4.8	0.974
1000	0	1.0
2000	−3.5	1.015
4000	−6.1	1.025
8000	−8.0	1.03

Table 14.6 Noise rating applications

Application	Noise rating	Comment
Acoustic laboratory	NR 15	Critical acoustics
Radio studio	NR 15	Critical acoustics
Concert hall	NR 15	Critical acoustics
TV studio	NR 20	Excellent listening
Large conference room	NR 25	Very good listening
Hospital, home, hotel	NR 30	Sleeping, relaxing
Library, private office	NR 35	Good listening
Office, restaurant, retail	NR 40	Fair listening
Cafeteria, corridor, workshop	NR 45	Moderate listening
Commercial garage, factory	NR 50	Minimum speech interference
Manufacturing	NR 55	Speech interference
Heavy engineering to industrial	NR 60 to NR 80	Sound levels judged on merits, leading to risk of hearing damage

Table 14.7 Noise spectrum in Example 14.8

Frequency	31.5 Hz	63 Hz	125 Hz	250 Hz	500 Hz	1 kHz	2 kHz	4 kHz	8 kHz
Room SPL dB	39	44	48	52	55	49	36	33	28

A manually calculated example for one noise rating curve is shown. The reader should use the spreadsheet graph or chart facilities to plot the whole figure. Calculate the SPL values for NR 55.

$$\mathrm{NR_f} = 55$$

$$\mathrm{SPL} = \mathrm{NR_f} \times B_f + A_f \qquad \mathrm{dB}$$

Calculate the SPL at each frequency for the values of B_f and A_f from Table 14.5. For 31.5 Hz,

$$\mathrm{SPL} = 55 \times 0.681 + 55.4 \qquad \mathrm{dB}$$

$$= 92\ \mathrm{dB}$$

For 63 Hz,

$$SPL = 55 \times 0.79 + 35.5 \quad dB$$
$$= 78\ dB$$

For 125 Hz,

$$SPL = 55 \times 0.87 + 22 \quad dB$$
$$= 69\ dB$$

For 250 Hz,

$$SPL = 55 \times 0.93 + 12 \quad dB$$
$$= 63\ dB$$

For 500 Hz,

$$SPL = 55 \times 0.974 + 4.8 \quad dB$$
$$= 58\ dB$$

For 1000 Hz,

$$SPL = 55 \times 1.0 + 0 \quad dB$$
$$= 55\ dB$$

For 2000 Hz,

$$SPL = 55 \times 1.015 - 3.5 \quad dB$$
$$= 52\ dB$$

For 4000 Hz,

$$SPL = 55 \times 1.025 - 6.1 \quad dB$$
$$= 50\ dB$$

For 8000 Hz,

$$SPL = 55 \times 1.03 - 8.0 \quad dB$$
$$= 48\ dB$$

These sound pressure levels are compared to the room data in Table 14.8.

Table 14.8 Noise spectrum in Example 14.8

Frequency	31.5 Hz	63 Hz	125 Hz	250 Hz	500 Hz	1 kHz	2 kHz	4 kHz	8 kHz
NR 55 SPL dB	92	78	69	63	58	55	52	50	48
Room SPL dB	39	44	48	52	55	49	36	33	28

The closest approach to the SPL limit for NR 55 occurs at 500 Hz. Check that NR 50 is exceeded. For 500 Hz,

$$\text{SPL} = 50 \times 0.974 + 4.8 \qquad \text{dB}$$

$$= 53 \text{ dB}$$

It should now be possible to check manually any sound pressure level against noise rating. Once the frequency that produces the greatest sound pressure level from the sound source is identified, other SPL values can be obtained for the peak frequency to check which NR is not exceeded.

The room does not exceed the NR 55 curve data but it does exceed NR 50. Plot the chart with the spreadsheet functions for NR 35 to NR 55 and with the room noise SPL. The spreadsheet will produce curves for six sets of data, so five NR curves and the one room curve can be displayed on one chart. The results are shown in Fig. 14.3.

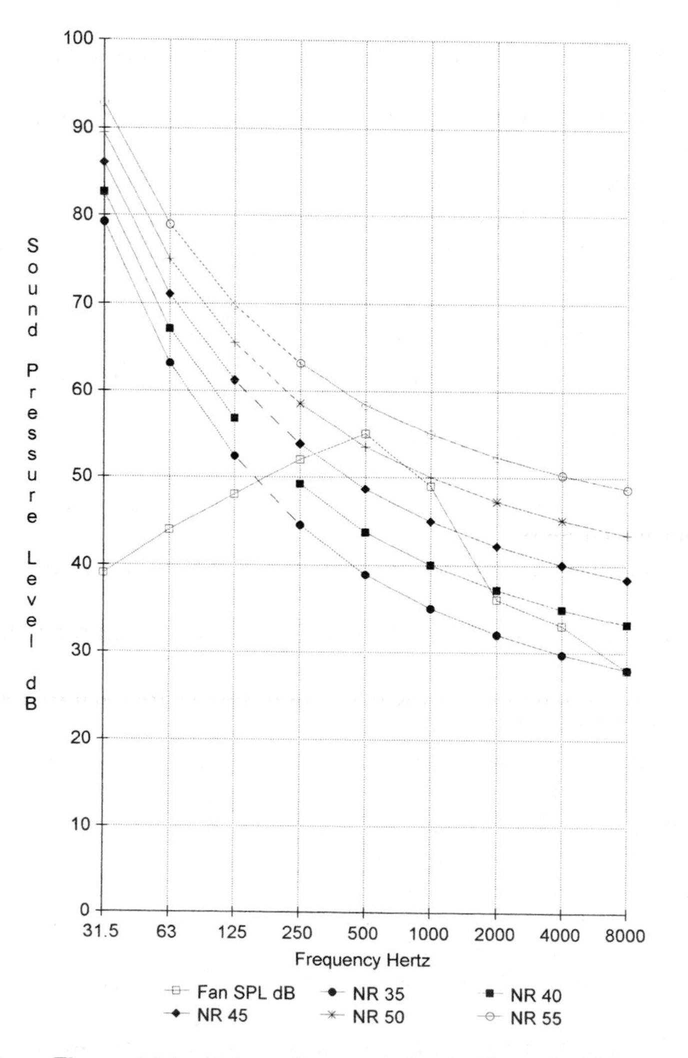

Figure 14.3 Noise ratings, solution to Example 14.8

Data requirement

The following new data should be entered:

1. user's name in cell B4;
2. user's job title in cell B5;
3. job reference number in cell B6;
4. file name in cell B10;
5. description of the plant item in cell B24;
6. plant room volume in cell B25;
7. plant room directivity index in cell B26;
8. distance from noise producing plant to the recipient in cell B27;
9. sound absorption coefficient for the plant room floor in the cell range D28 to I28;
10. sound absorption coefficient for the plant room ceiling in the cell range D29 to I29;
11. sound absorption coefficient for the plant room walls in the cell range D30 to I30;
12. sound absorption coefficient for the plant room windows in the cell range D31 to I31;
13. area of the plant room floor in cell B32;
14. area of the plant room ceiling in cell B33;
15. area of the plant room walls, excluding openings for doors and windows, in cell B34;
16. area of the plant room windows, or external doors, in cell B35;
17. description of the plant sound source in cells B44 to D44;
18. the plant manufacturer's sound power level data in the cell range B50 to I50;
19. the area of the exterior surface of the plant room that is projecting sound to an external recipient in cell B64;
20. the distance from the plant room surface, wall, window, door or roof to the recipient in cell B65;
21. the attenuation of the sound barrier between the plant and the recipient in cell B66;
22. the directivity index of sound projected to outdoors from the plant room in cell B67;
23. the sound reduction index of the plant room surface for each frequency in the cell range C71 to I71;
24. the attenuation of a sound barrier or screen between the plant room and the recipient at each frequency in the cell range C72 to I72;
25. description of the intermediate space in cell B84;
26. intermediate space volume in cell B85;
27. intermediate space directivity index in cell B86;
28. distance from the sound source surface to the recipient, person or location in the intermediate space in cell B87;
29. sound absorption coefficient for the floor of the intermediate space in the cell range D88 to I88;
30. sound absorption coefficient for the ceiling of the intermediate space in the cell range D89 to I89;

31. sound absorption coefficient for the walls of the intermediate space in the cell range D90 to I90;
32. sound absorption coefficient for the windows or doors of the intermediate space in the cell range D91 to I91;
33. area of the intermediate space floor in cell B92;
34. area of the intermediate space ceiling in cell B93;
35. area of the intermediate space walls, excluding openings for doors and windows, in cell B94;
36. area of the intermediate space windows, or external doors, in cell B95;
37. the area of the surface of the intermediate space that is common with the plant room in cell B104;
38. the distance from the plant room common surface in the intermediate space to the recipient in cell B105;
39. the attenuation of the sound barrier between the plant room surface and the recipient in the intermediate space in cell B106;
40. the directivity index of sound projected into the intermediate space in cell B107;
41. the sound reduction index of the common plant room surface for each frequency in the cell range C111 to I111;
42. description of the target occupied space in cell B124;
43. target room volume in cell B125;
44. target room directivity index in cell B126;
45. distance from the sound source surface to the recipient, person or location in the target room in cell B127;
46. sound absorption coefficient for the floor of the target room in the cell range D128 to I128;
47. sound absorption coefficient for the ceiling of the target room in the cell range D129 to I129;
48. sound absorption coefficient for the walls of the target room in the cell range D130 to I130;
49. sound absorption coefficient for the windows or doors of the target room in the cell range D131 to I131;
50. area of the target room floor in cell B132;
51. area of the target room ceiling in cell B133;
52. area of the target room walls, excluding openings for doors and windows, in cell B134;
53. area of the target room windows, or external doors, in cell B135.
54. area of noise source surface in the target room in cell B144;
55. distance from the noise source surface in the target room to the recipient or location in cell B145;
56. the attenuation of a barrier between the noise source and the recipient in cell B146;
57. directivity index of the target room in cell B147;
58. the sound reduction index of the noise source surface, wall, ceiling, floor or door in the target room in the cell range from D151 to I151.

When there is no intermediate space between the plant and target rooms, the calculations for the intermediate space need to be bypassed. This is done by: changing the formulae for the sound pressure levels SPL_3 in row 152; moving the cursor to cell D152 and entering the new formula +D56; copying this cell into cells E152 to I152 so that they contain the formulae +E56 to +I56. This can be achieved from manual entry or by using the COPY function, which will automatically enter the correct cell references. Remember that when there is an intermediate space, cells D152 to I152 must contain the formulae +D117 to +I117 as in the original file.

Output data

The data that will be given out are as follows:

1. current date in cell B7;
2. current time in cell B8;
3. mean absorption coefficient of the plant room surfaces at each frequency in the cell range D36 to I36;
4. total surface area of the plant room in cell B37;
5. plant room sound absorption constant at each frequency in the cell range D38 to I38;
6. reverberation time of the plant room at each frequency in the cell range D39 to I39;
7. reduction in sound pressure level with distance, *r* m, from the sound source at each frequency in the cell range D40 to I40;
8. sound pressure level in the plant room in the cell range C56 to I56;
9. noise rating in the plant room in cell B58;
10. sound pressure level at the stated distance from the sound source in the cell range C60 to I60;
11. sound pressure level experienced by the external recipient, due to the noise radiated from the plant room, at each frequency in the cell range C79 to I79;
12. noise rating at the recipient's location, due to the plant room sound radiation to outdoors, in cell B80;
13. mean absorption coefficient of the intermediate space surfaces at each frequency in the cell range D96 to I96;
14. total surface area of the intermediate space in cell B97;
15. intermediate space sound absorption constant at each frequency in the cell range D98 to I98;
16. reverberation time of the intermediate space at each frequency in the cell range D99 to I99;
17. reduction in sound pressure level with the stated distance, *r* m, from the sound source surface in the cell range D100 to I100;
18. sound pressure level in the intermediate space in the cell range D117 to I117;
19. noise rating in the intermediate space in cell B119;
20. mean absorption coefficient of the target room surfaces at each frequency in the cell range D136 to I136;
21. total surface area of the target room in cell B137;
22. target room sound absorption constant at each frequency in the cell range D138 to I138;
23. reverberation time of the target room at each frequency in the cell range D139 to I139;

24. sound pressure level in the target room in the cell range D157 to I157;
25. noise rating in the target room in cell B159.

Formulae

Representative samples of the formulae are given here. Each formula can be read on the spreadsheet by moving the cursor to the cell. The equation is presented in the form that it would normally be written and in the format that is used by the spreadsheet.

cell B7: @TODAY

This function produces the serial number of the current day and time. The cell is formatted to display the date.

cell B8: @TODAY

This function produces the serial number of the current day and time. The cell is formatted to display the time.

cell D36: (D28*B32 + D29*B33 + D30*B34 + D31*B35)/
(B32 + B33 + B34 + B35)

The mean sound absorption coefficient of the plant room is found from

$$\overline{\alpha} = \frac{(\alpha_1 \times A_1 + \cdots + \alpha_4 \times A_4)}{A_1 + \cdots + A_4}$$

α = sound absorption coefficient for the surface
A = surface area m^2
$\overline{\alpha}$ = mean absorption coefficient for the room

This is repeated in the cell range E36 to I36 for the frequency range 125 Hz to 4000 Hz.

cell B37: +B32 + B33 + B34 + B35

The total area of the surfaces in the plant room is

$$\Sigma A = A_1 + A_2 + A_3 + A_4 \text{ m}^2$$

cell D38: ($B$37*D36)/(1–D36)

The plant room sound absorption constant is found from the total room surface area, S_1, and the mean absorption coefficient at that frequency, $\overline{\alpha}$:

$$R = \frac{S_1 \times \overline{\alpha}}{1 - \overline{\alpha}}$$

R = room sound absorption coefficient m^2
S_1 = total surface area of the plant room m^2

This is repeated for the range of frequencies from 125 Hz to 4000 Hz in cells E38 to I38.

cell D39: 0.161*$B$25/$B$37/D36

The reverberation time of the plant room at 125 Hz is calculated from

$$T_1 = \frac{0.161 \times V_1}{S_1 \times \overline{\alpha}}$$

T_1 = reverberation time of the plant room	s
V_1 = volume of the plant room	m^3

This is repeated for the range of frequencies from 125 Hz to 4000 Hz in cells E39 to I39.

cell D40: 10*@LOG(($B$26/(4*@PI*$B$27*$B$27)) + (4/D38))

The effect of distance from the sound source on the received sound pressure level is found from

$$\text{SPL} = 10 \times \log\left(\frac{Q}{4 \times \pi \times r^2} + \frac{4}{R}\right) \text{ dB}$$

SPL = reduction in sound pressure level with distance	dB
Q = directivity of sound source	
r = distance from sound source to recipient	m

This is repeated for the range of frequencies from 125 Hz to 4000 Hz in the cell range E40 to I40.

cell C52: 10*@LOG($D$39)

Calculates the reverberation time effect on the plant room sound pressure level from

$$\text{SPL} = 10 \times \log(T_1)$$

This is repeated for the range of frequencies from 63 Hz to 4000 Hz in the cell range D52 to I52.

cell C53: −10*@LOG($B$25)

Calculates the effect of the room volume on the plant room sound pressure level from

$$\text{SPL} = 10 \times \log(V_1)$$

This is repeated for the range of frequencies from 63 Hz to 4000 Hz in the cell range from D53 to I53.

cell C56: @SUM(C50..C54)

Calculates the plant room sound pressure level from

$$\text{SPL} = \text{SWL} + 10 \times \log(T_1) - 10 \times \log(V_1) + 14 \quad \text{dB}$$

SWL = sound power level of the source	dB

This is repeated in the cell range D56 to I56 for the frequency range 63 Hz to 4000 Hz.

cell N46: @IF($C$56>C168, $A169, $A168)

The user does not need to look at the cells to the right of column L. Cell N41 contains the heading Plant Room NR Calculation. The cells below N41 and to the right of column L contain the comparison formulae to discover the highest noise rating for the plant room. This formula compares the plant room sound pressure level at 63 Hz with the maximum sound pressure level at 63 Hz that corresponds to a noise rating of NR 20. If the noise rating criteria is exceeded, the next NR number, 25, is held in cell N46. If the NR 20 criteria is not exceeded, then 20 is held in cell N46. This comparison is repeated for all the other frequencies and noise ratings until the appropriate maximum NR criteria are displayed in the matrix of cells. If all the noise rating data is used and the formulae are unable to find a peak NR criteria, then the text 'End of NR data' is displayed. This occurs at NR 80. The user could add further lines of NR data. Such applications would not normally be found in building services systems, with the exception of within a large fan plenum chamber, diesel engine, gas turbine, air or refrigeration compressor acoustic enclosure. Ear defenders would be worn prior to entry to such enclosures.

cell 059: @IF(N58>058, N58, O58)

The peak noise rating at each frequency is compared with that at the next frequency to find the larger noise rating. This comparison is repeated for the frequencies from 63 Hz to 4000 Hz in cells O59 to I59.

cell N60: +T59

The maximum noise rating that is not exceeded at any frequency is displayed in cell T59 and copied into cell N60.

cell B58: +T59

The maximum noise rating that is not exceeded at any frequency in the plant room is displayed in cell B58.

cell C60: +C50 + $D$40

The sound pressure level at the stated distance from the noise source, r m, is calculated from

$$\text{SPL} = \text{SWL} + \text{distance sound reduction} \qquad \text{dB}$$

cell C73: +C56

The plant room sound pressure level at 63 Hz is copied from cell C56 and displayed in cell C73. This is repeated for the frequency range 125 Hz to 4000 Hz in cells D73 to I73.

cell C74: 10*@LOG($B$64)

The external wall element of the noise radiation from the plant room is calculated from

$$\text{SPL} = 10 \times \log(S_2) \qquad \text{dB}$$

S_2 = surface area of the external plant room wall $\quad$ m^2

This is repeated in cells D74 to I74.

cell C75: –20*@LOG($B$65)

The distance component of the noise radiation from the plant room wall is calculated from

$$\mathrm{SPL} = -20 \times \log(d) \qquad \mathrm{dB}$$

This is repeated in cells D75 to I75.

cell C76: –$B$66

The attenuation of the sound barrier, if there is one, between the plant room external wall and the outdoor recipient is copied from the input data cell:

$$\mathrm{SPL} = -B \qquad \mathrm{dB}$$

This is repeated in cells D76 to I76.

cell C77: +$B$67

The directivity index of the sound leaving the plant room external wall is copied from the input data cell:

$$\mathrm{SPL} = -\mathrm{DI} \qquad \mathrm{dB}$$

This is repeated in cells D77 to I77.

cell C78: –17

The constant is included in the overall formula as

$$\mathrm{SPL} = -17 \qquad \mathrm{dB}$$

This is repeated in cells D78 to I78.

cell C79: +C73 – C71 – C72 + C74 + C75 + C76 + C77 + C78

The sound that is radiated from the external surface of the plant room is received as a sound pressure level at the specified distance from the wall by

$$\mathrm{SPL}_2 = \mathrm{SPL}_1 + 10 \times \log(S_2) - 20 \times \log(d) - B + \mathrm{DI} - 17 \qquad \mathrm{dB}$$

This is repeated in cells D79 to I79.

cells N66 to T79:

The selection of the peak noise rating at the distant recipient's location is carried out for each frequency in the same manner as for the noise rating in the plant room.

cell B80: +T79

The peak noise rating that is found at the outdoor recipient's position is displayed in cell B80.

cell D96: (D88*$B92 + D89*$B93 + D90*$B94 + D91*$B95)/ ($B92 + $B93 + $B94 + $B95)

The mean absorption coefficient of the intermediate room at a frequency of 125 Hz, is calculated from

$$\overline{\alpha} = \frac{(\alpha_1 \times A_1 + \cdots + \alpha_4 \times A_4)}{A_1 + \cdots + A_4}$$

cell B97: +B92 + B93 + B94 + B95

The total surface area of the intermediate room is found from

$$S_3 = A_1 + A_2 + A_3 + A_4 \text{ m}^2$$

cell D98: ($B97*D96)/(1–D96)

The intermediate room absorption constant is found from

$$R = \frac{S_3 \times \overline{\alpha}}{1 - \overline{\alpha}}$$

S_3 = total surface area of the intermediate room m^2

This is repeated in the cell range E98 to I98.

cell D99: 0.161*$B$85/$B$97/D96

The reverberation time of the intermediate space at a frequency of 125 Hz is found from

$$T_2 = \frac{0.161 \times V_2}{S_3 \times \overline{\alpha}}$$

T_2 = reverberation time of the intermediate room   s
V_2 = volume of the intermediate room m^3

This is repeated in the cell range E99 to I99.

cell D100: 10*@LOG(($B$26/(4*@PI*$B$27*$B$27)) + (4/D98))

The effect of distance from the sound source on the received sound pressure level in the intermediate space is

$$\text{SPL} = 10 \times \log\left(\frac{Q}{4 \times \pi \times r^2} + \frac{4}{R}\right) \text{ dB}$$

This is repeated in the cell range E100 to I100.

cell D112: +D56

The sound pressure level within the plant room is copied into cell D112. This is repeated in the cell range E112 to I112.

cell D113: 10*@LOG($B$104)

The sound pressure level effect of the common wall with the plant room is calculated for the intermediate space from

$$SPL = 10 \times \log(S_4)\ dB$$

S_4 = area of the sound source surface in the intermediate room m^2

This is repeated in the cell range from E113 to I113.

cell D114: +10*@LOG(D99)

The effect of the reverberation time on the sound pressure level in the intermediate space is

$$SPL = 10 \times \log(T_2)\ dB$$

This is repeated in the cell range of E114 to I114.

cell D115: −10*@LOG(0.16*$B$85)

The effect of the room volume on the sound pressure level in the intermediate room is

$$SPL = -10 \times \log(0.16 \times V_2)\ dB$$

This is repeated in the cell range of E115 to I115.

cell D117: +D112 − D111 + D113 + D114 + D115

The sound pressure level that is produced in the intermediate space is found from

$$SPL_3 = SPL_1 - SRI + 10 \times \log(S_4) + 10 \times \log(T_2) - 10 \times \log(0.16 \times V_2)\ dB$$

SRI = sound reduction index of the structure which allows the transfer of sound from the plant room into the intermediate space dB

This is repeated in cells E117 to I117.

cells N106 to T117:

The selection of the peak noise rating in the intermediate room at each frequency is carried out in the same manner as for the noise rating in the plant room

cell B119: +T117

The peak noise rating for the intermediate room or space is displayed in cell B119.

cell D136: (D128*$B132 + D129*$B133 + D130*$B134 + D131*$B135)/($B132 + $B133 + $B134 + $B135)

The mean absorption coefficient of the target room at a frequency of 125 Hz is calculated from

$$\overline{\alpha} = \frac{(\alpha_1 \times A_1 + \cdots + \alpha_4 \times A_4)}{A_1 + \cdots + A_4}$$

This is repeated in cells E136 to I136.

cell B137: +B132 + B133 + B134 + B135

The total surface area of the target room is

$$S_5 = A_1 + A_2 + A_3 + A_4 \qquad \text{m}^2$$

cell D138: ($B137*D136)/(1–D136)

The absorption constant of the target room is

$$R = \frac{S_5 \times \overline{\alpha}}{1 - \overline{\alpha}}$$

This is repeated in the cell range E138 to I138.

cell D139: 0.161*$B$125/$B$137/D136

The reverberation time of the target occupied room at a frequency of 125 Hz is found from

$$T_3 = 0.161 \times V_3/(S_5 \times \overline{\alpha})$$

This is repeated in the cell range E139 to I139.

cell D152: +D117

The sound pressure level that is projected into the target room from the intermediate space is copied from cell D117. This is repeated in the cell range from E152 to I152.

cell D153: 10*@LOG($B$144)

The sound pressure level effect of the common wall with the intermediate space is calculated for the target room from

$$\text{SPL} = 10 \times \log(S_5) \text{ dB}$$

This is repeated in the cell range from El53 to I153.

cell D154: +10*@LOG(D139)

The sound pressure level effect of the reverberation time is calculated for the target room from

$$\text{SPL} = 10 \times \log(T_3) \text{ dB}$$

This is repeated in the cell range from E154 to I154.

cell D155: −10*@LOG(0.16*$B$125)

The sound pressure level effect of the target room volume is calculated from

$$SPL = -10 \times \log(0.16 \times V_3) \text{ dB}$$

This is repeated in the cell range E155 to I155.

cell D157: +D152 − D151 + D153 + D154 + D155

The sound pressure level that is produced in the target room is found from

$$SPL_4 = SPL_3 - SRI + 10 \times \log(S_5) + 10 \times \log(T_3) - 10 \times \log(0.16 \times V_3) \text{ dB}$$

This is repeated for the frequency range in cells E157 to I157.

cells N146 to T157:

The selection of the peak noise rating at each frequency is carried out in the same manner as for the noise rating in the plant room.

cell B159: +T157

The peak noise rating that is found in the target room is copied from cell T157.

cell B168: +$A168*0.681 + 55.4

The sound pressure level at a frequency of 31.5 Hz that corresponds to a noise rating of NR 20 is found from

$$SPL = 0.681 \times NR\ 20 + 55.4 \text{ dB}$$

cell C168: +$A168*0.79 + 35.5

The sound pressure level at a frequency of 63 Hz that corresponds to a noise rating of NR 20 is found from

$$SPL = 0.79 \times NR\ 20 + 35.5 \text{ dB}$$

cell D168: +$A168*0.87 + 22

The sound pressure level at a frequency of 125 Hz that corresponds to a noise rating of NR 20 is found from

$$SPL = 0.87 \times NR\ 20 + 22 \text{ dB}$$

cell E168: +$A168*0.93 + 12

The sound pressure level at a frequency of 250 Hz that corresponds to a noise rating of NR 20 is found from

$$SPL = 0.93 \times NR\ 20 + 12 \text{ dB}$$

cell F168: +$A168*0.974 + 4.8

The sound pressure level at a frequency of 500 Hz that corresponds to a noise rating of NR 20 is found from

$$SPL = 0.974 \times NR\ 20 + 4.8 \text{ dB}$$

cell G168: +$A168

The sound pressure level at a frequency of 1000 Hz that corresponds to a noise rating of NR 20 is found from

$$SPL = 1 \times NR\ 20 + 0\ dB$$

cell H168: +$A168*1.015–3.5

The sound pressure level at a frequency of 2000 Hz that corresponds to a noise rating NR 20 is found from

$$SPL = 1.015 \times NR\ 20 - 3.5\ dB$$

cell I168: +$A168*1.025–6.1

The sound pressure level at a frequency of 4000 Hz that corresponds to a noise rating of NR 20 is found from

$$SPL = 1.025 \times NR\ 20 - 6.1\ dB$$

cells B169 to I180:

These repeat the noise rating calculations for the frequency range from 31.5 Hz to 4000 Hz at NR 20 to NR 80.

Questions

Questions 1 to 31 do not need to be evaluated on the worksheet. The worksheet is to be used for Questions 32 onwards. The solution to Question 32 is shown on the original file DBPLANT.WKS. Solutions to descriptive questions are to be found within the chapter, except where specific answers are provided.

1. List the sources of noise that could be found within an air conditioned building.
2. What is meant by noise?
3. State which items of mechanical services plant, equipment and systems within an occupied building are not likely to create noise?
4. Explain how sound travels from one location to another.
5. Explain what is meant by the term sound pressure wave.
6. Why is sound important?
7. Explain how we 'hear' sounds.
8. State what is meant by sound power and sound pressure level. State the units of measurement for sound power, sound pressure, sound power level and sound pressure level.
9. Explain why any decimal fraction of a decibel is not used in engineering design.
10. List the ways in which mechanical and electrical services plant, equipment and systems generate sound.
11. Explain, with the aid of sketches and examples, how sound is transferred, or can be, through a normally serviced multi-storey occupied building.
12. Discuss the statement: 'Turbulent flows in building services systems create a noise nuisance.'
13. State how the building structure transfers sound.
14. Explain, with the aid of sketches, ways in which the noise and vibration produced by the mechanical and electrical services of a building can be reduced before they become a nuisance for the building's users.

15. Explain how sound energy is dissipated into the environment.

16. State the range of frequencies that are detectable by the human ear and the frequencies that are used in acoustic design calculations. State the reasons for these two ranges being different, if they are.

17. Define the terms 'sound power level' and 'sound pressure level'.

18. Explain what is meant by direct and reverberant sound fields.

19. A plant room for a refrigeration compressor is $6\ m \times 4\ m$ in plan and 3 m high. It has four brickwork walls, a concrete floor and a concrete roof. Select the surface absorption coefficients for the frequency range 125 Hz to 4000 Hz from Table 14.1. Calculate the room absorption constant and the reverberation time for the plant room at each frequency. Do the calculations manually and then enter the same data onto the worksheet to validate the results.

20. An air-conditioning centrifugal fan has an overall sound power level SWL of 75 dBA. The fan is to be installed centrally within a plant room that has a room absorption constant R of 12 m^2. Calculate the sound pressure level that will be produced close to the fan, in the plant room at 1000 Hz when the fan is operating, and also generally within the room.

21. A 900 mm diameter axial fan is to be installed on the concrete floor of an $8\ m \times 4\ m \times 3\ m$ high plant room. The fan sound power level at 1000 Hz is 89 dB. The room absorption constant R at 1000 Hz is 8 m^2 and the reverberation time is 0.4 s. Calculate the room sound pressure level at a radius of 300 mm from the fan, and the reverberant room sound pressure level.

22. A reciprocating water chilling refrigeration compressor has an overall sound power level of 92 dBA. It is to be located within a concrete-and-brick plant room that has a reverberation time of 1.8 s and a volume of 250 m^3. Calculate the plant room reverberant sound pressure level.

23. An air-handling plant has an overall sound power level of 81 dB. The plant room has an external wall of 10 m^2 that has an acoustic attenuation of 35 dB and ventilation openings having a free area of 3 m^2. The windows of residential and office buildings are at a distance of 12 m from the plant room wall. Calculate the external sound pressure level at the windows and recommend what, if any, attenuation is needed at the plant room.

24. A forced draught gas-fired boiler has an overall sound pressure level of 96 dB. The boiler plant room has an external wall of 60 m^2 that has an acoustic attenuation of 25 dB and two louvre doors to admit air for combustion. Calculate the external sound pressure level at a distance of 20 m from the plant room wall. State your recommendations for the attenuation of the boiler and the plant room.

25. A single-storey office building has floor dimensions of $40\ m \times 30\ m$ and a height of 3 m to a suspended acoustic tile ceiling. The average height of the ceiling void is 1.5 m. A plant room is adjacent to the roof void. There is a common plant room wall of $10\ m \times 1.5\ m$ high in the roof void. The sound pressure level in the plant room is expected to be 61 dB. The reverberation time of the roof void is 0.6 s. The plant room wall adjoining the roof void has a sound reduction index of 13 dB. Calculate the sound pressure level that is produced within the roof void as the result of the plant room noise. Comment on the resulting sound pressure level.

26. A hospital waiting area has floor dimensions of $8\ m \times 12\ m$ and a height of 3 m to a plasterboard ceiling. A packaged air conditioning unit is housed in an adjacent room. There is a common wall of 15 m^2 and sound reduction index of 35 dB to the two rooms. The sound pressure level in the plant room is expected to be 72 dB. The reverberation time of the waiting room is 1.3 s. Calculate the sound pressure level that will be produced in the waiting room.

27. A meeting room has floor dimensions of $8\ m \times 6\ m$ and a height of 2.7 m to a suspended tile ceiling. The reverberation time of the room is 0.7 s. A fan coil heating and cooling unit creates a sound pressure level of 43 dB in the ceiling space. The acoustic tile ceiling has a sound reduction index of 8 dB. Calculate the sound pressure level in the meeting room.

28. An hotel bedroom is 6 m long, 5 m wide and 2.8 m high and it has a reverberation time of 0.4 s. The

air-conditioning plant room generates a sound pressure level of 56 dB in the service space above the ceiling of the bedroom. The plasterboard ceiling has a sound reduction index of 16 dB. Calculate the sound pressure level in the bedroom.

29. Explain how noise rating curves relate to the response of the human ear and are used in the design of mechanical services plant and systems.

30. The centrifugal fan in an air-handling plant produces the noise spectrum shown in Table 14.9 within an office. Calculate the sound pressure levels for noise ratings NR 35, NR 40, NR 45, NR 50 and NR 55 and plot the noise rating curves for the frequency range 31.5 Hz to 8 kHz. Plot the room sound pressure levels on the same graph and find which noise rating is not exceeded.

31. A model XT45 water chiller is to be located within a plant room on the roof of an hotel in a city centre. The plant room is 12 m long, 10 m wide and 3 m high. The room directivity index is 2. The plant operator will normally be 1 m from the noise source. The floor is concrete, the roof is lined internally with a 50 mm polyester acoustic blanket with a metallized film surface. The plant room walls are 115 mm brickwork. There are no windows. The water chiller manufacturer provided the sound power levels as 100 dB overall, 74 dB at 63 Hz, 89 dB at 125 Hz, 95 dB at 250 Hz, 97 dB at 500 Hz, 99 Hz at 1 kHz, 97 dB at 2 kHz and 90 dB at 4 kHz.

(a) Check that the correct data is entered onto the working copy of the original worksheet file DBPLANT.WKS and find the noise rating that is not exceeded within the plant room.

(b) The plant room has three external walls. The nearest openable window in nearby buildings is at a distance of 15 m from a plant room wall. There is no acoustic barrier between a plant room wall and the recipient's window. The directivity index for the outward projection of sound is taken as 3 dB. Find the noise rating at the recipient's window and state what the result means.

(c) A corridor adjoins the plant room. The target sound space, an office, is on the opposite side of the corridor. The corridor is 10 m long, 1 m wide and 3 m high. It has a room directivity index of 2, a carpeted concrete floor, plastered brick walls and a plasterboard ceiling. The common wall between the plant room and the corridor is 10 m long, it is constructed with 115 mm plastered brickwork and it does not have a door. There are no windows. There is no other sound barrier. Find the noise rating which would be found at a distance of 0.5 m from the plant room wall while within the corridor.

(d) The target office is 10 m long, 10 m wide and 3 m high. The room directivity index is 2. The nearest sedentary occupant of the office will be 1 m from the corridor wall. The floor has pile carpet, the walls are plastered brick and there is a suspended ceiling of 15 mm acoustic tile and 50 mm glass fibre-matt. The office has four 2 m × 2 m single-glazed windows on two external walls. The office wall that adjoins the corridor is 115 mm plastered brickwork and it has one 2 m^2 door into the corridor. Find the noise rating, NR, and sound pressure levels, SPL dB, that are experienced in the target office. State what effect the office and plant room doors will have on the noise rating in the target room. Recommend appropriate action to be taken with these doors.

32. A centrifugal fan is located within the basement plant room of an office building. The plant room is 8 m long, 6 m wide and 3 m high. The room directivity index is 2 and the plant operator will normally be 1 m from the noise source. The floor and ceiling are concrete, there are four 230 mm brick walls and one acoustically treated door. There are no windows in the plant room.The sound power levels of the fan are, 86 dB overall, 64 dB at 63 Hz, 66 dB at 125 Hz, 72 dB at 250 Hz, 80 dB at 500 Hz, 86 Hz at 1 kHz, 82 dB at 2 kHz and 77 dB at 4 kHz.

(a) Find the noise rating that is not exceeded within the plant room.

Table 14.9 Noise spectrum in Question 30

Frequency	31.5 Hz	63 Hz	125 Hz	250 Hz	500 Hz	1 kHz	2 kHz	4 kHz	8 kHz
Room SPL dB	30	35	32	40	42	31	28	20	10

(b) A corridor and staircase connect the plant room to the Reception area of the building. The corridor is 6 m long, 1 m wide and 3 m high. It has a room directivity index of 2. The corridor has a concrete floor, plastered brick walls and a plasterboard ceiling. The common wall between the plant room and corridor is 2 m long. The sound reduction index of the plant room door is 20 dB at each frequency from 125 Hz to 4 kHz. There is no other sound barrier. Find the noise rating that would be found at a distance of 1 m from the plant room in the corridor.

(c) The Reception area is 12 m long, 8 m wide and 3 m high. The room directivity index is 2. There are 10 m^2 of single-glazed windows in Reception. There is a door at the top of the staircase down to the plant room. The stairs door is 1 m wide, 2 m high and it has a sound reduction index of 20 dB at each frequency from 125 Hz to 4 kHz. The nearest occupant will be 1 m from the stairs door. The floor has thermoplastic tiles on concrete, the walls are plastered brick and there is a plasterboard ceiling. Find the noise rating which is not exceeded in Reception.

33. Oil-fired hot water boilers are located in a plant room in the basement of an exhibition and trade centre building in a city centre. The plant room is 10 m long, 10 m wide and 5 m high. The room directivity index is 2. The floor, walls and ceiling are concrete. There are no windows. The reference sound power level of the boiler plant is 88 dBA.

(a) Find the anticipated spectral variation in the sound power level for the frequency range from 63 Hz to 4 kHz from Table 14.4 and Fig. 14.1, enter the data onto the worksheet and find the noise rating that is not exceeded within the boiler plant room.

(b) The plant room has three 100 mm concrete external walls. The nearest recipient can be 1 m from the external surface of a boiler plant room wall. There is no acoustic barrier between a plant room wall and a recipient. The directivity index for the outward projection of sound is taken as 3 dB. Find the noise rating at the nearest recipient's position and state what the result means.

(c) A hot-water pipe and electrical cable service duct connects the boiler plant room to other parts of the building. The concrete-lined service duct is 30 m long, 2 m wide and 1 m high. Both ends of the service duct have a 100 mm concrete wall. Calculate the noise rating within the service duct at its opposite end from the boiler plant room.

(d) A conference room 115 mm brick wall adjoins the service duct at the furthest end from the boiler plant room. The conference room is 12 m long, 10 m wide and 4 m high. The room directivity index is 2. The nearest sedentary occupant will be 0.5 m from the service duct wall. The floor has pile carpet, the walls are plastered brick and there is a suspended ceiling of 15 mm acoustic tile and 50 mm glass fibre matt. There are no windows. Find the noise rating that is produced in the conference room by the boiler plant.

34. A four-pipe chilled- and hot-water fan coil unit is located within the false ceiling space above an office in an air-conditioned building. Conditioned outdoor air is passed to the fan coil unit through a duct system. The office is 5 m long, 4 m wide and 3 m high. The room directivity index is 2. The office has a concrete floor with thermoplastic tiles and 115 mm plastered brick walls. The 700 mm deep suspended ceiling has 12 mm fibreboard acoustic tiles, recessed fluorescent luminaires, ducted supply and return air with a supply air diffuser, a return air grille and a concrete ribbed slab for the floor above. The office has a double-glazed window of 2 m × 2 m. The reference sound power level of the fan coil unit is 85 dBA. Enter the ceiling space as the plant room and bypass the intermediate space data as directed.

(a) Find the anticipated spectral variation in the sound power level of the fan coil unit for the frequency range from 63 Hz to 4 kHz from Table 14.4 and Fig. 14.2. Enter the data onto the worksheet and find the noise rating that is not exceeded within the ceiling space above the office.

(b) Find the noise rating that is not expected to be exceeded within the office at head height. Assume that the sound reduction of the acoustic tile ceiling is maintained across the whole ceiling area.

(c) Sketch a cross-section of the fan coil unit installation and identify all the possible noise paths into the office.

(d) List the ways in which the potential noise paths into the office can be, or may need to be, attenuated.

15 Fire protection

Key terms and concepts

Learning objectives

Study of this chapter will enable the reader to:

1. classify fire hazards;
2. identify the necessary ingredients for a fire;
3. describe the development of a fire;
4. recognize the hazards of smoke;
5. apply the correct fire-fighting system, or combination of systems, to different fire classifications;
6. understand the principles of portable fire extinguishers;
7. know the criteria for the use of hose reel, dry riser, wet riser, foam, sprinkler, carbon dioxide (CO_2), vaporizing liquid, dry powder and deluge fixed fire-fighting systems;
8. know the sources of water used in fire-fighting;
9. identify how fire development can be detected;
10. recognize the importance of smoke ventilation;
11. identify the locations for fire dampers in air ductwork and know how they operate.

Introduction

The systems required to meet the needs of tackling small fires, evacuation and major fire-fighting both by the occupants and then the Fire Service are outlined. Building management systems under computer monitoring and control will incorporate such systems, together with security functions. Integration of such equipment with the architecture, decor and other services is planned from the earliest design stage.

Fire classification

A building's fire risk is classified according to its occupancy and use. Table 15.1 gives representative information (CIBSE, 1986).

A fire is supported by three essential ingredients: fuel, heat and oxygen. The absence of any one of these causes an established fire to be extinguished. The fire-fighting system must be appropriate to the location of the fire and preferably limited to that area in order to minimize damage to materials, plant and the building structure. Radiation from a fire may provoke damage or combustion of materials at a distance. Structural fire protection can include water sprays onto steelwork to avoid collapse, as used in the Concorde aircraft production hangar.

The system of fire-fighting employed depends upon the total combustible content of the building (fire load), the type of fire risk classification and the degree of involvement by the occupants. Fire escape design where children, the elderly or infirm are present needs particular care so that sufficient time is provided in the fire resistance of doors and partitions for the slower evacuation encountered.

Smoke contains hot and unpleasant fumes, which can be lethal when produced from certain chemicals and plastics. Visual obstruction makes escape hazardous and familiar routes become confused. Packaging materials, timber, plastics, liquefied petroleum gas cylinders and liquid chemicals must not be stacked in passageways or near fire exits in completed or partially completed buildings. Each working site or building needs a safety officer responsible for general oversight.

Fires are classified in Table 15.2.

Table 15.1 Classification of occupancies

Category	Group	Hazard occupancy
Extra light	—	Public buildings
Ordinary	1	Restaurant
Ordinary	2	Motor garage
Ordinary	3	Warehouse
Ordinary	3 (special)	Woodwork
Extra high	—	Paint manufacture
Extra high (storage)	1	Electrical appliance
Extra high (storage)	2	Furniture
Extra high (storage)	3	Wood, plastic or rubber
Extra high (storage)	4	Foamed plastics or rubber

Table 15.2 Fire classifications

Classification	Fire type	Fire-fighting system
A	Wood and textiles	Water, cools
B	Petroleum	Exclude oxygen
C	Gases	Exclude oxygen
D	Flammable metals	Exclude oxygen
E	Electrical	Exclude oxygen, non-conducting

Regular fire drills are conducted by the safety officer and employees are clearly notified of their responsibilities in an emergency. Staff duties will be to shepherd the public, patients or students out of the building to the rendezvous, while maintenance personnel may be required to operate fire-fighting equipment while awaiting the fire brigade.

Portable extinguishers

Portable extinguishers are manually operated first-aid appliances to stop or limit the growth of small fires. Staff are trained in their use and the appliances are regularly maintained by the suppliers. Table 15.3 summarizes their types and applications. Fire blankets are provided in kitchens where burning pans of oil or fat need to be covered or personnel need to be wrapped to smother ignited clothing.

Water

A 9 l water extinguisher is installed for each 210 m^2 floor area, with a minimum of two extinguishers per floor. A high-pressure CO_2 cartridge is punctured upon use and a 10 m jet of water is produced for 80 s. Water must not be used on petroleum, burning liquids or in kitchens as it could spread the fire.

Dry powder

Dry powder extinguishers contain from 1 to 11 kg of treated bicarbonate of soda powder pressurized with CO_2, nitrogen or dried air. A spray of 2–7 m is produced for 10–24 s depending on size. The powder interrupts the chemical reactions within the flame, producing rapid flame knockdown. The powder is non-conducting and does little damage to electric motors or appliances. A deposit of powder is left on the equipment.

Foam

Portable foam extinguishers may contain foaming chemicals that react upon mixing or a CO_2 pressure-driven foam. They cool the combustion, exclude oxygen, and can be applied to wood, paper, textile or liquid fires. Garages are a

Table 15.3 Type of portable fire extinguisher

Group	Extinguishing agent	Fire type	Action	Colour
1	Water	Class A	Cools	Red
2	Dry powder	All	Flame interference	Blue
3	Foam	Class B	Excludes oxygen	Cream
4	Carbon dioxide (CO_2)	Classes B, E	Excludes oxygen	Black
5	Vaporizing liquid	Small fires, motor vehicles, class E	Flame interference	Green

particular application. Sizes range from 4.5 to 45 l. A 7 m jet is produced for 70 s with a 9 l capacity model.

Vaporizing liquid

Vaporizing liquid extinguishers use bromochlorodifluoromethane (BCF) or bromotrifluoromethane (BTM). These are 1–7 kg extinguishers containing a nitrogen-pressurized liquefied halogen gas, which is highly efficient at interrupting the flames of chemical reactions and producing rapid knockdown without leaving any deposit. They are more powerful than CO_2 extinguishers and are used on electrical, electronic and liquid fires. Halogen is used for outdoor fires and motor vehicles, where the toxic vapour given off is adequately ventilated. They are not suitable for enclosed areas because of the danger to occupants. These are CFCs (p. 151) and are part of the international agreement to cease their use. A suitable replacement for fire-fighting is being sought.

Carbon dioxide

Pressurized CO_2 extinguishers leave no deposit and are used on small fires involving solids, liquids or electricity. They are recommended for use on delicate equipment such as electronic components and computers. The CO_2 vapour displaces air around the fire and combustion ceases. There is minimal cooling effect, and the fire may restart if high temperatures have become established. Water-cooling backup is used where appropriate.

Fixed fire-fighting installations

Various fire-fighting systems are employed in a building so that an appropriate response will minimize damage from the fire and the fire-fighting system itself. Backup support for portable extinguishers may be provided by a hose reel installation and this can be used by staff while the fire brigade is called.

Some public buildings, shops and factories are protected by a sprinkler system, which only operates directly over the source of fire. This localizes the fire to allow evacuation. Where petroleum products are present, a mixture of foam and water is used. The Fire Officers' Committee (FOC) rules should be consulted for further information.

Hose reels

Hose reels are a rapid and easy to use first-aid method, complementary to other systems and used by the building's occupants. They are located in clearly visible recesses in corridors so that no part of the floor is further than 6 m from a nozzle when the 25 mm bore flexible hose is fully extended.

The protected floor area is an arc 18 m to 30 m from the reel, depending on the length of hose. A minimum water pressure of 200 kPa is available with the 6 mm diameter nozzle. This produces a jet 8 m horizontally or 5 m vertically. Minimum water flow rate at each nozzle is 0.4 l/s, and the installation should be designed to provide not less than three hose reels in simultaneous use: a flow rate of 1.2 l/s. Figure 15.1 shows a typical installation.

The local water supply authority might allow direct connection to the water main, and there may be sufficient main pressure to eliminate the need for pressure boosting. Pump flow capacity must be at least 2.5 l/s. The stand-by pump can be diesel-driven. Flow switches detect the operation of a hose and switch on the pump.

Dry hydrant riser

A dry hydrant riser is a hydrant installation for buildings 18–40 m high where prompt attendance by the fire brigade is guaranteed. A dry riser pipe 100 mm or 150 mm in diameter is sited within a staircase enclosure with a 65 mm instantaneous valved outlet terminal at each landing. All parts of the building floor are to be within 60 m of the hydrant, measured along the line on which a hose would be laid. A test hydrant is fitted at roof level, and also a 25 mm

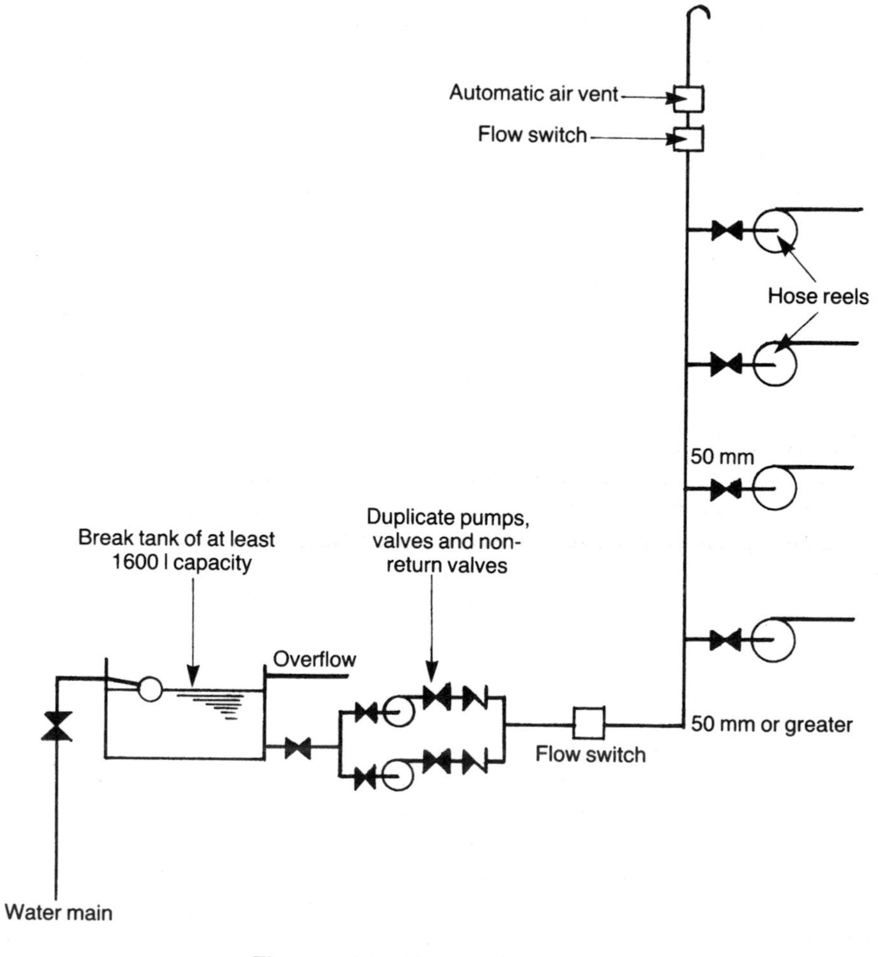

Figure 15.1 Hose reel installation

automatic air vent. A double inlet breeching piece with two 65 mm instantaneous terminals is located in a red-wired glass box in an external wall, 760 mm above ground level and not more than 12 m from the riser.

The inlet point is within 18 m of an access road suitable for the fire brigade pumping appliance. A brass blank cap and chain is fitted to each landing valve. The riser is electrically earthed. Landing valves are 1 m above floor level and are used by the fire brigade for their own hoses.

Wet hydrant riser

A permanently charged rising pipe 100 mm in diameter or greater supplies a 65 mm instantaneous valved outlet terminal at each floor at a pressure of between 410 and 520 kPa. The upper pressure limit is to protect the fire brigade hoses from bursting and is achieved by fitting an orifice plate restriction before the landing valve on the lower floors of a tall building. The maximum static pressure in the system when all the landing valves are shut is limited to 690 kPa by recirculating water to the supply tanks through a 75 mm return pipe.

Each hydrant valve is strapped and padlocked in the closed position. They are 1 m above floor level and are only used by the fire brigade for buildings over 60 m high which extend out of the reach of turntable ladders. The maximum normally permitted height is 60 m for a low-level break tank and booster set. Higher buildings have separate supply tank and pump sets for each 60 m height.

Pressure boosting of the water supply is provided by a duplicate pump installation capable of delivering at least 23 l/s. Pumps are started automatically on fall of water pressure or water flow commencement. Audible and visual alarms are triggered to indicate booster plant operation.

A break tank capacity of 11.4–45.5 m^3 is required and mains water make-up rate is 27 l/s or 8 l/s for the larger tank. Additionally, four 65 mm instantaneous fire brigade inlet valved terminals are provided at a 150 mm breeching fitting in a red wired-glass box in an external wall, as described for the dry hydrant riser. The box is clearly labelled.

A nearby river, canal or lake may also be used as a water source with a permanently connected pipe from a jack well and duplicate pumps.

Pneumatic pressure boosting is used to maintain system pressure in a similar manner to that shown in Fig. 6.4. The standby pump may be driven by a diesel engine fed from a 3–6 h capacity fuel storage tank providing a gravity feed to the engine.

Foam inlets

Oil-fired boiler plant rooms and storage tank chambers in basements or parts of buildings have fixed foam inlet pipework from a red wired-glass foam inlet box in an outside wall as for the dry hydrant riser.

A 65 mm or 75 mm pipe runs for up to 18 m from the inlet box into the plant room. The fire brigade connect their foam-making branch pipe to the fixed inlet and pump high-expansion foam onto the fire. The foam inlet pipe terminates above the protected plant with a spreader plate. A short metal duct may be used as a foam inlet to a plant room close to the roadway. Vertical pipes cannot be used and the service is electrically bonded to earth.

On-site foam-generation equipment is available and may be used for oil-filled electrical transformer stations. In the event of a fire, the electricity supply is automatically shut off, a CO_2 cylinder pressurizes a foam and water solution and foam spreaders cover the protected equipment. Figure 15.2 shows a typical fire brigade inlet box.

Automatic sprinkler

High-fire-risk public and manufacturing buildings are protected by automatic sprinklers. These may be a statutory requirement if the building exceeds a volume of 7000 m^3. Loss of life is very unlikely in a sprinkler-protected building. Sprinkler water outlets are located at about 3 m centres, usually at ceiling level, and spray water in a circular pattern. A deflector plate directs the water jet over the hazard or onto walls or the structure.

Each sprinkler has a frame containing a friable heat-sensing quartz bulb, containing a coloured liquid for leak detection, which seals the water inlet. Upon local overheating, the quartz expands and fractures, releasing the spray. Water flow is detected and starts an alarm, pressure-boosting set and automatic link to the fire brigade monitoring station.

Acceptable sources of water for a sprinkler system are as follows:

1. a water main fed by a source of 1000 m^3 capacity where the correct pressure and flow rate can be guaranteed;
2. an elevated private reservoir of 500 m^3 or more depending on the fire risk category;
3. a gravity tank on site, which can be refilled in 6 h, with a capacity of 9–875 m^3 depending on the fire risk category;
4. an automatic pump arranged to draw water from the main or a break tank of 9–875 m^3 capacity;

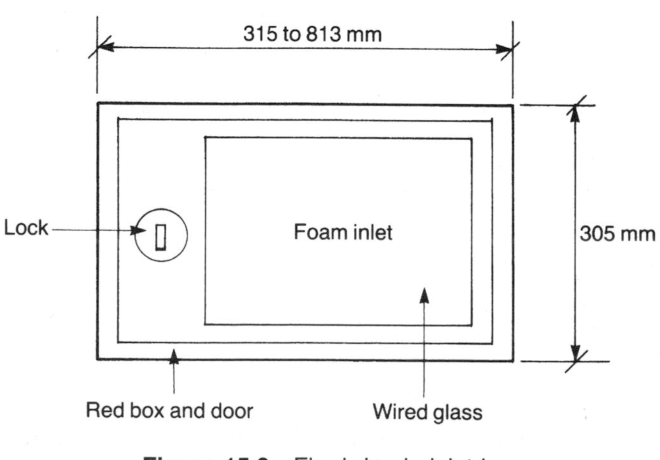

Figure 15.2 Fire brigade inlet box

5. a pressure tank: a pneumatic pressure tank source can be used for certain light fire risk categories or as a back-up facility to some other system.

Sprinkler installations are classified under four principal types.

1. Water-filled pipes are permanently charged with water.
2. Dry pipework: pipes are filled with compressed air and used where pipework is exposed to air temperatures below 5 °C or above 70 °C.
3. Alternate system: pipes are filled with water during the summer and air in the winter.
4. A pre-action system is a dry pipe installation but has additional heat detectors which pre-empt the opening of sprinkler heads and admit water into the pipework, converting it to wet-pipe operation.

Different types of sprinkler head are used depending on the hazard protected, their object being to produce a uniform density of spray.

Fusible link: a soldered link in a system of levers holds the water outlet shut. At a predetermined temperature of 68 °C or greater, the solder melts and water flow starts

Chemical: similar to the fusible link but using a block of chemical, which melts at 71 °C or greater, depending on the application

Glass bulb: a quartz bulb containing a coloured fluid with a high coefficient of expansion, which fractures at 57 °C or more

Open sprinkler heads (deluge system): these are used to combat high-intensity fires and protect storage tanks or structural steelwork. They are controlled by a quick-opening valve actuated from a heat detector or a conventional sprinkler arrangement. A drencher system provides a discharge of water over the external openings of a building to prevent the spread of fire

Each sprinkler installation must be provided with the following:

1. main stop valve, which is strapped and padlocked in the open position to enable the water flow to be stopped after the fire is extinguished;
2. alarm valve: differential pressure caused by water flow through the valve opens a branch pipe to the alarm gong motor;
3. water motor alarm and gong: water flow through a turbine motor drives a rotary ball clapper within a domed gong to give audible warning of sprinkler operation and commence evacuation of the building.

A satisfactory pipework installation serving an automatic sprinkler distribution and hose reels is shown in Fig. 15.3. Two 65 mm instantaneous fire brigade inlet pipes are provided at a clearly marked access box.

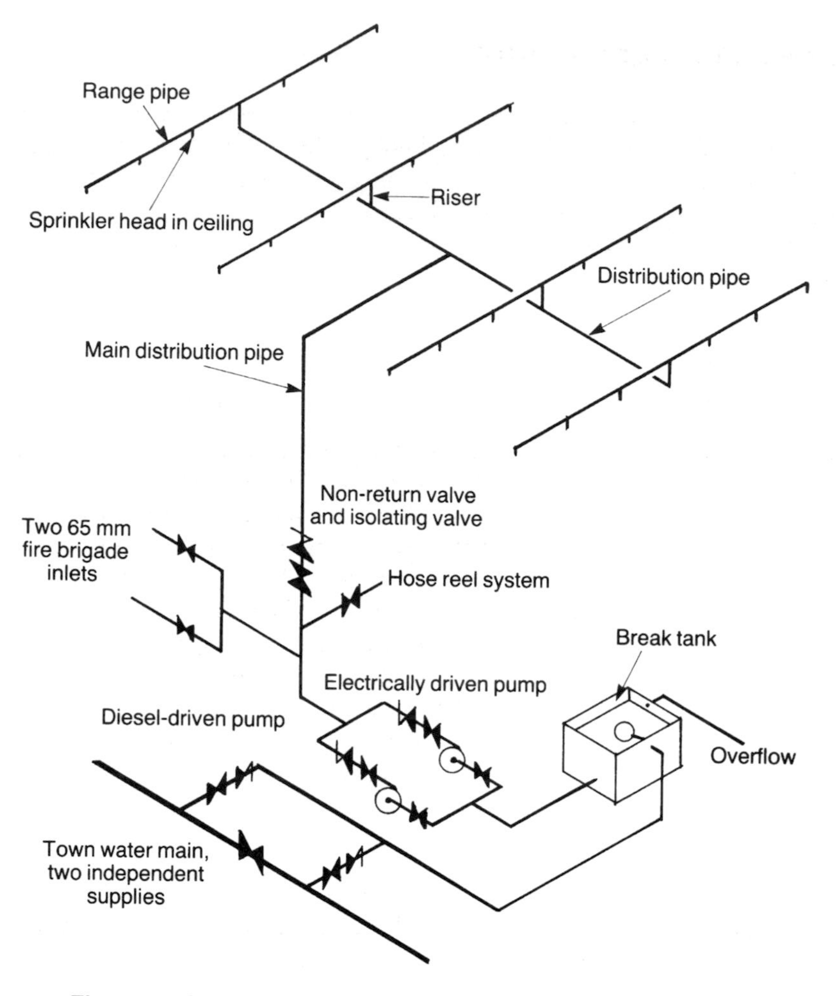

Figure 15.3 Water supply to hose reel and sprinkler installation

Carbon dioxide

Carbon dioxide is used in fixed installations protecting electrical equipment such as computer rooms, transformers and switchgear. Heat or smoke detectors sound alarms and CO_2 gas floods the room from high-pressure storage cylinders. Pipework transfers the CO_2 to ceiling and underfloor distributors. System initiation can be manual or automatic but complete personnel evacuation is essential before CO_2 flooding is allowed.

Fixed BCF, BTM and dry powder

Extinguishers are installed within rooms or false ceilings and are operated from a manual push-button or automatic fire detector. Personnel evacuation is followed by the release of halogen gas to flood the room with a 5% concentration in air, which is sufficient to inhibit fire.

Fire detectors and alarms

Detection of a potentially dangerous rise in air temperature or pressure or the presence of smoke is required at the earliest possible moment to start an alarm. Evacuation of the building and manual or automatic contact with the fire brigade monitoring switchboard should take place before people are at risk. Means of detection can be combined with security surveillance. Fire detection takes the following forms.

Hazard detectors

Hazard detectors give an early warning of the risk of a fire or explosion.

Temperature rise:	a local rise in temperature leads to the melting of a fusible link in a wire holding open a valve on a fuel pipe to a burner, thermal expansion of a fluid-filled bellows or capillary tube or movement of a bimetallic strip to make an alarm circuit
Flammable vapour detector:	gas, oil, petrol or chemical vapour presence is detected by a catalytic chemical reaction
Diffusion:	butane and propane vapour diffusion through a membrane is detected
Explosion:	rise of local atmospheric pressure above a set value, or at a fast rate, is detected

Ionization smoke detector

Ionization smoke detectors contain a radioactive source of around 1 microcurie, typically americium-241, which bombards room air within the detector with alpha particles (ionization). Electrical current consumption is 50 μA. The presence of smoke reduces the flow of alpha ions; the electric current decreases and at a pre-set value an alarm is activated.

Visible smoke detector

A source of light is directed at a receiving photocell. Smoke obscures or scatters the light and an alarm is triggered.

Laser beam

A laser beam is refracted by heat or smoke away from its target photocell and an alarm is initiated. A continuous or pulsed infrared beam can be transmitted up to 100 m and can be computer-controlled to scan the protected area. It can also serve as an intruder alarm.

Closed-circuit television

Manned security monitoring also acts as fire and smoke detection. Infrared imaging cameras reveal overheating of buried pipes and cables and can detect heat sources unseen by visual techniques.

Fire alarms are a statutory requirement. Audible bells, sirens, klaxons, hooters and buzzers are arranged so that they produce a distinctive warning. A visual alarm should also be provided throughout a building. Breakable glass call points are located 1.4 m above floor level within 30 m of any part of the premises.

The electrical system for fire detectors comprises alarms, a central control panel, an incoming supply and distribution board, emergency batteries, a battery charger and fire-resistant cable. A permanent cable or telephone line connection is made to the fire brigade and computer-controlled monitoring indicates any system faults.

Smoke ventilation

Positively removed smoke through automatically opened roof ventilators can greatly aid escape and reduce smoke damage, often localizing a fire that would otherwise spread.

The spread of smoke through ventilation ductwork is arrested by fire dampers where fire compartments within the building are crossed. Fire dampers may be motorized or spring-loaded multileaf, eccentrically pivoted flaps, sliding plate or intumescent-paint-coated honeycombs, which swell and block on heating. A typical arrangement of a pivoted flap damper is shown in Fig. 15.4.

An air pressurization ductwork and fan system is switched on at the commencement of a fire to inject outdoor air into escape routes, corridors and staircases. The staircase static air pressure is maintained at 50 Pa above that of adjoining areas to overcome the adverse force caused by wind, mechanical ventilation and the fire-produced stack effect ventilation pressure. This ensures that clear air is provided in the escape route and smoke movement is controlled.

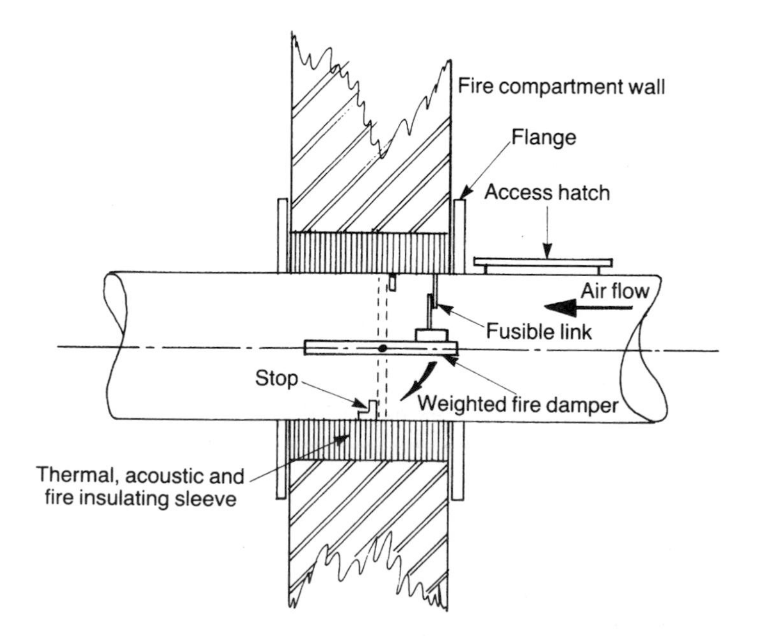

Figure 15.4 Hinged deadweight single blade fire damper in a ventilation duct

Questions

1. List the sources of fire within a building and describe how they may develop into a major conflagration. State how the spread of fire is expected to be limited by good building and services practice.

2. List the ways in which fire and smoke are detected and fire-fighting systems are brought into action.

3. Describe the methods and equipment used to fight fires within buildings in their likely order of use.

4. State the principal hazards faced by the occupants of a building during a fire. How are these hazards overcome? Give examples for housing, shops, cinemas, office blocks, single-storey factories and local government buildings.

5. Sketch and describe the fire-fighting provisions necessary in large industrial oil-fired boiler plant.

6. How are water and foam systems used to protect building structures from fire damage?

7. Compare a fixed sprinkler installation with other methods of fire-fighting. Give three applications for sprinklers.

8. Explain how sprinkler systems function, giving details of the alternative operating modes available. State the suitable sources of water for sprinklers.

9. Tabulate the combinations of fire classification and types of extinguisher to show the correct application for each. State the most appropriate fire-fighting system for each fire classification and show which combinations are not to be used.

16 Plant and service areas

Key terms and concepts

Learning objectives

Study of this chapter will enable the reader to:

1. identify the actions necessary prior to commencement of a construction, in order to facilitate the correct provision of utility services;
2. coordinate utility services under public footpaths;
3. design suitable routes for utility services;
4. calculate the areas of buildings needed for services plant;
5. find the sizes of plant room needed for all services from preliminary building information at the design stage;
6. allocate routes for services through the building structure;
7. understand the need for fire barriers in service ducts and correctly locate them;
8. choose suitable sizes for pipe service ducts;
9. understand the need to allow space for pipes crossing over each other;
10. estimate sizes for service shafts carrying air ducts;
11. know the requirements for walkway and crawlway ducts;
12. understand the need for expansion provision in pipework;
13. identify the ways in which pipes can be supported with allowance for thermal movement;
14. understand the ways in which thermal movement is accommodated;
15. know how thermal, fire, support and vibration measures are applied to pipes and air ducts passing through fire barriers;
16. apply flexible connections to plant;
17. appreciate the need for coordinated drawings;
18. understand the use of services zones;
19. be able to design boilerhouse ventilation.

Introduction

The building design team needs information on the size and location of services plant spaces and their interconnecting ducts before the engineer has sufficient data on which to base calculations. There may be only general definitions of the spaces to be heated or air conditioned and preliminary design drawings. Past experience of similar constructions reveals the likely plant and service duct requirements.

The use of such data is explained and worked examples are used. The planning of external utility supplies is shown, together with typical arrangements for internal multiservice ducts.

Support and expansion provision for distribution services is demonstrated, as is the design of combustion air ventilation for boilers.

Mains and services

The planning and liaison with public utilities (National Joint Utilities Group, 1979) must be included in the initial application made by the developer for planning permission. Each utility requires detailed information at the estate design stage in order to facilitate the following:

1. siting of plant or governor houses, substations, service reservoirs, water towers and other large items of apparatus and also early completion of associated easements and acquisition of and early access to land in order to ensure service to the development by the programmed date;
2. design of mains and service layouts;
3. location of and requirements for road crossings;
4. provision and displacement of highway drainage;
5. programming of cut-offs from existing premises that are to be demolished;
6. arrangements for protecting and/or diverting existing plant and services;
7. provision of supplies to individual phases of the development, including temporary works services;
8. acquisition of materials and manpower resources;
9. siting of service termination and/or meter positions in premises and service entry details;
10. provision of meter-reading facilities;
11. provision of public lighting.

Developers must provide information on the following:

1. the intended position of public carriageways, verges, footpaths, amenity areas and open spaces;
2. existing and proposed ground levels;
3. the position and level of proposed foul and surface-water sewers and any underground structures.

The utility will inform the developer of the need to close or restrict roadways, and these matters will then be discussed with the local authorities.

All main services to more than one dwelling should be located on land adopted by the Highway Authority. The location on private property of a main designed to serve a number of dwellings can lead to friction between residents if excavation for repairs or maintenance is needed, and also makes it difficult for the utilities to gain ready access in an emergency.

With the exception of road crossings, mains and services other than sewers should not be placed in the carriageways. The routes chosen should be straight and on the side of the carriageway serving most properties. Any changes of slope should be gradual. The prior approval of the utilities must be sought if landscaping will alter the levels of underground services.

Public sewers must be laid to appropriate levels and gradients in straight lines between manholes, usually under the carriageway.

An underground clear width of 1.8 m is needed between a private boundary and the kerb foundations but extra allowance should be agreed for the following:

1. fire hydrants;
2. inspection covers and manholes;
3. large-radius bends for pipes and cables;

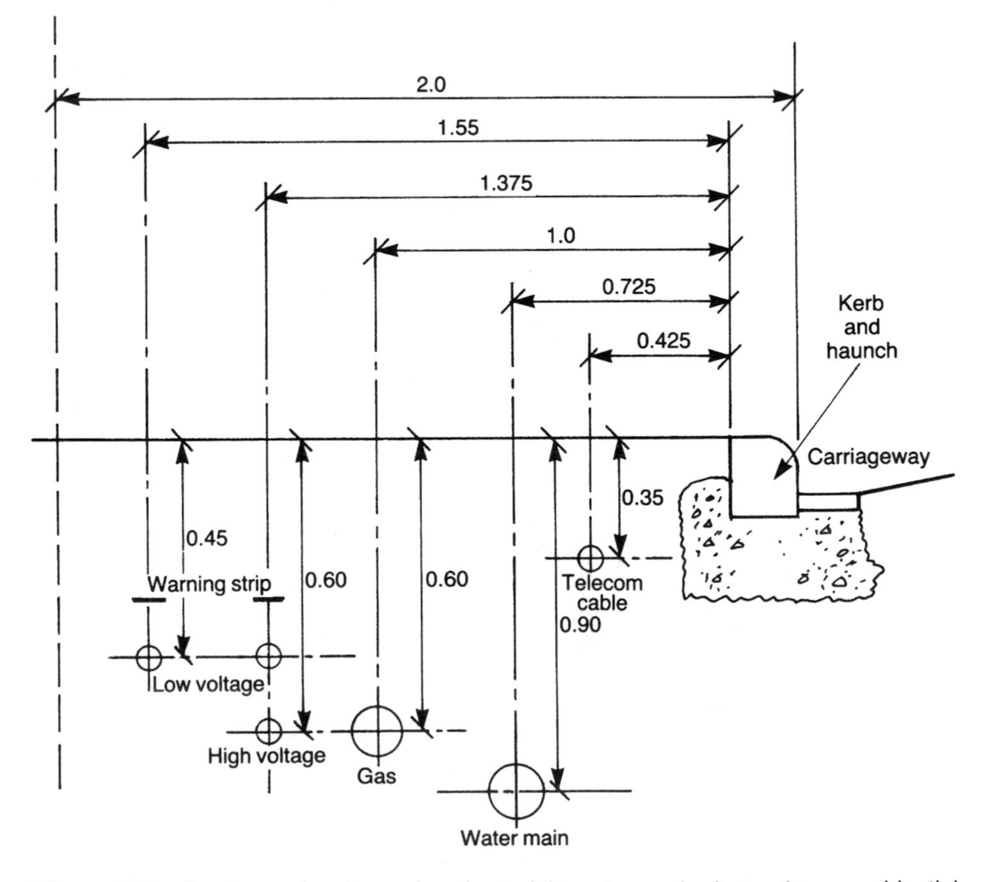

Figure 16.1 Positions of main services in straight routes under footpaths on residential estates

4. fuel oil distribution pipes;
5. district heating pipework and manholes;
6. through-services not connected to the development;
7. cross-connections between services to form ring mains rather than dead ends;
8. imposed loads from adjacent buildings – medium-pressure gas mains must be 2 m from the building line.

Protective measures are taken where there is a risk to pipes or cables from vehicles that may park on soft ground. Where special paving is used, early consultation with the utilities will help to avoid subsequent defacement due to maintenance work.

Footpaths should be used for the utilities. Sewers need to be laid in conjunction with the early stages of road building. Utilities operate under statutory powers and will not carry out work as a subcontractor. On completion of site construction, a copy of the plans showing the installed routes and details of the mains and services is sent to each utility to enable permanent records to be established.

The minimum dimensions for the locations of mains and services under a pavement are shown in Fig. 16.1. Brick tiles, concrete covers or yellow marker tapes are put over 11 kV and 415 V electricity cables. The 11 kV cables have a red PVC oversheath and 415 V cables have a black PVC oversheath.

The mains and services are surrounded with selected back-fill that is free of sharp or hard objects, and the trench is filled and compacted with earth that is free of rubble or site debris.

Plant room space requirements

Coordination of the services with architectural and structural design is required at the earliest possible moment during conception of the project. This stage is too early for heating and cooling loads, plant sizes and system types to be known with any certainty, but reliable information is required to form the basis for decisions. Building Services Research and Information Association (BSRIA) surveys (Boyer, 1979) of existing buildings have shown that their plant room requirements can be expressed as a percentage of floor area, as given in Table 16.1.

Some of the outline requirements for services plant rooms quoted by Boyer (1979) are as follows.

Table 16.1 Services plant room space requirements

Building type	Plant room floor area as a percentage of building floor area, excluding the plant room
Simple factory or warehouse	4
Most types	9
Small, well-serviced hospital	15

Cold-water storage

The volume of cold water to be stored to cover a 24 h period is calculated from the building's occupancy and type. An incoming break tank may be required at ground or basement level for pneumatically boosted systems. Fire-fighting services may need water storage at ground level.

Tanks can be 1, 2 or 3 m high, with 1 m clearance allowed around them for insulation, pipework and access.

Hot-water storage

The space needed for vertical indirect hot-water storage cylinders, secondary pumps, pipework, valves, controls and heater battery withdrawal is given by

$$\text{plant room floor area m}^2 = 1.7 \times \text{cylinder volume m}^3 + 10$$

Room heights of 3–4.8 m are needed depending on cylinder height.

Boilers

For buildings constructed to the Building Regulations, an assessment of the boiler power in watts can be made by multiplying the heated volume in metres cubed by 30.

The required boiler plant room floor area is

$$\text{plant room floor area m}^2 = 80.99 + 31.46 \times \ln\text{ (boiler capacity MW)}$$

Plant room heights are up to 5 m. Domestic and small commercial boilers are accommodated within normal ceiling heights and their floor areas are not predicted with this equation. The area calculated allows for two equally sized boilers, pipework and water treatment, pressurization and pumping equipment.

Fuel storage and metering

Electricity and gas meters are part of the incoming services accommodated within the plant room space calculated for other equipment. Partition walls and access for reading and removal are required.

Two equally sized oil storage tanks, supports, tanked catch pit and access are accommodated in

$$\text{oil storage tank room area m}^2 = 22.52 + 0.64\text{ (oil storage volume m}^3\text{)}$$

Plant room height is up to 4.5 m. Tanks are frequently located externally and stood on three brick or concrete block piers so that oil will flow by gravity to the burners. They are of mild steel and protected with bitumastic paint.

Air handling

The air-handling plant room size is assessed by assuming that the mechanically ventilated parts of the building have between 6 and 10 air changes per hour. The expected supply air volume flow rate is

$$Q = \frac{NV}{3600} \text{ m}^3/\text{s}$$

where V is the volume of ventilated space (m^3) and N is the number of air changes per hour (6–10).

The plant room will be 2.5–5 m high depending on the sizes of fans, ducts, filters, heater and cooler batteries, humidifiers and control equipment. The floor area is

$$\text{plant room floor area m}^2 = 6.27 + 7.8 \times Q \text{ m}^3/\text{s}$$

A fresh-air-only system, such as an induction system, is sized on one air change per hour.

Cooling plant

Refrigeration plant capacity may be as high as 30 W/m^3 of building volume, and an early estimate of heat gains should be made. A Building Energy Estimating Programme (BEEP) is available through the electricity supply authority, and other computer packages are in use. Plant room height is 3–4.3 m and the floor area is

$$\text{plant room floor area m}^2 = 80.49 + 35.46 \times \ln \text{(cooling load MW)}$$

The area is for two refrigeration machines, pumps, pipework and controls. Additional space on the roof is needed for the cooling tower.

Telephones

Space requirements will change with advances in technology.

Lifts

An early assessment of requirements is made in conjunction with the lift engineer. Boyer (1979) gives further information.

Electrical substation

The incoming high-voltage supply is located in a substation, which may be external or on an external wall of the building. The floor area needed is 35 m^2 for a 200 kVA load and up to 48 m^2 for a 2000 kVA load.

Standby diesel electric generator

A standby diesel electric generator supplies emergency electrical power of up to 100% of the connected load from the mains. The plant room will be adjacent to the substation and will be up to 4 m high. The floor area required is 18 m^2 for

50 kVA up to 37 m^2 for 600 kVA, plus a diesel oil storage tank for 7 days of continuous running.

Service ducts

Service ducts are passageways vertically and horizontally throughout a building, or between buildings, large enough to permit the satisfactory installation of pipes, cables and ducts, together with their supports, thermal insulation, control valves, expansion allowance and access for maintenance. Each service duct might be constructed as a fire compartment, and BS Code of Practice 413: 1973 should be consulted. An example of current practice is given in Fig. 16.2.

Casings and chases of 100 mm diameter or less are fire stopped to the full thickness of the wall or floor. The passage of a service must not reduce the resistance of the fire barrier. Plastic pipes can soften and collapse during a fire and allow the passage of flames and smoke. A galvanized steel sleeve with an intumescent liner can be used to surround the plastic pipe where a fire stop is needed. When its temperature rises to 150 °C, the intumescent liner expands inwards to close the softened pipe and seal the wall aperture.

Service duct sizes can be found from an estimate of ventilation air supply rate Q m^3/s, doubled to allow for the recirculation duct, with an assumed air velocity

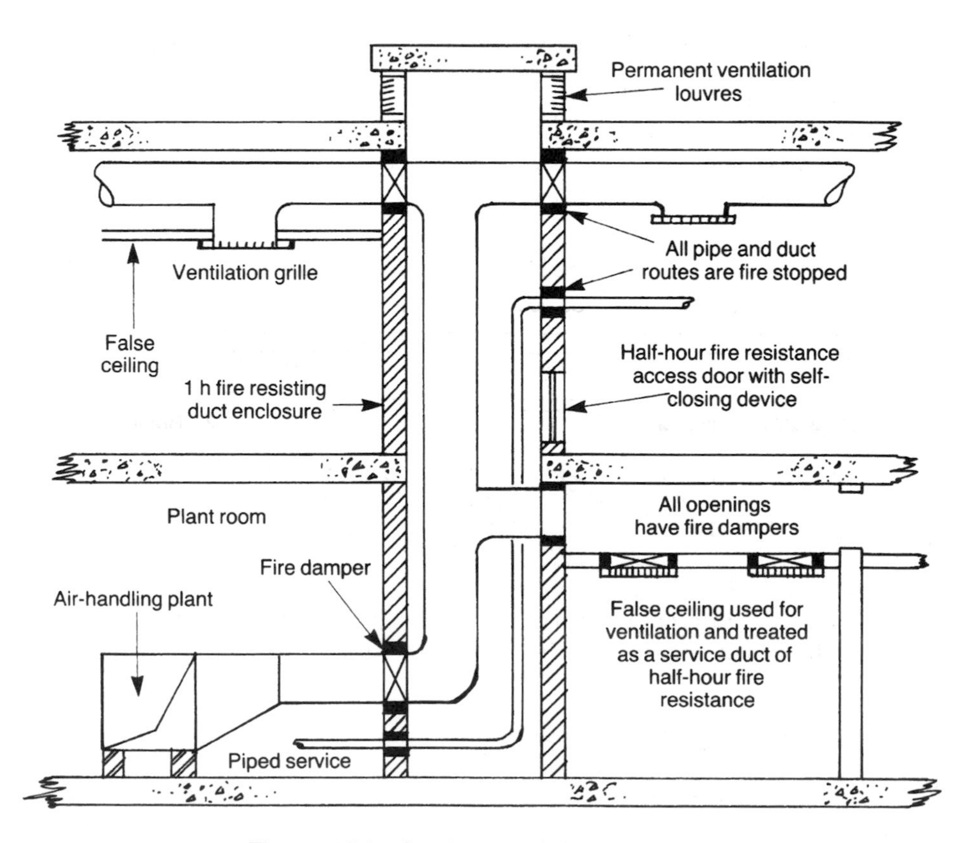

Figure 16.2 Service duct fire compartment

of 10 m/s for vertical ducts within brick or concrete enclosures or 5 m/s in false ceilings where quiet operation is important. At least 150 mm clearance is allowed between ducts and other surfaces for thermal insulation, jointing, supports and access.

Fan noise is contained within the air-handling plant room by acoustic attenuators, anti-vibration machine mountings and heavy concrete construction.

The total floor space taken by vertical pipe and cable routes will be up to 1% of the gross floor area. Horizontal service ducts and false ceilings 500 mm deep are used for air-conditioning ducts and other services. Recessed luminaires and structural beams encroach into the nominally available spaces.

Underfloor service ducts should be constructed to allow access for jointing and maintenance. The minimum standard for an underfloor duct is shown in Fig. 16.3. The duct route is accurately marked on installation drawings and access is gained by breaking the screed. Hot-water pipes are insulated with 50 mm thick rigid glass fibre and wrapped with polyethylene sheet, sealed with waterproof tape. The duct is filled with dry sand. Sufficient depth is allowed for branches to cross over the other pipes. Recommended sizes are given in Table 16.2 (Butler, 1979b).

Vertical and other service ducts can be sized in a similar manner by allowing a 50 mm gap between thermal insulation and other surfaces. Additional smaller pipes run in the same duct will require an increase in the width.

When builder's work holes for services are specified, the dimensions of the structural opening required should always be used, rather than the nominal pipe diameter.

An underfloor, or underground, crawlway or walkway has the following features:

1. crawlway duct height 1.4 m;
2. walkway duct height 2 m;
3. 750 mm clear width between fixtures;
4. reinforced concrete construction;

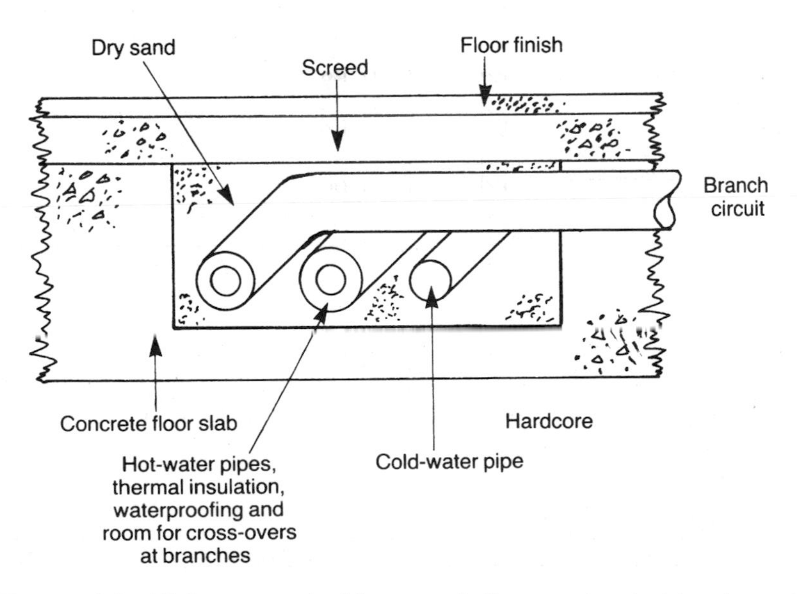

Figure 16.3 Minimum standard for an underfloor service duct for pipework

Table 16.2 Minimum underfloor duct sizes for pipework

Hot-water flow and return nominal diameter (mm)	Underfloor duct dimensions width and depth (mm)
15	294
22	304
28	346
35	364
42	376
54	400

5. floor laid to fall to a drainage channel along the length of the duct, with connections to the surface drainage system;
6. watertight access manhole covers at intervals with built-in galvanized steel stepladders;
7. watertight lighting fittings and power sockets;
8. services are painted to appropriate British Standards colours and clearly labelled, and control valves are numbered with an explanation list provided;
9. services branching into side ducts do not block through-access.

EXAMPLE 16.1

A four-storey office block of 20 m × 15 m is to be air conditioned with an induction system using gas and electricity fuel. The main plant room is to be built on the roof. There will be 300 occupants. Calculate the plant room and service duct space requirements for the preliminary design stage.

It is expected that the plant rooms will require 9% of the floor area of $(20 \times 15 \times 4)\ \mathrm{m}^2$, and so a first estimate is $108\ \mathrm{m}^2$. This will be mainly on the roof; thus an oblong room of dimensions l and $2l$ could be used.

$$\text{area} = l\,(2l) = 108\ \mathrm{m}^2$$

$$2l^2 = 108\ \mathrm{m}^2$$

$$l = \left(\frac{108}{2}\right)^{0.5} = 7.35\ \mathrm{m}$$

A plant room of 7.35 m × 14.7 m could be accommodated on the roof.

A further estimate of the requirements can be made through consideration of each service.

1. Cold-water storage of 45 l per person:

$$\text{volume} = 300\ \text{people} \times 45\ \frac{1}{\text{person}} \times \frac{1\ \mathrm{m}^3}{10^3\ \text{litre}} = 13.5\ \mathrm{m}^3$$

$$\text{tank dimensions} = 2.6\ \mathrm{m} \times 5.2\ \mathrm{m} \times 1\ \mathrm{m}$$

Add 1 m all round the tanks. Thus the plant room floor space is 33.12 m^2. This could be reduced by stacking the tanks if there is sufficient headroom. The water main pressure will be sufficient to reach the roof; ground-level break tanks and pumps will not be needed.

2. Hot-water storage of 5 l per person:

$$\text{volume} = 300 \text{ people} \times 5 \frac{\text{l}}{\text{person}} \times \frac{1 \text{ m}^3}{10^3 \text{ litre}} = 1.5 \text{ m}^2$$

$$\text{volume of an indirect cylinder} = \frac{\pi d^2}{4} \times l \text{ m}^3$$

For a cylinder 1 m in diameter, its length l is

$$l = 1.5 \times \frac{4}{\pi} = 1.91 \text{ m}$$

The floor area required is 1 m^2.

3. The boiler power is given by

$$\text{boiler power} = (20 \times 15 \times 4 \times 3) \text{ m}^3 \times 30 \frac{\text{W}}{\text{m}^3} \times 1 \frac{\text{MW}}{10^6 \text{ W}}$$

$$= 0.108 \text{ MW}$$

$$\text{plant room floor area} = 80.99 + 31.46 \times \ln 0.108 \text{ m}^2$$

$$= 10.972 \text{ m}^2$$

4. Gas and electricity meters will be housed either in the roof plant room or in cubicles at the rear of the building on the ground floor.
5. The induction system air-handling plant will pass only the fresh air supply, say one air change per hour.

$$Q = \frac{1 \times (20 \times 15 \times 4 \times 3)}{3600} \text{ m}^3/\text{s}$$

$$= 1 \text{ m}^3/\text{s}$$

$$\text{plant room area} = 6.27 + 7.8 \times 1 \text{ m}^2$$

$$= 14.07 \text{ m}^2$$

6. The cooling plant capacity is 30 $W/m^3 = 0.108$ MW

$$\text{plant room floor area} = 80.49 + 35.46 \times \ln 0.108 \text{ m}^2$$

$$= 1.569 \text{ m}^2$$

This is unrealistically small, as the data in Boyer (1979) do not cover plant smaller than 0.3 MW. An estimate of plant room space of 20 m^2 will be made.

Thus the total plant room space requirements are estimated to be

$$(33.12 + 1 + 10.972 + 14.07 + 20)\ m^2 = 79.162\ m^2$$

These two methods of estimation show that plant room space requirements are of the order of 79–108 m^2.

A vertical service duct is needed from the roof plant room to ground level carrying supply and exhaust air ducts, drainage and water pipework and cables. If the maximum air velocity in the air ducts is 6 m/s, their sizes will be

$$\text{duct cross-sectional area} = \frac{Q\ m^3/s}{V\ m/s}$$

$$= \frac{1}{6}\ m^2$$

$$= 0.167\ m^2$$

If square ducts are used, they will be 408 mm × 408 mm. An estimated service duct arrangement is shown in Fig. 16.4. This allows for thermal insulation and access to all the services.

False ceilings provide space for the horizontal distribution of services. The induction units will be located along the external perimeter under the windows or within the false ceiling. Holes 150 mm in diameter are needed in the floor slab, one for each unit, for the air-duct and, close by, two holes 50 mm in diameter for pipes.

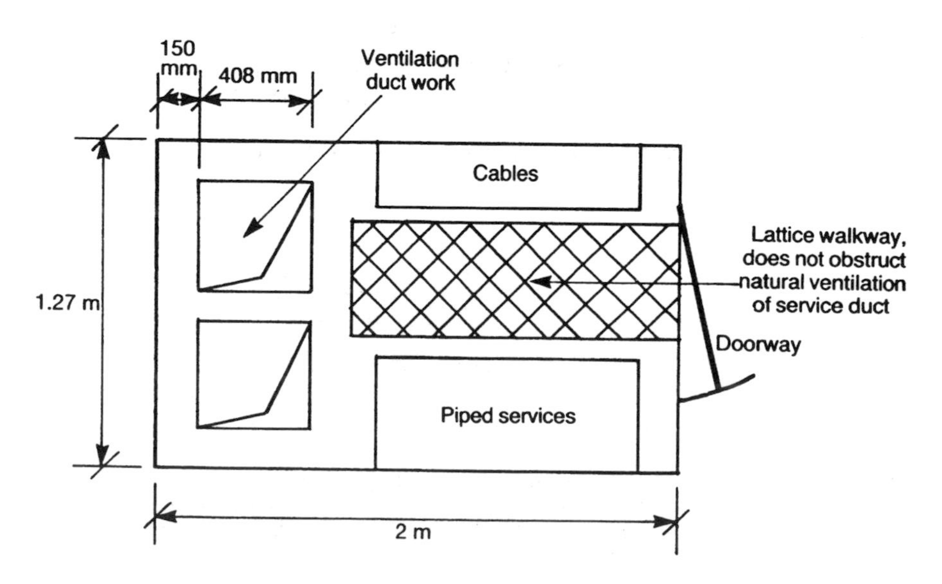

Figure 16.4 Layout of the vertical service duct in Example 16.1

Pipe, duct and cable supports

Hot-water pipework can be supported with hardwood insulation rings clamped in mild steel brackets, as shown in Fig. 16.5. Saddle pipe and cable clips, shown in Fig. 16.6, are extensively used because of their low cost. They should be made of a material that is compatible with that of the service. A row of services may be bolted to a mild steel angle iron whose ends are built into the structure. The longitudinal spacing of supports depends on pipe size, material and whether the service is horizontal or vertical. Rollers, as shown in Fig. 16.7, allow pipes to move freely during thermal movement.

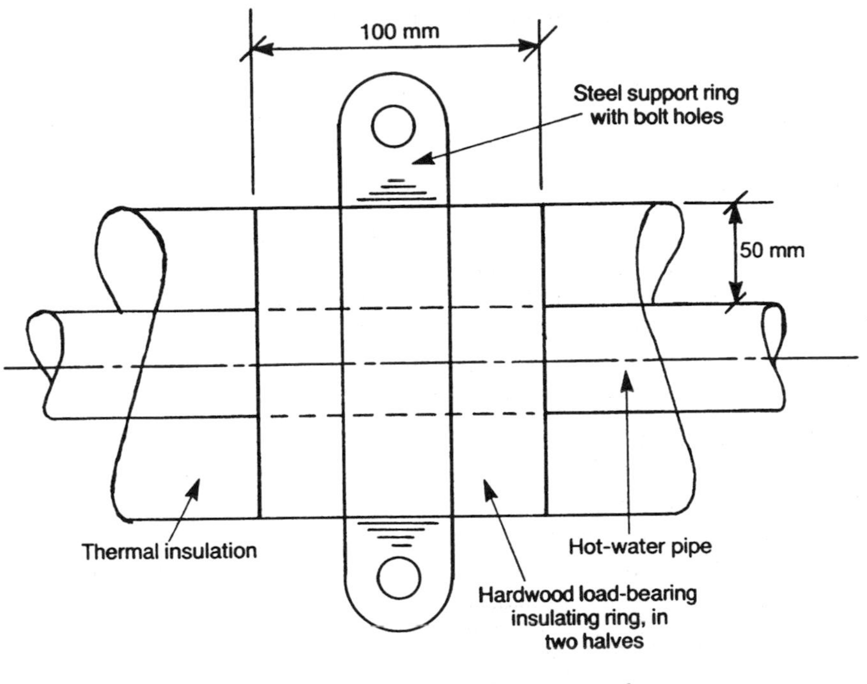

Figure 16.5 Insulated pipe support ring

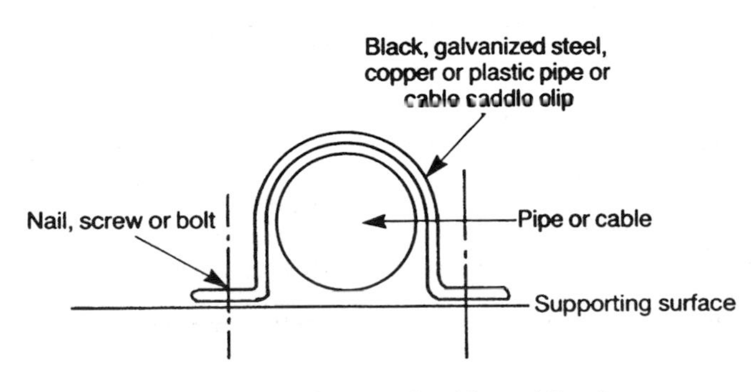

Figure 16.6 Pipe and cable saddle clip

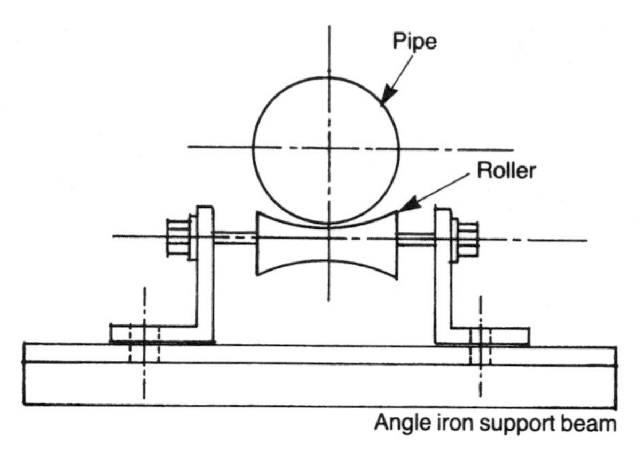

Figure 16.7 Roller pipe support

Expansion and contraction of short pipe runs is accommodated at frequent bends and branches, the pipes moving within their thermal insulation and non-rigid brackets. Spaces between pipes are sufficient to avoid contact. Long pipe runs need expansion devices, anchors and guides.

1. With a pipe anchor the pipe is rigidly bolted or welded to a steel bracket which is firmly built into a brick or concrete structure.
2. Pipes can be supported by tubular guides, as shown in Fig. 16.8, which allow longitudinal movement with minimum metal contact.
3. Several types of thermal expansion device are used, depending on application, space available and fluid pressure.

   (a) Pipe loops are least expensive in some cases and can be formed where external pipes pass over a roadway. They can be prefabricated with pipe fittings, welded, bent or factory formed, as shown in Fig. 16.9.
   (b) Bellows are made of thin copper or stainless steel and have hydraulic pressure limitations. A complete installation is shown in Fig. 16.10.
   (c) Articulated ball joints take up pipe movement at a change in direction, as shown in Fig. 16.11.

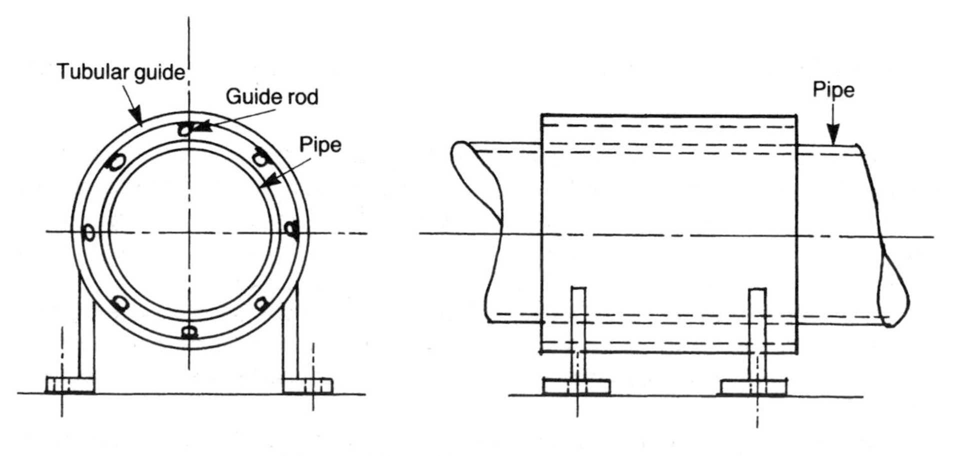

Figure 16.8 Tubular guide support

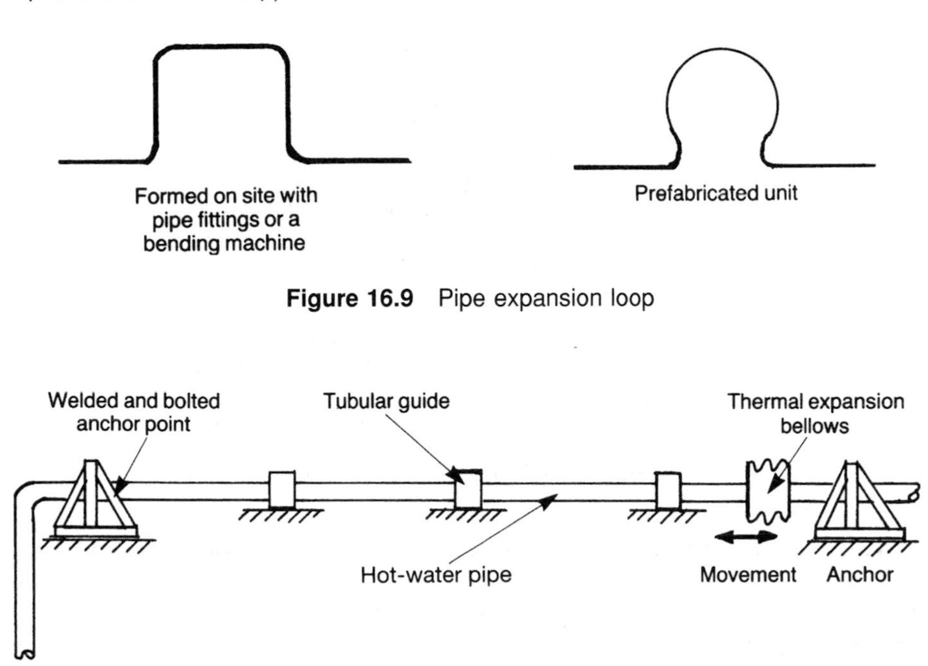

Figure 16.9 Pipe expansion loop

Figure 16.10 Complete pipework installation for thermal expansion provision

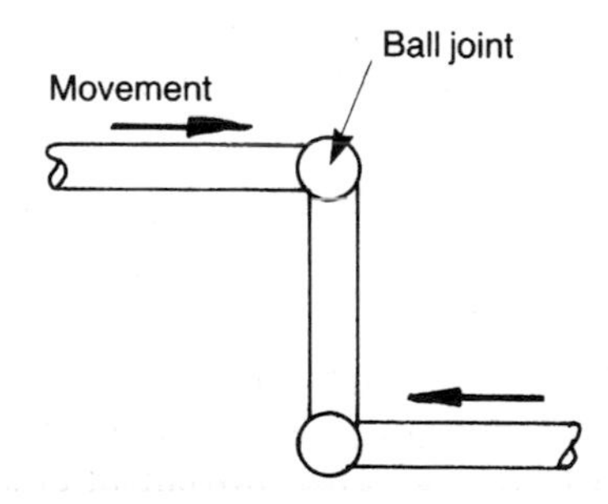

Figure 16.11 Articulated expansion joint

(d) Sliding joints are packed with grease and the pipe slides inside a larger diameter sleeve.

A fire stop unit is used where pipes pass through a fire compartment wall or floor and incorporates structural support, vibration insulation and fire resistance within a steel flanged sleeve which is in two halves, as shown in Fig. 16.12. Silicone fire stop foam is used to seal the space around pipes. When it is exposed to heat, the foam chars to form a hard flame-resistant clinker.

Ventilation ducts are fixed to the building with galvanized mild steel saddle clips for up to 300 mm diameter light-gauge metal; larger ducts have flanged joints, which are suspended with rods from angle brackets. Figure 16.13 shows a typical fixing.

Cables are supported along their entire length by the conduit or a perforated metal tray.

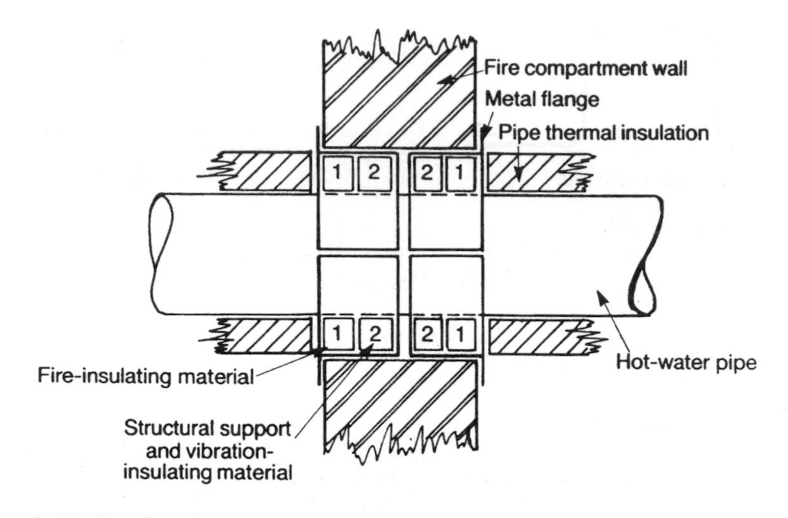

Figure 16.12 Insulated pipe sleeve (reproduced by courtesy of Stuart Forbes (Grips Units) Ltd, Woking)

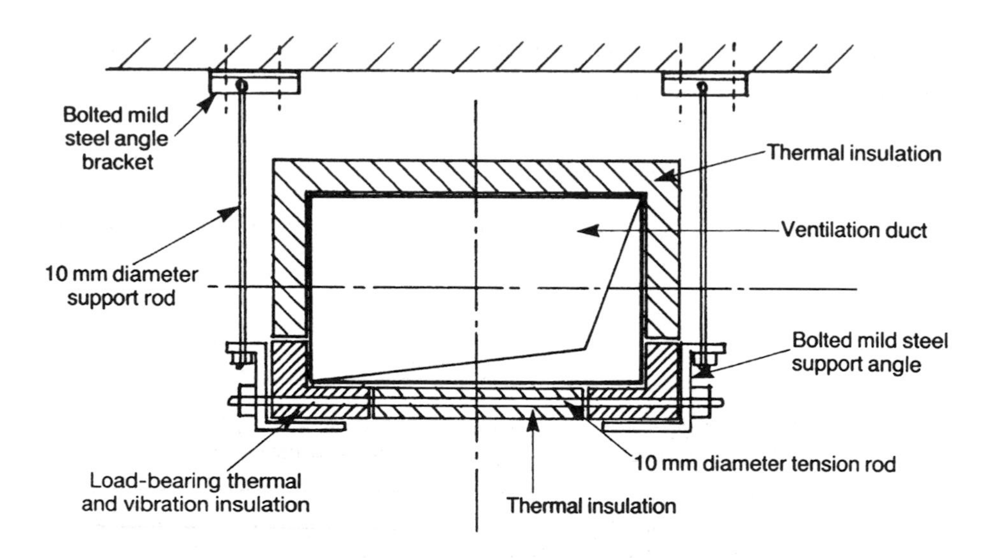

Figure 16.13 Insulated duct support

Plant connections

Connections to plant are made in flexible materials to reduce the transmission of vibration from fans, pumps and refrigeration compressors to the distribution services, or to allow greater flexibility in siting the equipment. A fan installation is shown in Fig. 16.14. The discharge air duct has a sound attenuator to absorb excess fan noise. Polyurethane foam held in place by perforated metal sheet is used in the attenuator.

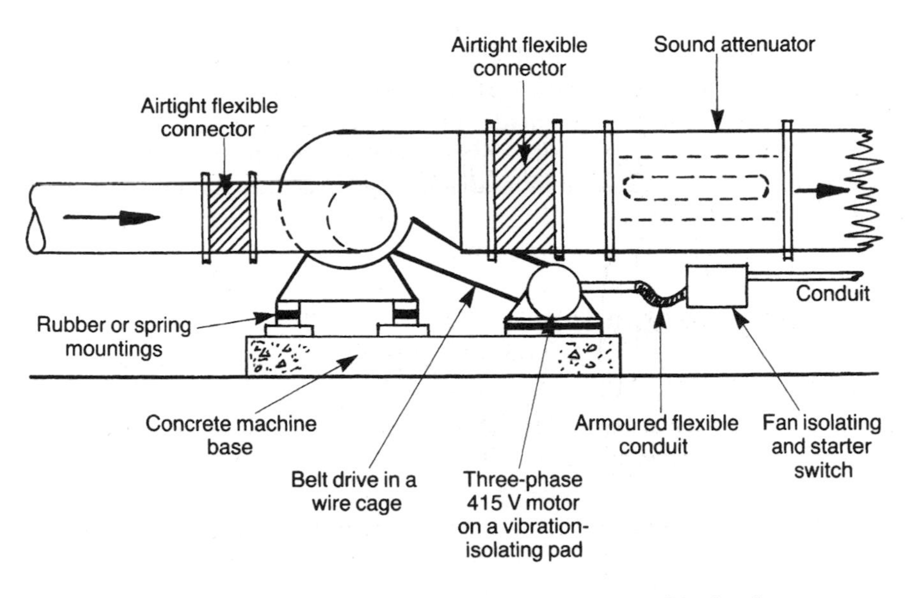

Figure 16.14 Flexible connections to an air-conditioning fan

Coordinated service drawings

A master set of all service drawings is maintained under the overall charge of a coordinating engineer for the project (Crawshaw, 1976). Drawings and a schedule of builder's work associated with the services are circulated round the

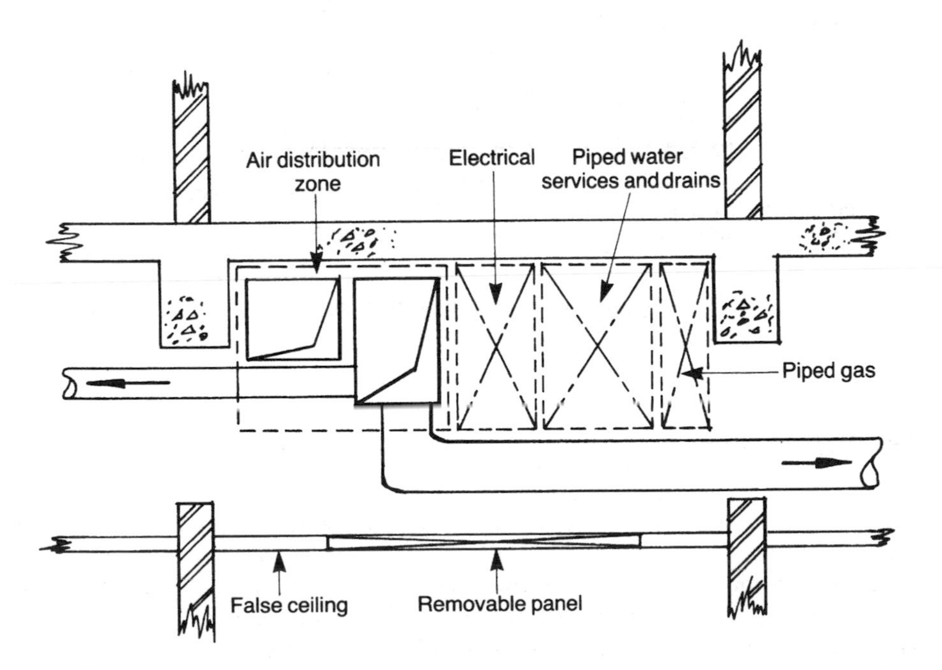

Figure 16.15 Service zones in a false ceiling over a corridor

construction team. The structural engineering implications of holes through floors and beams are checked at an early stage.

Service space allocation is made on the basis of zones for particular equipment within structural shafts and false ceilings. Each engineering service is restricted to its own zone. Common areas are provided for branches, as shown in Fig. 16.15.

Boiler room ventilation

Combustion appliances must have an adequate supply of outdoor air, otherwise the fuel will not burn properly and carbon monoxide will be produced. Under down-draught conditions, this will be a danger to occupants. Fatalities have occurred through improper appliance operation.

Good installation practice is to introduce combustion air so that it does not cause a nuisance through draught, noise or poor appearance. Heat and fumes produced by the appliance are ventilated to outdoors through high-level openings.

Any room containing a fuel-burning appliance may be positively supplied with ventilation air by a fan if this is needed for some other purpose. An extract fan must not be used, as combustion products can be drawn into the occupied room. Natural inlet and outlet ventilation is predominant.

The combustion air inlet for a domestic kitchen or living area can be (CIBSE, 1973):

1. through an external wall, just below ceiling level, to enable the incoming cold air to diffuse with the room air above head height (this avoids most draughts and occasional blockage by snow or debris);
2. through an external wall at low level behind a hot-water radiator or other heat emitter (a frost thermostat switches on the heating system at an internal air temperature of 5 °C);
3. by direct connection of the combustion air from outside to the appliance casing, locality or enclosing cupboard with an underfloor duct. Two suitably sized air bricks are fitted into the external walls of a suspended timber floor on opposite sides of the building. Either a duct connection between the appliance casing and the floor space or a ventilation grille is put into the floor by the heater. A drain pipe or galvanized steel duct can be cast into a concrete floor slab for this purpose.

Table 16.3 Minimum free areas of ventilation openings for combustion appliances for outdoor air per kilowatt of heat output

	Minimum free area (mm^2)	
Position of opening	Conventionally flued	Balanced flue
High level	550	550
Low level	1100	550

An appliance of up to 30 kW heat output may be fitted within a compartment, which is ventilated to an adjoining room, which in turn is ventilated to outdoors by the stated openings. Recommended ventilation openings are given in Table 16.3.

Questions

1. List the principal information and activities involved in the provision of main services throughout a housing estate.

2. Sketch a suitable arrangement for the services beneath the public highway and leading into a dwelling. Show the recommended dimensions and explain how the ground is to be reinstated.

3. Estimate the plant room and service duct space requirements of the following buildings, using the preliminary design information given.

   (a) A naturally ventilated hotel with a hot-water radiator heating system. Roof and basement plant rooms are available. The hotel dimensions are 50 m × 30 m, with ten storeys 3 m high. Total occupancy is 750. An oil-fired boiler plant is to be used.
   (b) A single-storey engineering factory of dimensions 100 m × 40 m, using overhead gas-fired radiant heating. The roof height slopes from 3 m to 5 m at the central ridge. There are 300 occupants. Mechanical ventilators and smoke extractors will be fitted in the roof. A standby diesel electricity generator and an electrical substation are required.
   (c) A 12-storey city centre educational building of 40 m × 20 m, 3 m ceiling height, with a single-storey workshop block and laboratory area 40 m × 60 m × 4 m. The whole complex is to be mechanically ventilated with 4 air changes per hour. Hot-water radiators and fan convectors provide additional heating. Gas and electricity are to be used.

   The tower building has a basement with ramp access to ground level. A refectory is located at ground level. The total building occupancy is 2000.

4. Draw the installation of services in a vertical duct through a three-storey office building. The duct is 2.5 m × 1.2 m. Boiler and ventilation plant are in the basement. There are false ceilings on all floors.

5. Sketch and describe how the spread of fire through a building is limited by the services installation.

6. A false ceiling over a supermarket contains recessed luminaires, sprinkler pipework and a single-duct air-conditioning system. The false ceiling is 400 mm and has structural steel beams 250 mm deep. Extract air from the shop passes through the luminaires. Draw the installation to scale.

7. A concrete floor with a wood block finish houses a service duct carrying two 35 mm heating services, two 28 mm hot-water services, a 42 mm cold-water service and 54 mm gas pipework. Side branches are required to carry a maximum of three 22 mm pipes. Continuous access covers are to be provided. The hot-water pipes are to have 50 mm thick thermal insulation, and at least 25 mm clearance is needed around the pipes. Draw a suitably detailed design showing dimensions, materials, pipe support, cover construction and pipe routes at the branch.

8. Describe, with the aid of sketches, how successful coordination between all the services can be achieved within builders' work ducts.

9. Explain how fuel-burning appliances fitted in kitchens, living rooms, cupboards and domestic garages can be adequately ventilated. Illustrate an example of each location and state the areas of ventilation openings required for appliances of 3 kW, 18 kW and 40 kW heat output.

Appendix: Answers to questions

Chapter 1

7. t_r 15.0 °C; t_{ei} 16.3 °C. The room condition is below the comfort zone shown in Fig. 1.14. Surface temperatures are to be increased by adding thermal insulation, such as double glazing, and the air temperature should be raised.
8. Site A WCI 862, EWCT −5.8 °C; Site B WCI 757, EWCT −1 °C; site A has the more severe conditions.
9. HSI −26.8.
10. HSI 96.0; this is the maximum 8 h exposure for a fit, acclimatized young person; conditions vary through the day.
11. t_r 14.3 °C; t_{ei} 17.2 °C. The room condition is outside the sedentary comfort zone.
16. 345 W.
21. C_i 0.977 decipol, C_o 0.3 decipol for vitiated outdoor air, Q 1770 l/s total, N 3.5 air changes/h.
22. Answers in litre/s. First figure is based on floor area, second is per person: Fanger 6250, 5000; ASHRAE 875, 700; BS 1625, 1300; DIN 2375, 1900; CIBSE 1625, 1300; maximum N 5.1 air changes/h.
23. C_i 0.612 decipol, G 0.517 olf/m^2, C_o 0.05 decipol, Q 9.2 l/s m^2, total Q 2760 l/s, N 6.6 air changes/h.

Chapter 2

11. U_e 0.51 W/m^2 K, L 53 mm.
12. A_f 2640 m^2, A_W 1776 m^2, B 1.035, T_T 8.46 W/m^2, T_E 24.85 W/m^2, T 33.3 W/m^2.

Chapter 3

4. 2330.5 W.
5. 20.112 kW.
6. 12.4 °C.

7. Allowed heat loss per degree Celsius difference inside to outside is 3746.8 W/K; thus the proposal complies. Proposed heat loss 3407 W/K.
8. 83.14 kW.
9. 43%.
13. R_{si} 0.1 m^2 K/W, Q 50 W, U 2.78 W/m^2 K, R_n 6.67 m^2 K/W, 221 mm.
14. R_{si} 0.12 m^2 K/W, Q 19.2 W, 114 mm, 120 mm used, U_n 0.29 W/m^2 K, 17.4 °C.
15. R_{si} 0.1 m^2 K/W, Q 20 W, U 1.82 W/m^2 K, extra R_a 0.18 m^2 K/W; 81.75 mm, 90 mm used, U_n 0.23 W/m^2 K, new Q 4.83 W, 15.5 °C.

Chapter 4

11. 0.95 litre/s.
12. 2.4 m long × 700 mm high.
13. X 42 mm, Y 35 mm, Z 28 mm, radiator 1 22 mm, radiator 2 28 mm.
14. Expected internal temperature 26.5 °C, system performance is satisfactory.

Chapter 5

1. From $Q = \dfrac{\text{SH kW}}{t_r - t_s} \times \dfrac{(273 + t_s)}{357}\ \text{m}^3/\text{s}$

$$357Q\ (t_r - t_s) = \text{SH}(273 + t_3)$$

$$357Qt_r - 357Qt_s = 273\text{SH} + \text{SH} \times t_s$$

$$357Qt_r - 273\text{SH} = \text{SH} \times t_s + 357Qt_s$$

$$357Qt_r - 273\text{SH} = t_s(\text{SH} + 357Q)$$

and

$$t_s = \frac{357Qt_r - 273\text{SH}}{\text{SH} + 357Q}$$

$$= \frac{357 \times 5 \times 23 - 273 \times 50}{50 + 357 \times 5} = 14.94\ ^\circ\text{C}$$

2. 0.793 m^3/s.
3. 0.007469 kg H_2O/kg air.
4. (a) no, (b) 21.2 °C w.b., 0.877 m^3/kg, (c) 6.186 kW.
5. (a) 4.25 m^3/s, (b) 4.86 m^3/s, (c) 87.45%, (d) 4.13 m^3/s, (e) 0.61 m^3/s, (f) 3.52 m^3/s.
6. 20 air changes/h.
13. 1680 mm × 930 mm.
15. 2.68 m^3/s, 10.72 air changes/h, 0.0076 kg H_2O/kg air.

16. t_S 28.6 °C d.b., reduce supply air quantity to 1.7 m^3/s and use t_s 30 °C d.b. if the room air change rate will not be less than 4 changes/h.
20. 14.45 air changes/h.
21. 15 air changes/h, 710 mm × 710 mm, 2 m^3/s fresh air, 2 m^3/s recirculated air, 3.6 m^3/s extract air, 4 m^3/s supply air-duct 0.4 m^3/s natural exfiltration.

Chapter 6

12. 13.44 kW.
15. 0.56 h.
16. 0.05 kg/s, 3.15 m head, pump C.

Chapter 7

7. The furthest WC can only be 12.353 m from the stack. This produces a branch gradient of 0.812° and this complies with Fig. 7.8.
8. 53.

Chapter 8

2. 50 litre/s, at least three.
3. 5.04 litre/s.
4. 1.337 litre/s.
5. 1.45 litre/s.
6. 35.53 litre/s.
7. 2.921 litre/s. Note that this is less than the figure given in Table 8.2 because of the simplifying assumptions made here.
8. Either 125 mm with centre outlet or 150 mm with end outlet.
9. Width W 120 mm, depth D 71.876 mm.
10. Yes, flow load 2.1 litre/s, gutter capacity 2.936 litre/s.
12. Storage volume 2.4 m^3, one pit diameter 1.25 m.

Chapter 9

3. 100 mm.
4. 150 mm.
5. 1.235 m.
6. 225 mm, approximately 614.
8. Yes.
9. Yes.

Chapter 10

15. (a) 6.79 °C, 6.11 °C; (b) 13.89 °C, 13.55 °C, 2.89 °C, 2.55 °C; (c) 18.47 °C, 17.83 °C, 6.71 °C, 2.91 °C, 0.27 °C; (d) 19.45 °C, 19.06 °C, −0.38 °C.
16. 2.72 °C.
17. −7.46 °C.
18. 28.1 mm.
19. *U* value 0.46 W/m^2 K, heat flow 9.26 W/m^2. Thermal temperature gradient is 22 °C, 21.07 °C, 20.38 °C, 18.71 °C, 11.76 °C, 3.08 °C, 2.9 °C, 2.7 °C, 2 °C. Indoor dew-point 11.3 °C, vapour pressure 1300 Pa, outdoor air −0.8 °C and 568 Pa. Vapour resistance R_v 6.265 GN s/kg, mass flow of vapour *G* 1.168×10^{-7} kg/m^2 s. Dew-points at the same interfaces as the thermal temperatures are 11.3 °C, 11.3 °C, 10.5 °C, 10.5 °C, 8.8 °C, −0.7 °C, −0.8 °C, −0.8 °C, −0.8 °C. Condensation does not occur.
20. *U* value 0.6 W/m^2 K, heat flow 7.74 W/m^2. Thermal temperature gradient is 14 °C, 13.07 °C, 12.84 °C, 11.32 °C, 3.58 °C, 2.19 °C, 1.22 °C, 1 °C. Indoor dew-point 6.5 °C, vapour pressure 936 Pa, outdoor air −1.8 °C and 531 Pa. Concrete blockwork resistivity taken as 200 GN s/kgm. Vapour resistance R_v 25.35 GN s/kg, mass flow of vapour *G* 1.6×10^{-8} kg/m^2 s. Dew-points at the same interfaces as the thermal temperatures are 6.5 °C, 6.5 °C, 6.3 °C, 0.09 °C, −0.06 °C, −0.06 °C, −1.8 °C, −1.8 °C. Condensation does not occur.

Chapter 11

9. 59%.
11. Room index 3, UF 0.73, MF 0.9, 36 luminaires in 3 rows of 12 along the 20 m dimension, 16.8 W/m^2, 21 A.
22. Lighting 3750 h/yr, 3.81×10^6 lm, tungsten, 1814 lamps, replace 3401 per year, total annual cost £80 255 per year, fluorescent 569 lamps, replace 178 per year, total annual cost £14 403 per year, sodium 139 lamps, replace 22 per year, total annual cost £10 917 per year.
23. Lighting 1200 h/yr, 350 685 lm, tungsten 352 lamps, replace 211 per year, total annual cost £3559 per year, fluorescent 95 lamps, replace 16 per year, total annual cost £772 per year, halogen 37 lamps, replace 2 per year, total annual cost £423 per year.

Chapter 12

1. 1.026 litre/s, 125 mm H_2O.
2. 5.394 mb, 3.5 mb, 0.75 mb, 15 mb, 1050 mb.
3. 3.333 Pa/m.
4. 27.17 m.
5. 1.47 litre/s, 2.609 Pa/m, 32 mm.
6. 72 Pa.

Chapter 13

5. 0.00172 ohm.
6. 0.2867 ohm.
7. 10.7%.
10. 12.6 kVA.
11. 19.2 ohm.
12. 0.2857 mA.
14. 9716.8 kW.
15. 28.6 m.
16. (i) 18.25 kVA, 25.4 A, (ii) £691.95.
26. Three earth rods give a total system resistance of 8.937 ohm.

Chapter 14

19. reverberation time T 2.901 s at 125 Hz, 3.462 s at 250 Hz, 3.462 s at 500 Hz, 3.157 s at 1 kHz, 2.752 s at 2 kHz and 3.253 s at 4 kHz.
20. r 100 mm SPL 87 dB; r 1 m SPL 71 dB.
21. Directivity Q 2, r 0.5 m SPL 92 dB, reverberant SPL 79 dB.
22. 84 dBA.
23. Through the wall SPL_2 19 dB; through air vent 49 dB; open air vent causes noise to bypass the attenuation of the wall and may need acoustic louvres or an acoustic barrier.
24. Through the wall SPL_2 47 dB; through air vents in doors 59 dB; open air vent causes noise to bypass the attenuation of the wall; burner needs an acoustic enclosure.
25. SPL in roof is 32 dB; the large volume and short reverberation time assist in attenuating the plant room noise.
26. 33 dB.
27. 37 dB.
28. 39 dB.
29. See chapter explanation.
30. NR 40 is not exceeded in the room.
31. (a) NR 80; (b) NR 25, no intrusive noise from the chiller; (c) NR 45; (d) NR 20 when doors have equal sound reduction to the walls, have air-tight seals and are closed.
32. (a) NR 80; (b) 65 dB due to sound escape through door; (c) NR 35.
33. (a) NR 75; (b) NR 45, equivalent to the background noise level in a corridor; (c) NR 35; (d) NR 20, there is no intrusive noise.
34. (a) NR 60; (b) NR 40; (c) through the supply and return air ducts, noise radiation from the outer case of the fan coil unit, from the ceiling space through ceiling tiles, light fittings, noise break-in from the ceiling space into the supply and return air ducts and then into the office, structurally transmitted vibration from the fans, main air-handling plant noise through the outside air duct to the fan coil unit; (d) acoustic lining in the outdoor air, supply air and return air ducts, anti-vibration rubber mounts for the fan coil unit and the fan within it, acoustic lining within the fan coil unit, acoustic blanket above the recessed luminaires and above the ceiling tiles.

References

ASHRAE (1985) *ASHRAE Fundamentals Handbook*, American Society of Heating, Refrigeration and Air Conditioning Engineers, Atlanta, GA.

Awbi, H.B. (1991) *Ventilation of Buildings*, E & FN Spon.

Belding, H.S. and Hatch, T.F. (1955) Index for evaluating heat in terms of resulting physiological strains. *Heating, Piping and Air Conditioning*, **27** (8), 129.

Boyer, A. (1979) Space allowances for building services – outline design stage, *Technical Note TN4/79*, Building Services Research and Information Association.

BRE Digest 205: 1977. *Domestic Water Heating by Solar Energy*, September, Building Research Establishment, Garston, Watford, UK.

BRE Digest 206: 1977. *Ventilation Requirements*, Building Research Establishment, Garston, Watford, UK.

BRE Digest 369: February 1992. *Interstitial Condensation and Fabric Degradation*, Building Research Establishment, Garston, Watford, UK.

British Gas (1980) *The Effect of Thermal Insulation in Houses on Gas Consumption.*

BS 848: Part 1: 1980. *Fans for General Purposes*, Part 1: *Methods of Testing Performance.*

BS 5572: 1978. *Sanitary Pipework.*

BS Code of Practice CP 326: 1965. *The Protection of Structures Against Lightning.*

BS Code of Practice 413: 1973. *Ducts for Building Services.*

BS Code of Practice CP 1013: 1965. *Earthing.*

Butler, H. (1979a) Lightning protection of buildings, *Technical Note TN1/79*, Building Services Research and Information Association.

Butler, H. (1979b) Space requirement for building services distribution systems – detail design stage, *Technical Note TN3/79*, Building Services Research and Information Association.

Chadderton, D.V. (1997a) *Air Conditioning, A Practical Introduction*, 2nd edn, E & FN Spon.

Chadderton, D.V. (1997b) *Building Services Engineering Spreadsheets*, E & Spon.

Chartered Institution of Building Services Engineers (1973) *Practice Notes*, Number I, *Combustion and Ventilation Air for Boilers and Other Heat Producing Appliances: Installations Not Exceeding 45 kW*,

Chartered Institution of Building Services Engineers (1981) *Building Energy Code*, Parts 1, 2 and 3.

Chartered Institution of Building Services Engineers (1986) *CIBSE Guide.*

Chartered Institution Building Services Engineers (1995) Building Services. *CIBSE Journal*, **17** (4), 39.

Coffin, M.J. (1992) *Direct Digital Control for Building HVAC Systems*, Chapman & Hall.

Courtney, R.G. (1976a) An appraisal of solar water heating in the UK, *Building Research Establishment Current Paper CP 7/76*, January.

Courtney, R.G. (1976b) Solar energy utilisation in the UK: current research and future prospects, *Building Research Establishment Current Paper 64/76*, October.

Crawshaw, D.T. (1976) Coordinated working drawings, *Building Research Establishment Current Paper CP 60/76*.

Department of Energy (1983) *Fuel Efficiency*, booklets.

Department of Energy (1988) *Energy Efficiency in Buildings*.

Diamant, R.M.E. (1977) *Insulation Deskbook*, Heating and Ventilating Publications.

Ellis, P. (1981) Listening to the user. *Journal of the Chartered Institution of Building Services*, **3** (4), 40.

Fanger, P.O. (1972) *Thermal Comfort*, McGraw-Hill.

Fanger, P.O. (1988) Introduction of the olf and decipol units to quantify air pollution perceived by humans indoors and outdoors. *Energy in Buildings*, **12**, 1–6.

Haines, R.W. and Hittle, D.C. (1983) *Control Systems for Heating, Ventilating and Air Conditioning*, Chapman & Hall.

Horlock, J.H. (1987) *Cogeneration: Combined Heat and Power Thermodynamics and Economics*, Pergamon.

Jenkins, B.D. (1991) *Electrical Installation Calculations*, Blackwell Scientific.

Jones, W.P. (1985) *Air Conditioning Engineering*, 3rd edn, Edward Arnold (Publishers) Limited.

Kut, D. (1993) *Illustrated Encyclopedia of Building Services*, E & FN Spon.

Levermore, G.J. (1992) *Building Energy Management Systems*, E & FN Spon.

McKenna, G.T., Parry, C.M. and Tilic, M. (1981) What future for task lighting? *Journal of the Chartered Institution of Building Services*, **3**, 39.

McVeigh, J.C. (1977) *Sun Power. An Introduction to the Applications of Solar Energy*, Pergamon.

Moss, K.J. (1996) *Heating and Water Services Design in Buildings*, E & FN Spon.

Moss, K.J. (1997) *Energy Management and Operating Costs in Buildings*, E & FN Spon.

National Joint Utilities Group (1979) *Provision of Mains and Services by Public Utilities on Residential Estates*, Publication Number 2, November.

Palz, W. and Steemers, T.C. (1981) *Solar Houses in Europe, How They Have Worked*, Pergamon.

Payne, G.A. (1978) *The Energy Manager's Handbook*, IPC Science and Technology Press.

Saunders, L.H. (1981) Thermal insulation and condensation. *Building Research Establishment News*, 55.

Smith, B.J., Peters, R.J. and Owen, S. (1985) *Acoustics and Noise Control*, Longman.

Southampton Geothermal booklet, Southampton Geothermal Heating Company, Pelham House, Broadfield Barton, Broadfield, Crawley, West Sussex, RH11 9BY.

Swaffield, J.A. and Wakelin, R.H.M. (1976) Observation and analysis of the parameters affecting the transport of waste solids in internal drainage systems. *Public Health Engineer*, November.

Szokolay, S.V. (1978) *Solar Energy and Building*, 2nd edn, The Architectural Press.

Threlkeld, J.L. (1962) *Thermal Environmental Engineering*, Prentice-Hall International.

Uglow, C.E. (1981) The calculation of energy use in dwellings. *Building Services Engineering Research and Technology*, **2** (1), 1–14.

Wise, A.F.E. (1979) *Water, Sanitary and Waste Services for Buildings*, Batsford.

Index